THIRD CALTECH CONFERENCE ON

VERY LARGE SCALE INTEGRATION

THIRD CALTECH CONFERENCE ON
VERY LARGE SCALE INTEGRATION

Editor
RANDAL BRYANT

California Institute of Technology

COMPUTER SCIENCE PRESS

RECID = 7350.1

Computer Science Press
11 Taft Court
Rockville, Maryland 20850

1 2 3 4 5 6   88 87 86 85 84 83

Preface

The papers in this book were presented at the *Third Caltech Conference on Very Large Scale Integration,* held March 21-23, 1983 in Pasadena, California. The conference was organized by the Computer Science Department, California Institute of Technology, and was partly supported by the Caltech Silicon Structures Project.

This conference focused on the role of systematic methodologies, theoretical models, and algorithms in all phases of the design, verification, and testing of very large scale integrated circuits. The need for such disciplines has arisen as a result of the rapid progress of integrated circuit technology over the past 10 years. This progress has been driven largely by the fabrication technology, providing the capability to manufacture very complex electronic systems reliably and at low cost. At this point the capability to manufacture very large scale integrated circuits has exceeded our capability to develop new product designs quickly, reliably, and at a reasonable cost. As a result new designs are undertaken only if the production volume will be large enough to amortize high design costs, products first appear on the market well past their announced delivery date, and reference manuals must be amended to document design flaws.

Recent research in universities and in private industry has created an emerging science of very large scale integration. This science covers a wide variety of subjects, including formal models of the behavior and performance of digital systems, systematic and algorithmic approaches to computer-aided design, and new architectures and interconnection strategies to exploit the capabilities of VLSI. By providing a forum for this research, we hope this conference and these proceedings can contribute to the field and stimulate further research and new ideas.

The book is divided into seven sections, according to the sessions at the conference.

Invited Papers

Four distinguished researchers from industry and academia were invited to present their work and the insights they have gained from designing large chips. Hörbst, Sandweg, and Wallstab describe the problems faced by their research group at Siemens in designing two VLSI chips and the solutions

v

they adopted. Anceau presents a silicon compiler to generate chip designs automatically from a specification of the algorithm to be implemented. Hennessy, et al present the design of the MIPS processor, a microprocessor with a novel approach to instruction pipelining that promises to yield higher performance than traditional designs. Dobberpuhl discusses some of the issues faced by circuit designers in designing VLSI chips.

Circuit Timing

Traditionally, circuit designers estimate the speed of a circuit manually or by extensive circuit simulation, and find ways to improve its performance using guesswork and intuition. The papers in this session describe algorithmic approaches to performance analysis and optimization that potentially reduce the effort and amount of expertise required to design high performance chips. Both Ousterhout of U.C. Berkeley and Jouppi of Stanford describe programs that estimate the speed of circuits fabricated in nMOS technology. These programs can be used to identify regions of the circuit requiring further optimization to meet the desired clock rate. The similarities and differences of their two programs provide insight into the range of possible solutions to the problem. Leiserson, Rose, and Saxe present an algorithm for improving the performance of a circuit by finding an optimal rearrangement of the combinational logic and the registers.

Routing and Interconnection

The authors in this section describe some new algorithms and fundamental results for connecting together the different components of a VLSI system. Both Chan and Pinter address the subject of automatic wire routing, in which a computer program determines the layout of the interconnecting wires based on a specification of which terminals are to be connected. Chan presents a new algorithm for channel routing in which connections are made from one set of components to another set across a rectangular wiring channel. Pinter presents some fundamental results and algorithms for river routing, in which all wires run on the same layer (no crossovers). Greene and El Gamal, on the other hand, investigate the problem of interconnecting components in a system which is so large (wafer-scale), that some percentage of the fabricated components are defective and hence must be bypassed after fabrication. Their paper describes methodologies for designing restructurable systems and some statistical analyses of their reliability.

Formal System Models

Models of the behavior of digital systems form the basis of techniques for synthesizing and verifying VLSI systems. For example, a logic simulator has some model of the operation of a digital system as its basis. A variety of models have been proposed, differing greatly in generality, accuracy, practicality, and mathematical rigor. The papers in this section present research on the more mathematical side of system models. Shostak describes a method using temporal logic to verify the correct operation of a digital system based on assertions about the behavior of the individual components. Chen and Mead describe a simulator that can model a system at several levels of abstraction from the transistor level description up to a functional description. Rem, van de Snepscheut, and Udding present trace theory as a formal model for describing the behavior of concurrent systems. And then van de Snepscheut describes a method for synthesizing a circuit based on a specification of its desired behavior in terms of its trace structure.

System Building Blocks

The construction of a large system is greatly facilitated by a design methodology in which components obeying a set of interface conventions are connected according to a set of composition rules to yield a subsystem with well-defined behavior. The papers in this section illustrate two such methodologies: self-timed systems in which the components communicate according to an asynchronous protocol, and systolic systems in which the components operate in tight synchrony. Hayes presents a design system for realizing self-timed, finite state controllers using a cellular design method called PPL's. Frank and Sproull describe the design of a static RAM chip using self-timed protocols. Finally, Fisher, et al describe a chip which can be used to construct highly parallel architectures based on the systolic approach, with the functionality of the architecture determined both by the interconnection structure and the programming of the individual chips.

Special-Purpose Chip Architectures

Many algorithms can be performed at a much higher speed when implemented directly in hardware rather than as software on conventional computers. Such algorithms form attractive applications for computer-aided custom VLSI design, because neither high chip density nor carefully optimized circuits are required to produce a system with higher performance than would otherwise be available. Furthermore, the design of special-purpose chips often leads to new architectures with applications beyond the original problem. In this section Truong, et al describe the design of a chip for encoding data to be sent from a deep space satellite to Earth. Such an application clearly has performance, weight, and power constraints that only a custom VLSI chip can provide. By careful selection of the encoding and multiplication algorithms, they were able to fit the entire encoder on a single chip. Schaeffer, Powell, and Jonkman describe several approaches to the design of special-purpose hardware for generating the set of legal moves from a particular board position in chess. This problem consumes a large portion of the computation time in chess-playing programs and lends itself well to some unusual architectures. Ja'Ja' and Owens describe ways to reduce the amount of processor hardware in systems consisting of many interconnected processors operating in parallel. It is shown that for certain problems a relatively small number of processors accessing a large number of memory elements can achieve high performance.

Silicon Compilation

The term "silicon compiler" is applied to a large variety of computer programs that synthesize chip designs directly from a specification of either the functionality, the logic design, or the circuit schematic. The building blocks used by the compiler to generate the layout can range from arbitrary geometry to a library of parameterized standard cells. Furthermore, silicon compilers differ greatly in their range of applications and in the flexibility of their floorplans. Wolf, et al describe a very general program that transforms a circuit schematic into a stick diagram which can then be compacted to yield a layout. Both Pope and Broderson as well as Bergmann present silicon compilers specifically designed to produce chips for signal processing, with one using parallel arithmetic and the other using bit-serial. By focusing on a particular class of applications, these compilers can produce designs with high performance and density.

Acknowledgments

This conference and proceedings have benefited from the efforts of many people. Clearly the authors deserve most of the credit, because their efforts to carry out research and report the results form the basis of the conference. The members of the program committee served to define the direction of the conference and reviewed many papers in a short amount of time. The other reviewers also aided greatly in the paper selection process. Chuck Seitz of Caltech and both Paul Penfield and Barbara Lory of MIT shared much of their wisdom gained from organizing previous VLSI conferences. Linda Getting and Pam Hillman handled most of the administrative aspects of organizing and running the conference. Finally, the staff of the Caltech Computer Science Department including Phyllis Weiss, Vivian Davies, Dianne Hahn, and Helen Derevan spent many hours handling the large amount of paperwork required to put on such a major event.

Cover Illustration

The illustration on the cover shows part of the layout for the tree machine processor designed at Caltech by Chuck Seitz, Don Speck, Steve Rabin, and Peter Hunter. This design was written in the layout language EARL in which the design is specified as a hierarchical composition of cells. The lowest level (leaf) cells consist of a set of geometric objects that are required to satisfy a set of programmer-defined spacing constraints. These cells are composed by stretching the geometry subject to the constraints so that the ports of adjacent cells align. Close inspection of the layout will show examples of wires which have been specified as paths which must miss other objects by minimum radii, and cells which have been composed by stretching. This layout also illustrates the use of "Boston" geometry in which lines at arbitrary angles and circular arcs are permitted. Recent extensions to EARL enable the designer to produce "Manhattan" geometry in which only lines parallel to the X and Y coordinate axes are permitted, providing better compatibility with existing pattern generators and design rule checkers.

R.E. Bryant, January, 1983

Program Committee

Randal Bryant, Caltech
Lynn Conway, Xerox PARC
John Hennessy, Stanford
Lennart Johnsson, Caltech
H.T. Kung, CMU
Alain Martin, Caltech and Philips Research
Carver Mead, Caltech
Richard Newton, Berkeley
Paul Penfield, MIT

Referees

Martin Buehler, JPL
Clement Leung, Patil Systems, Inc.
Bill Athas, Caltech
Marina Chen, Caltech
Gary Clow, Caltech
Erik De Benedictis, Caltech
Tzu-Mu Lin, Caltech
Ricky Mosteller, Caltech
Mike Schuster, Caltech
Chuck Seitz, Caltech
John Tanner, Caltech
Steve Trimberger, Caltech
John Wawrzynek, Caltech
Dan Whelan, Caltech
Doug Whiting, Caltech
Telle Whitney, Caltech

Authors

Contents

System Building Blocks

Special-Purpose Chip Architectures

Silicon Compilation

Invited Papers

Practical Experience with VLSI Methodology
E. Hörbst, G. Sandweg and S. Wallstab

CAPRI: A Design Methodology and a Silicon Compiler for VLSI Circuits Specified by Algorithms
F. Anceau

Design of a High Performance VLSI Processor
J.L. Hennessy, N.P. Jouppi, S. Przybylski, C. Rowen and T. Gross

Fundamental Issues in the Electrical Design of VLSI Circuits
D.W. Dobberpuhl

Practical Experience With VLSI Methodology

Egon Hörbst, Gerd Sandweg and Stefan Wallstab

This work was supported by the Technological Program
of the Federal Department of Research and Technology
of the Federal Republic of Germany.

The authors are with the Cooperate Laboratories for
Information Technology, Siemens AG, Munich, Germany.

ABSTRACT: Some VLSI typical problems and their solutions
with the help of architectural concepts are presented.
These principles are demonstrated on two experimental
chips fabricated in our research process line. Regular
structures for the control part of a VLSI processor
are described in more detail.

Typical Problems for VLSI

VLSI circuits comprise more than 100 000 transistors
(including places for transistors in ROMs and PLAs).
The density and low power consumption needed for such
a large number of transistors is only achievable with
MOS technology. The characteristics of the MOS technology influence the design style as shall be shown later
on.

1

There are typical VLSI problems, some of which might
be solved by suitable architectural concepts.

The most evident problem is managing the complexity.
This is the reason why first VLSI chips were memories.
From a logic point of view these chips are very simple.
You have "only" to solve circuit and process design
problems, whereas for logic oriented chips the problem
is how to reduce complexity. One method is to use as
much as possible regular modules like RAMs, ROMs, PLAs
or slice structures. Additionally it is essential to
employ CAD tools extensively.

As a result of the high packaging density one can run
into power consumption problems. A way to reduce this
problem is to use dynamic techniques e.g. precharged
busses. But the dynamic techniques are time critical
and prone to trouble (testing!). For VLSI design it
is better to use static techniques and solve the power
problem with a low power consuming technology e.g. CMOS.

A severe problem is the limitation of pins. We are able
to put a whole system with tremendous computing power
on a chip but we have difficulties to get the inputs
to and the results from the chip. The transfer to the
outside word of the chip is approximately one magnitude
slower than the transfer inside the chip. The solution
is to broaden up and separate the communication paths
but this again is limited by the cost and the mechanical
problems of big packages. Architectural solutions might
be serial structures like systolic arrays or concepts
like pipelining and distributed processing.

Another consequence of the narrow communication channel

is that the controllability and observability of chip
components decrease with increasing integration. This
leads to testing problems unless special design techniques for testability and selftest are used.

VLSI chips need much area. But the yield decreases very
significantly with increasing area. In memories there
is the possibility of adding redundand elements (spare
rows and columns). For logic circuits redundancy and
error correction is still a field of research. Therefore it is not true that we can waste chip area because
integration is so cheap. The relation is a little bit
more complicated. Concerning small chips it is better
to spend silicon area instead of development effort.
For large chips it is necessary to use area saving
structures and to find good floorplans.

The next typical problem is the wiring. In logic oriented circuits most area is not consumed by the active
elements but by the wiring between them. Since even
modern MOS technologies have only two metal layers for
wiring (the polysilicon and diffusion layers can be
used for short connections only because of their high
resistance) the designer must find flat layouts for
the functions to be realized.

For the VLSI typical case, that logical and geometric
structure have to be designed together, there is an
additional reason. Long wires result in large capacitances, large transistors to drive them and therefore
long signal delays. The designer therefore has to ar-
range function blocks that fit together and need only
few and short interconnections.

The last VLSI typical problem to be mentioned here is

the small production volume. Only memories and some
successful microprocessors reach quantities of more
than 100 000 a year. But the future VLSI market will
belong to the custom-ICs produced in small quantities.
One solution to this problem is to make low volume production economic, the other to reduce development time
and costs by extensive use of computer aided design
combined with architectural concepts suitable for automation.

Structures for VLSI Processors

Processor structures have proved to be very flexible
and well suited for the implementation of complex functions. Therefore most of the realized logic oriented
VLSI chips have processor structures. Typical examples
are microprocessors, peripheral controllers, signal
processors, graphic processors and communication chips.
The two chips we gained our VLSI experience from are
processors too.

The first chip is a 32-bit execution unit /1/. When
counting the number of transistors it is not really
a VLSI chip because it has only 25 000 tranistors. It
is rather a model for a VLSI chip. We tried to make
this execution unit as regular as possible without loos-
ing performance or wasting silicon area. Beside this
it was used as a test circuit for our research fabrication line developing a scaled NMOS single-layer poly-
technology with 2-μm minimum gate length and low-ohmic
polycide for gates and interconnections. The chip was
produced in 1981 and the test results were very satisfying. The execution unit performs logic and arithmetic
operations on two 32-bit operands in 125 ns (8 MHz).

Multiplication and division is supported by a special
control circuit to speed up the shift and add logic.
Thus multiplication and division on signed 32-bit operands need only 34 cycles.

This execution unit chip has an area of 16 mm^2 and is
mounted in a 64 pin DIP. Power dissipation is about
750 mW.

Our second chip is a real VLSI chip. It is a peripheral
processor with about 250 000 transistors and an area
of more than 100 mm^2. The biggest part of this chip
is a static 36-kbit RAM (200 000 transistors) for microprogram or data storage. The data word size is 16 or
8 bit. The instruction format is 32 bit. The instruction
set is tuned to the special task of this processsor
and rather irregular. Most of the instructions are re-
gister-register-operations but there are also three-
address-operations and operations between external operands.

To realize this chip we chose the same technology as
mentioned above except for an additional second metal
layer. The design of this chip is completed and we will
get first silicon in spring 83.

With both chips we could show that for VLSI processors
it is a good architectural concept to partition the
processor in an operative part, an I/O unit, a memory
and a control unit (Fig.1). These are the classic components of a computer.

For the operative part a slice architecture has proved
to be very effective concerning processing speed, area
consumption and design effort (Fig. 2).

The integration of memories on logic oriented circuits
is becoming more and more interesting. On-chip memories
may help to reduce transfers to and from the chip and
microprogramming is an established way to implement
complex functions with simple hardware. But there are
still other memory structures well suited for VLSI design like associative memories, stacks or queues.

The I/O units in our experimental chips are simple and
consist merely of I/O ports which were integrated easily
in the data path.

Architectures for the control unit are not so well known
as for the other processor components. Some of our experiences in this field is therefore reported in more
detail in the next chapter.

Control Unit

The control unit is assumed to be the most irregular
part of a processor. This is true if we look at older
microprocessors where the aim was to squeeze the instruction word. But if the VLSI designer has some influence on the instruction format and if he is allowed
to spend a little bit more area of silicon then there
are quite regular and effective solutions for the control unit as we are able to show with our experimental
chips.

In the execution unit chip the control part is rather
simple. Its only task is to decode the 8-bit wide opcode
into 40 control lines. This is done by a ROM with 208
40-bit words, one word for each opcode. The bit lines

run in aluminium parallel to the data path and it is
possible to pick up every bit line at any point and
connect it with a poly-line that crosses the bit lines.
The poly-line leads to an output buffer and from there
through some timing logic directly to the control input
of a function slice. With this technique a minimum area
for wiring is required (Fig. 3).

This approach was not suitable for the control unit
in the peripheral processor chip. One reason is the
32-bit wide instruction word, the other reason is the
large number of control lines, namely more than 200.
The obvious solution is to use several small ROMs or
PLAs. But the different sizes of these units would have
led to wasted areas and we would have to spend extra
area for wiring.

We chose another way. We found that in most decoder
PLAs the Or-plane is occupied very weakly. This is the
case when the instruction format is structured into
groups of bits that correspond to distinct function
units, e.g. a function code for the arithmetic logic
unit, addresses for the registers or a condition code.
In this case it is possible to reduce the Or-plane of
a decoder PLA just to one splitted Or-line. Each sum
term is generated at the location near to the control
line that needs this sum term. And if a sum term is
needed at different locations it is generated twice
to avoid global wiring.

In the peripheral processor chip we use such a "dege-
nerated" PLA with about 500 product terms (Fig. 4).
The pitch between the product terms is 10 μm. Here it
was necessary to use the second metal layer to achieve
a short transit time of 25 ns through such a large PLA.

The output signals of the decoder ROM or PLA are not
yet qualified for control lines. They have to be combined with clock signals for the exact timing. This
timing stage might be regarded as a second decoder,
as a decoder for the different phases of an instruction
cycle. This second stage adds some flexibility to the
decoding scheme. In the peripheral processor chip this
stage is used to combine step signals instead of clock
signals. These step signals are a mixture of timing
and functional signals.

The advantages of the described decoder concepts are
that they are adaptable to the data path with only local
wiring and that the basic structures (ROM and PLA) can
be generated automatically.

Another task of the control part is to calculate the
next instruction address and fetch the next instruction
word.

We find that this task can be done by a sequencer which
fits very well into the slice concept of a data path.
In many applications data word size and instruction
word size are the same and if the sequencer lies at
one end of the data path there is a good connection
to the data busses. The function of the sequencer can
be realized with parts of the arithmetic logic unit
(incrementer) and some registers. In the peripheral
processor chip we added an address stack for subroutine
calls to the sequencer. This stack is very similar to
a dual-port register file but instead of a decoder a
pointer in a shift register is used.

Conclusion

With both our experimental chips we were able to show
that is possible to design VLSI processors in a very
regular manner. The described architectural concepts
are well suited for computer aided design and we can
expect that there will soon be CAD tools which allow
the realization of custom oriented VLSI processors with
astonishingly small design effort.

References

/1/ Pomper, M.; Beifuss, W.; Horninger, K.; Kaschte, W.:
    A 32-bit Execution Unit in an Advanced NMOS Techno-
    logy. IEEE Journal of Solid-State Circuits, Vol.
    SC-17, No. 3, June 1982.

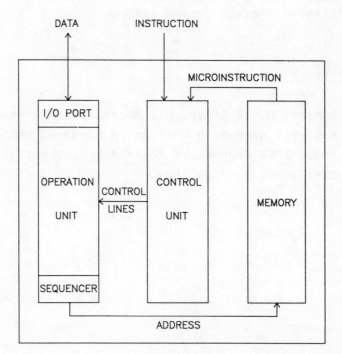

Fig. 1.   Simplified structure of a VLSI processor

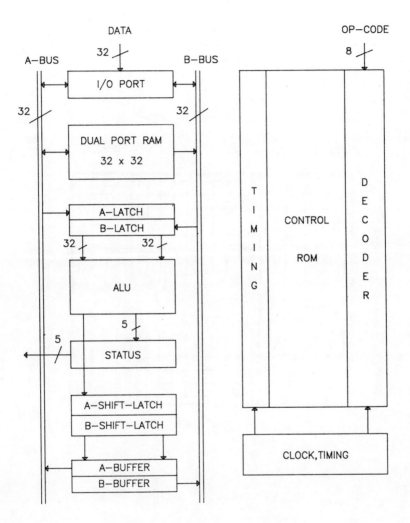

Fig. 2. Logic plan of the 32-bit execution unit

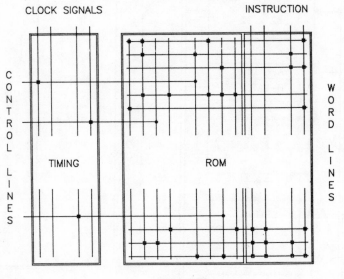

Fig. 3.   Instruction decoding with a ROM

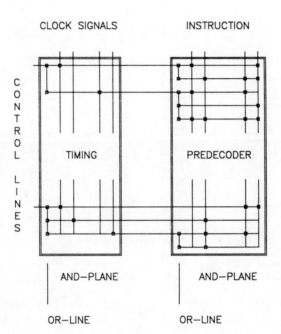

CLOCK SIGNALS   INSTRUCTION

CONTROL LINES

TIMING   PREDECODER

AND—PLANE   AND—PLANE

OR—LINE   OR—LINE

Fig. 4.   Instruction decoding with PLA structures

CAPRI: A Design Methodology and a Silicon Compiler for VLSI Circuits Specified by Algorithms

F. Anceau
Computer Architecture Group
BP 68
38042 St-Martin D'Heres cedex
France

ABSTRACT

This paper presents a design methodology and a silicon compiler for VLSI circuits specified by algorithms. The first part of this paper describes some principles to obtain optimized layout. The second part describes the design methodologies based on the use of predefined architectural templets selected for their efficiency. The last part describes the tools which are currently in development at the Computer Architecture group of IMAG. The efficiency of this methodology has been proved by its application in the design of industrial IC.

I - INTRODUCTION

The CAPRI system is a Silicon Compiler under development by the IMAG Computer Architecture Group. This tool implements the CAPRI design methodology developed by this group from its design experience and the analysis of the internal structure of existing microprocessors [10, 13].

The aim of this system is the reduction of the design cost and design duration and the increase of the design reliability for mass production and custom digital VLSI. These design improvements will be obtained without a large increase of the silicon area of these circuits or a big decrease in their performance.

The area of application of the CAPRI system is very extensive: it includes all digital VLSI circuits which are specified by their anticipated behaviour such as:
- microprocessors (one chip and one chip CPU),
- co-processors,
- peripherial controlers,
- communication controlers,
- automata,
- ...

II - CAPRI DESIGN METHODOLOGY

The design of a VLSI circuit from its behavioural specification using CAPRI design methodology is divided into four main steps:
- architectural design step,
- data path design step,
- control section design step,
- miscellaneous part design step.

Before detailling these design steps, it is useful to remember that the aim of designing a digital VLSI circuit is to draw the masks which will be used to build them. This drawing work is currently the most costly and critical step of the whole design process.

Research into increasing the integration density, synonymous with a good production yield, has to be done in parallel with the increasing of design reliability to limit errors. Experience from other domains has shown that in order to reach this goal, the search for a global optimization is more important than the search for local optimizations. This principle is expressed here in a few simple design rules [17].

III - GENERAL TOPOLOGY OPTIMIZATION

Unused area between cells must be reduced as much as possible. This goal is obtained by adjusting the shape of each cell to fit its destination place. This flexibility in the shape of the cells under dimensional constraints is shown by figure 1. It differs for each cell and corresponds to the scanning of different implementations of it.

It is preferable to design a cell not for minimal area but in order effectivelly to fit into the general topological structure. Thus, individual optimization of each cell area does not necessarily lead to area optimization of the whole circuit.

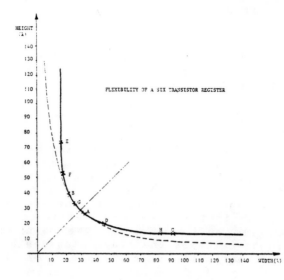

FLEXIBILITY OF A SIX TRANSISTOR REGISTER

Figure 1 - Deformability of a cell from [19].

III.1. Interconnexion optimization

The area occupied by the connection in an integrated circuit increases with its complexity. In the case of LSI/VLSI circuits this area is frequently at least equivalent to the area taken up by the logical block [1].

The classical approach, consisting of the placement and routing of functional blocks designed and optimized independently, gives a circuit area equal to the sum of the connection area and the gates area.

In order to get a better optimization of the circuit area, we suggest the use of the following rules:

- Suppression of pure interconnection zones: by the use of direct connections between cells which must be designed taking this constraint into account.

-Suppression of connections and gates: the existence of many conductive layers allows us to draw the gates under the connections "Wires first, gates under" (Craig MUDGE).

-Use of the "transparency" of the cells: it is often better that the connections pass through the cells rather than going around them. In fact, it is often possible to include a connection passing through a cell without a large modification of its size. Each cell will be characterized by a transparency factor in both directions, horizontal and vertical, which indicates how many lines may pass through this cell.

The application of these rules suggests that it is possible to characterize the design quality of an integrated circuit by the complement of the percentage of its area occupied only by pure interconnection lines without any transistors underneath.

III.2. - Floor plan

The topological design must be considered in parallel with the logical design and not as a final step. By using a top down design, the main topological data (floor plan) must be built as soon as possible from the functional specification. The refinement of the functional and topological descriptions must be metric to take into account the size and spacing of the main blocks. The dimension of the cells of the floor plan can be evaluated from their functional specification by heuristical algorithms.

It is well known in the microelectronic industry that the best way to start the design of an integrated circuit is to draw up its floor plan as soon as possible, where the shape, size, connections and power supply distribution lines of each block are predictable. This floor plan must be considered as giving the upper level of topological information. In the case of an integrated CAD system, it is used to define the organization of the data base which is itself used to manage the design of the whole circuit [8].

III.3. Interconnection layers reservation

The use of the above rules shows that the design and the optimization of the interconnection layers must be done before those of the topological gates which will be put under the interconnection lines. In practice, the application of this strategy is often affected by problems linked to the fact that gate implementation may sometimes also need the use of the upper layers which should be reserved for interconnection layers.

Such problems may occur in the following cases:
 - aluminium gate technology (now obsolete),
 - butting contacts used in some technologies to implement interconnections between deep layers (for example, in CMOS technology, butting contacts are used to connect p+ diffusion and n+ polysilicon),
 - topological problems may require the use of more conductive layers than those reserved for gate implementation. In this case, some local connection of the gates must be implemented in the upper layers reserved for interconnections (for example flip flop

feedback, or straps to put transistors in parallel into CMOS gates).

These disturbances of the interconnection layers may be so small that they can be predicted in the interconnection layer design by making reservations into these layers. The strategy of overlapping interconnections and gates may become impossible to use when this disturbance becomes larger and routing channels must be provided.

III.4. Layer utilisation strategy

The allocation of different conductive layers will be made according to their electrical properties. The better ones as regards resistance-capacitance will be used for the longest connections.

- nMOS/HMOS/CMOS technology using only one polysilicon layer (table 1). The double use of the polysilicon layer for both interconnections and gates reduces the transparency of the cells in this direction.

- nMOS/HMOS/CMOS technology using one supplementary interconnection layer (table 2). The two upper layers may almost only be used for interconnection, giving a high transparency of the cell in the both directions.

The power supply lines use metal layers (with few exceptions).

layers (by decreasing quality)	interconnection layer use	gate use
metal	x	reserved
polysilicon	x	x
diffusion		x

Table 1 - Layer allocation for nMOS/HMOS/CMOS single polysilicon layer

```
------------------------------------------------
layers (by    interconnection      gate use
decreasing    layer use
quality)
------------------------------------------------
metal                        x
metal or polysilicon         x         reserved
polysilicon                            x
diffusion                              x
------------------------------------------------
```

Table 2 - Layer allocation for nMOS/HMOS/CMOS double
          aluminium or polysilicon layers.

III.5. Crossed interconnection layers

In almost all cases, the main blocks of an integrated
circuit may be seen as the crossing area of two
information flows (for example, data and control flow).
This can be physically carried out by two straight wiring
layers running perpendicularly using the two best
conductive layers.

The metal layer must be used to implement the longer and
denser wiring layer and the power supply distribution.
The polysilicon layer must be shared between the
implementation of the gates and the second wiring layer.

The orientation of the wiring layers and their allocation
to the conductive layers must be done for the whole
circuit (from information given by the floor plan) and
must be the same for the largest possible areas.

III.6. Strip organization of the layout

Considering a given interconnection layer, a high
transparency of a set of cells for this direction allows
this interconnection layer to overlap above the cells, a
low transparency requires this interconnection layer to
be implemented   by using wiring channels.

The characteristics of the nMOS/HMOS technology using one
polysilicon layer suggest the   use of a strip approach
for layout. The high transparency of the cells in the
direction of the metal layer and the fact that the power
supply must be implemented as two interleaved metal
combs, leads this layer to be implemented   as a set of
parallel tracks at minimum spacing betweem the teeth of
the power supply combs. The pitches of these tracks only
allow the use of alternated contacts. Some free tracks
must be reserved into the metal wiring layer to implement

the local connections of the cells. The low transparency
of the cells in the direction of the polysilicon layer
leads interconnections to be implemented in this layer
as channels between the cells. The high transparency in
the both directions of the nMOS/HMOS technologies using
one supplementary interconnection layer suggests the use
of a strip approach for the both interconnection layers.

CMOS circuitery requires the gates of the n and p
transistors of a gate to be interconnected (except for
switches). This constraint suggests gates be realized
by using two parallel strips of n+ and p+ diffusion
crossed by straight polysilicon gates. Transistors are
put in parallel by straping over diffusion strip. The
separate gates of switches make exceptions which suggests
metal or second polysilicon be used in the direction of
the control flow and first polysilicon in the direction
of the data flow.

IV - ARCHITECTURAL DESIGN STEP

This design step starts from the behavioural
specification to produce a low level precise description
of the behaviour of the future circuit. This design step
is performed by:
   - introducing intermediate levels of interpretation
   into the algorithmic specification (microprogram,
   clock phases, etc...),
   - transformation of the algorithm:
   * by breaking down complex operations into simple
   operations. Example: a multiplication operation is
   replaced by a loop including addition and shifting.
   * by translating operative instructions into a pre-
   defined format corresponding to the potentialities
   of a standardized organization of the data path.
   - extraction of the data path and control section
   specifications from the accurate behavioural
   description.

The two first steps are common with computer design. The
influence on VLSI implementation starts by using a pre-
defined format for operative instructions.

V - DATA PATH DESIGN STEP

This design step consists in generating the layout of
the data path from its specification extracted from the
accurate behavioural description.

The use of a pre-defined format for operative instruction
corresponds to standardize the data path organization
from:

- Architectural point of view:
The data path of a circuit is organized as the assembly
of one or several sub-data paths performing parallel
operation (for example data and address computation)
(figure 2).
The architecture of each sub-data path is organized
around a double bus system carrying data from/to double
access registers, double access I/O registers (also
connected to I/O buses), ALU containing input registers
and shifters, and double access RAM to store immediate
values. The buses of each sub-data path are
interconnected by switches to allow to transmit data.

This data path organization is very powerful. Each
microcycle consists of two phases, each of them allowing
two bus transfers. The pre-defined format describes the
possibilities of this kind of data path.

⟨phi 1⟩ A ⟨= ... Reg2 ⟨= (Reg 1 or immediate value 1) ,
         B ⟨= ... Reg4 ⟨= (Reg 2 or immediate value 2) ;

⟨phi 2⟩ Reg6 ⟨= ... Reg 5 ⟨= A op B ,
         Reg9 ⟨= ... Reg 8 ⟨= Reg 7 ;

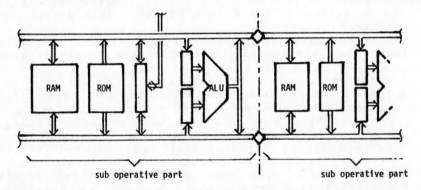

Figure 2 - Standardized data path organization.

- Electrical point of view:
All data sources and data sink connected on the bus
system have the same electrical behaviour. Two systems
are possible:
    - precharge/decharge single line bus where each
    register have separate input and output ;
    - static memory like bus system (MC 68000) where each
    bus has two differential lines and each register uses
    the same way for reading and writing.

In the first system, the size of the output transistors
of the cells must be adjusted to the length of the bus
system. In the second system data amplifiers allow the
same cells to be kept for a large range of data path

complexities.

- Layout point of view:
This kind of data path may be organized as bit slice
structure which obeys the topological optimization rules.
When using nMOS/HMOS technology, the metal layer is used
to carry data and power supply whereas polysilicon layer
is used to carry control lines. The pitch of the bit
slice directly comes from the bus and power supply
specification (including free tracks for intra-cell
connections).
When using CMOS technology, the metal layer is used to
carry control lines and power supply. The data buses are
implemented by the polysilicon layer. The pitch of the
bit slices is more difficult to fix. It must allow all
the cells to be drawn efficiently.
The properties of these kinds of data path allow their
layout to be obtained by an assembly relatively simple
of the layout of pre-designed cells taken from a library.

VI - CONTROL SECTION DESIGN STEP

The control section design is not as easy as the data
path design. Several internal organizations (styles) may
be used and the standardization of the control section
layout is not yet completely clear. Several proposals of
templets for control section layout already exist (figure
3) but they must be exercised on practical applications.
When not using a standardized layout for control section
design tools can only provide automatically the content
of repetitive blocks (ROM, PLA).

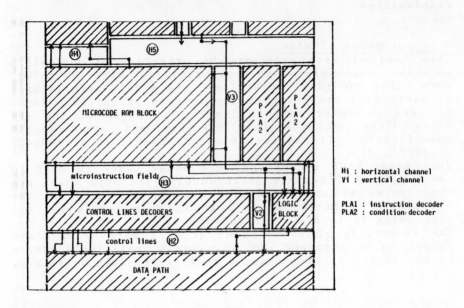

Figure 3 — Block Slice approach for control section organization (from M.Nemour).

The manner in which decisions are taken into the control structure of the algorithm has a high influence on the optimization of the control section. This influence depends on the style of control section which is used. The technique of decision taking by switching between several control sequences (MOORE approach) is suitable for control section styles where the sequencing mechanism and command operation mechanism are different (for example microprogrammed styles). The technique consisting to take decision by using conditional generation of commands (MEALEY approach) is suitable for control section styles where sequencing and command generation mechanisms are merged (for example single PLA style). The use of direct computation of commands from instruction register selected from sequencing mechanism is always a good approach.

This means that the form of the control structure of the algorithm must be adjusted to the selected style of the control section and to the balance of the size of its inner blocks.

VII - DESIGN OF THE MISCELLANEOUS PARTS

The preceding steps provide up to 80% of the layout of this kind of VLSI circuits. At this state of the

development of the methodology, the design must be finished by hand for:
- control section glue (if its layout is not standardized),
- interruption mechanism,
- I/O pads,
- clock system.

In order to avoid losing design reliability, the CAD tools which generate the layout of the data path and the control section will provide all necessary documents about their work (electrical and logical schemata) in order that they can be modified by hand.

VIII - EFFICIENCY

The efficiency of the CAPRI design methodology comes from the fact that it consists of the parametrization of pre-existing architectural solutions (architectural templets) which are proven to be efficient from architectural (performance) and implementation (layout) points of view.

Figure 4 shows the data path organization of an exercise of redesign of the I8048 microprocessor using this methodology. For the same performances (in terms of clock cycles) the layout becomes far more regular for the same size of chip area.

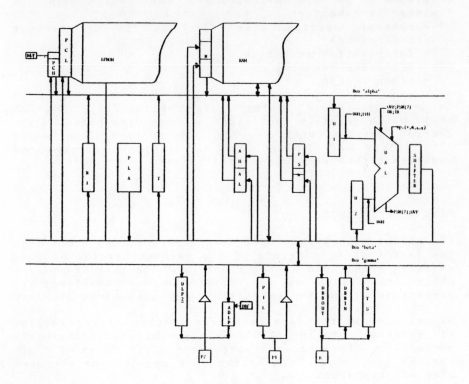

Figure 4 - Repetitive data path structure for I8048 microprocessor (from F.Bertrand).

The CAPRI design methodology has already been used on several industrial examples and is currently used for the design of a P.code microprocesssor by the research team as a pilot project.

IX - CAPRI MAIN TOOLS

The tools supporting CAPRI methodology are currently developed. The strategy of tools developement is bottom-up. Starting from the layout generation, each new layer of tools must satisfy criterium of design efficiency before being put in the CAPRI system.

IX.1. IRENE description language

The main support of the methodology is a description language able to describe the behaviour and/or the structure of a future circuit without any ambiguity.A Register Transfer Language (called IRENE), has been developed to support the design steps. The structure of this language is based on a one-to-one correspondence with hardware components. This feature is the key to the ability of representing any hardware organization.

IX.2. Extractor

The aim of this tool is to get the specification of the structure of the data path and of the control section of the future circuit from its behavioural description. This extractor collects separately all the operative instructions (assignments and connections) and the control structure from the behavioural description.

IX.3. Data path generator

The data path components are selected and sized from the syntactic analysis of the operative instructions by a grammar representing the pre-defined format of operative instructions. This tool must provide a relative assembly of cells to get the data path layout and the topology of the control lines on the side of the data path block.

IX.4. Control section generator

The more complex generating tool starts from the microprogram structure and the topology of the control lines on the side of the data path block to generate the regular elements of the control section. These tools are:

IX.4.1. Style selector - An expert program using techniques of the Artificial Intelligence will implement the strategy of the selection of the control section organization from the design experience of the research group.

IX.4.2. MOORE-MEALEY transformer - The possibility of balancing the size of the various components which constitute a control section is linked to the relative importance of the decision taken at the level of the sequencing mechanism (multiways branches, MOORE approach) and the decision taken at the level of the actions generation (conditional actions , MEALEY approach).
The possibility of transforming some sequences into a procedure using parameters, also allows another form of balancing between the relative size of the PLAs constituting the control section.

IX.4.3. Generators of the repetitive components - These tools generate the content of the repetitive components (ROM, PLA) constituting a given style of control section. These tools are called generators and are adapted to each type of control section we want to generate. These tools also provide the logical description of the repetitive components. A prototype of such a tool for the multi-PLA approach is in development .

IX.4.4. PLA optimizer - This tool generates and optimizes the PLA specified by the preceding program. A topological optimizer, called PAOLA [14] has been developed and may be used for this job.
The efficiency of the compaction C given by this tool on the OR matrix of a PLA is given by the formula

$$C = Ln(F)/0.033 - 25.22$$

where F indicates the filling ratio (in bits) of the unoptimized OR matrix.

This tool gives the optimized layout of a PLA and evaluates its performance. All the PAOLA computations are parametrized from the technological design rules (topological and electrical parameters).

IX.5. Topological evaluator

The topological organization of a future chip (floor plan) may be evaluated at each step by statistical methods, taking into account the relative deformability of the block and their ability to be crossed through by connection. A prototype as such a tool, called TESS, [12], [18], [19], is currently under development .

IX.6. Layout assembler

The layout of the circuit is obtained by a tool called LUBRICK which is an extension of the PASCAL language able to assemble cells from a topological description (relative positionning).

The cells are taken from a library which contains their layout, size, the description of their boundaries as lists of segments typed by material, and the description of their connectors where metal or poly strips may be connected in each direction.

The basic operations to assemble cells are called "up" and "right". They consist of putting a new cell up or right as close as possible to the previous one by an undirectional cell compaction. The connections between these cells are realized by straight metal or poly strips. Another function is used to place and to connect a strip of cells to an irregular topology of connectors like putting the drivers of the control lines along the side of a data path block.

The layout of non repetitive cells, or the exception into the repetition of cells, is directly generated by LUBRICK routines like routing channels used when straight connections are impossible. The structure of the cell assembly is directly controlled by the structure of the LUBRICK program.

The electrical performance of the cell asssembly may be automatically computed from the combination of the electrical properties of each cell.

X - CONCLUSION

The methodology consisting in designing complex VLSI or functional parts of ULSI by selecting pre-existing good engineering solutions and adjusting them to new problems by parametrization seems a very efficient engineering solution.

Such design strategies may be helped by the use of artificial intelligence techniques to select styles depending upon their application domain and to make choice between several options.

XI - BIBLIOGRAPHY

[1]   W.R.HELLER, W.F.MIKHAIL, W.E.DONATH
      Prediction of wiring space requirement for VLSI
      Journal of Design automation and fault-tolerant
      computing, pp. 117/144, 1978.

[2]   D.JOHANSEN
      Bristle blocks: a silicon compiler
      16th DAC, San Diego, June 1979.

[3]   H.DEMAN
      Computer aided design techniques for VLSI
      NATO Advanced summer Institute, Louvain la Neuve,
      Belgium, July 8/18, 1980.

[4]   J.M. COSTA ALVES MARQUES
      MOSAIC: a design methodology for VLSI systems
      circuits
      Docteur-ingenieur thesis, INPG, Grenoble, 24/9/80.

[5]   D.PATTERSON, C.SEQUIN
      Design considerations for single chip computers of
      the future
      IEEE Trans. on Comp., vol C 29, n.2, pp.108/115,
      February 1980.

[6]   C.MEAD, L.CONWAY
      Introduction to VLSI systems
      California, Addison Wesley, 1980.

[*7] Van CLEEMPUT
      A structural design automation environment for
      digital systems
      CAD and VLSI course, Linkoping, July 5/11, 1980.

[8]   H.de Man, W.SANSEN, J.CORNELISSEN, J.HURT, W.HEYNS
      An integrated CAD system supporting hierarchical
      design of nMOS VLSI circuits
      ESSCIRC 81, Freiburg, pp.94//97, 1981.

[9]   A.SUZIM
      Study of data-flow blocks with modular elements for
      monolithic processors
      Docteur-ingenieur thesis, INPG, Grenoble, 6/11/81.

[10] F.ANCEAU
      VLSI processor architecture
      ESSCIRC 81, Freibourg, pp.24/30, 22-24/9/81.

[11] J.M.SISKIND, J.R.SOUTHARD, K.W.CROUCH
      Generating custom high performance VLSI designs
      form succint algorithmic description
      Conf. on advance research in VLSI, MIT, January 82.

[12] R,REIS
      A topological evaluator as a first step in the
      VLSI design
      MICROELECTRONICS 82, Adelaide (Australia),
      May 12/14, 1982.

[13] F.ANCEAU
      LSI processor architecture and design
      MICROELECTRONICS 82, Adelaide (Australia),
      May 12/14, 1982.

[14] T.PEREZ SEGOVIA, S.CHUQUILLANQUI
      PAOLA: a tool for topological optimization of
      large PLAs
      19th Digital Automation conference,Las Vegas,June 82

[15] P.B..DENYER
      An introduction to bit-serial architecture for VLSI
      signal processing
      EEC CREST Summer school, Bristol, July 82.

[16] F.ANCEAU, J.P.SCHOELLKOPF
      CAPRI: a silicon compiler for VLSI circuits specified
      by algorithms
      EEC CREST Summer school, Bristol, July 82.

[17] F.ANCEAU, R.REIS
      Complex integrated circuit design strategy
      IEEE Journal of solid state circuit, vol.SC 17,n.3,
      pp. 458/464, june 82.

[18] R.REIS
      TESS: a topological evaluator tool
      ICCC 82, IEEE Intern. conf. on Circuits
      and Computers, New York, sept.29/octob.1st, 1982.

[19] R.REIS
      TESS: evaluateur topologique predictif pour la
      generation automatique des plans de masse de
      circuits VLSI
      Docteur-ingenieur thesis, INPG, Grenoble,
      January 11, 1983.

Design of a High Performance VLSI Processor

John L. Hennessy, Norman P. Jouppi, Steven Przybylski, Christopher Rowen and Thomas Gross
Computer Systems Laboratory
Stanford University
Stanford, California 94305

Abstract

Current VLSI fabrication technology makes it possible to design a 32-bit CPU on a single chip. However, to achieve high performance from that processor, the architecture and implementation must be carefully designed and tuned. The MIPS processor incorporates some new architectural ideas into a single-chip, nMOS implementation. Processor performance is obtained by the careful integration of the software (e.g., compilers), the architecture, and the hardware implementation. This integrated view also simplifies the design, making it practical to implement the processor at a university.

1 Introduction

MIPS [10] is a new 32-bit processor designed to execute compiled code for general purpose applications. MIPS is both a streamlined (or reduced) instruction set architecture and an implementation of that architecture as an nMOS chip. The MIPS implemetation has several goals. First, we wanted to determine whether a competitive 32-bit processor could be built using the concepts in the MIPS architecture and Mead-Conway style design. A number of benchmarks have been run on the 68000 (at 8MHz) and our MIPS simulator (assuming a clock period of 4MHz). These benchmarks show an average speed-up of 460% for C programs and 550% for Pascal programs using the same compiler technology for both processors. If we can build a 4MHz part, we will have succeeded in demonstrating the architecture and implementation.

Other experimental goals were just as important. We wanted to simplify the VLSI implementation as much as possible: at the heart of our strategy was to replace hardware with simple software wherever there was no significant performance advantage to the hardware implementation; specific examples of the types of tradeoffs we made are discussed in [11]. Last, we wanted to experience the problems of designing a large chip, to further focus our architectural and design aid research activities.

In this paper, we give an overview of the architectural design and VLSI implementation of MIPS. We will attempt to show how a set of new ideas in architecture was implemented in a VLSI processor, to yield a high performance CPU. We will discuss the limitations we encountered in our efforts to increase performance, the design principles that simplified our efforts and made a fairly ambitious design possible, and the problems we have yet to overcome before we embark upon another project of similar scope.

2 Architectural Design

One overall goal of the MIPS architecture design was to explore the extent to which instruction set design can be made into a scientific process. This can be done by evaluating the design choices on the basis of two data points: quantative data about the usefulness of an instruction and careful estimates of the hardware cost. For MIPS, a C compiler, derived from the Portable C compiler, and a simulator formed a measurement system that provides firm, realistic data. This provided a fundamental workbench to test out new instructions. The proposer of the new instruction had to get the compiler to generate the instruction, collect data for a number of benchmarks, and examine what instructions, if any, were displaced by the new instruction. These measures are then combined to give a reasonably accurate estimate of the potential of the new instruction. Unfortunately, determining the hardware cost of an instruction that does not fit into the model of other instructions in the architecture is at best an ad hoc process.

MIPS is a reduced or streamlined instruction set machine, like the Berkeley RISC processor [13] and the IBM 801 [15]. The fundamental philosophy of such architectures is to concentrate on the most frequently–used simple instructions and to build more complex instructions from a customized series of simpler instructions. Because the simpler instructions, e.g., simple loads and stores, arithmetic operations, etc., are used with much higher frequency than more complex instructions, eliminating the complex instructions will allow the architecture to optimize its performance for the simple instructions. Such a philosophy becomes even more applicable in a VLSI implementation, because power and area constraints force the designer to choose among fundamental alternatives.

We also chose to design a load-store architecture, i.e., a machine in which only the load and store operations access memory and all ALU instructions are register-register format. This structure is very compatible with a streamlined instruction set; it also simplifies the implementation of page faults, traps, and interrupts, which can be particularly difficult to implement in a pipelined machine. Lastly, given reasonable register allocation algorithms [1], using a register-based machine is a good match for VLSI technology. Since the cost of off-chip communication is high, getting data onto the chip and operating on it in the register set is a cost–effective strategy. An as alternative to the software register allocator, the RISC project provides the hardware for a register stack [13].

We observed, like the 801 project, that a simplified instruction set can expose all internal machine cycles and states at the instruction set level. In contrast, a more complex machine hides many internal cycles within a single instruction. With modern compiler technology, exposing all machine operations and making them subject to optimization provides obvious benefits. The simpler instructions also have consistent running times so that the code generator can more easily choose between candidate implementations of a function.

In MIPS, we decided to attempt to expose all the internal implementation details affecting performance. Primary among these implementation issues is the pipeline structure. In choosing to expose this structure, we also decided that the synchronization function performed by pipeline interlocks could be moved into the software thus substantially simplifying of the hardware and losing little or no performance [9]. Therefore, the MIPS hardware architecture was defined without pipeline interlocks and with delayed branches (see 4.2).

Performance in executing a single sequential instruction stream comes from the ability to execute portions of that stream in parallel. This parallelism can be obtained from both pipelining and using of multiple function units. The internal micromachine of a pipelined CPU is inherently parallel. One goal that we imposed was to exploit internal microengine parallelism, to the degree that it makes sense, by projecting it into the instruction set. VLSI technology reinforces this goal; on-chip parallelism is cheap compared to off-chip, sequential communication. Thus, the MIPS instruction set is an attempt to carefully blend the instruction set requirements with the capabilities of a high–performance, pipelined, VLSI microengine.

Many of our goals required that implementation–dependent features, such as the pipeline length and the parallel activity within instructions, be visible in the instruction set. This can lead to significant complexities in the user's view of the machine and in the construction of code generators. It also makes it difficult to alter the instruction set design specification during the actual hardware design, which we wanted to be free to do. Our implementation dependent optimization includes three primary code improvements:

1. reordering instructions to avoid pipeline dependencies,
2. reordering instructions so that the instruction following a branch can always be executed, and
3. packing together separate, simple instructions (called instruction pieces) into a single parallel instruction word.

These optimizations, similar to those performed by a microcode compactor [3, 6], can be performed with a high degree of success in the final phase of assembling instructions. Thus, we decided to define the instruction set at two different levels. The first level, called the user–level or assembly language instruction set, defines instructions that are unpacked and have no pipeline dependencies or branch delays. The assembly language instruction set is comprehensible and easy to generate code for. The machine-level instruction set is the low level instruction set actually run by the processor. This instruction set is generated only by a single program, the *reorganizer*. The reorganizer does all implementation–dependent optimization, and isolates the user level instruction set from these implementation details.

A related goal of the actual implementation process was to simplify and regularize the hardware wherever possible, subject to reasonable performance constraints. The unavoidable irregularities in the processor design often became software responsibilities. This shifting of responsibility included not only management of the pipeline but also:

• saving and restoring the pipe contents during a fault or interrupt,
• restoring operands after arithmetic overflows,
• specifying all program counter related operations (e.g. saving the PC on procedure call and relative jumps) so as to simplify the actual hardware implementation rather than conform to a predefined software model.

This division of the architectural definition into two parts and the freedom of giving the software responsibility for irregularities in the organization increased performance significantly. This strategy also simplified the processor design and thereby made the implementation of the MIPS chip possible with limited manpower and tools.

3 VLSI Implementation

The MIPS chip is a monolithic VLSI implementation of the MIPS processor architecture. The chip is implemented in standard, one-level metal nMOS using Mead-Conway design rules with buried contacts. The total dimensions are 3750λ by 4220λ. With $\lambda = 2\mu m$, we expect to run with a clock period of 250ns (4 MHz clock); this will give an execution rate of two million instructions per second.

The MIPS chip consists of four logical blocks. The 32-bit data path contains two bidirectional buses linking a fast ALU, a barrel shifter, a file of 16 registers, a complex program counter, and an address masking unit. The data path is controlled by an encoded control bus that distributes register transfer commands to the data path. The Instruction Decode Unit latches and interprets all the components of the instruction in parallel and drives the appropriate control information onto the control bus at the appropriate time. The Master Pipeline Control communicates with the outside world, responds to all internal and external exceptional conditions, and controls the basic sequencing of the pipeline. Figure 1 shows the major parts of the chip.

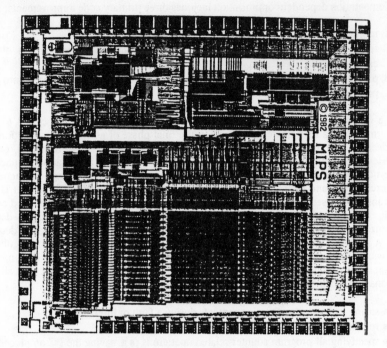

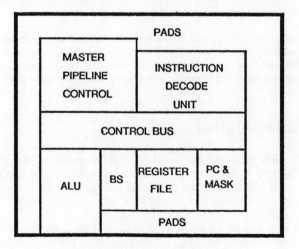

Figure 1:   MIPS layout and floorplan

Maximum performance of the MIPS machine requires a high–bandwidth memory interface. MIPS memory is word-addressable and all memory references are 32 bits wide. Separate address and data pins permit partial overlapping of memory references, and also eliminate delays often incurred in multiplexing the address and data pins. To execute at full speed, the instruction access time is bounded by 250ns. A simple read-only instruction cache is possible, since the architecture forbids writes into the instruction stream. Data memory may be slower, with a maximum access time of 450ns.

The bulk of the implementation was done by three designers over the course of about 18 months. These designers were also involved in other significant efforts on design aids and architectural refinements. They were assisted occasionally by other logic and circuit designers. The total time for logic and circuit design and layout for the MIPS processor and the test chips has been approximately 2.3 man-years; the chronology and effort expenditures are given in Section 5.1.

3.1 Timing Methodology

The timing methodology for MIPS uses a two-phase, non-overlapping clock without precharging. Both the data bus and the control bus are potentially active during each clock phase of every cycle. The control bus is actively driven on both cycles by the Instruction Decode Unit. Each data bus writer incorporates bus pulldowns, but the bus itself includes the depletion pullups. Although precharging is normally a powerful tool, it does not mesh well with high bus utilization and overlapped execution of the MIPS pipeline. A simple non-overlapping precharge scheme would require a doubling of the clock rate and possibly an increase in clock skew overhead. A more intelligent precharge scheme with overlap between the precharge and the following bus transfer would be more difficult to design. Either precharge technique is susceptible to erroneous bus discharge. The MIPS method offers good performance and greater tolerance to timing errors, for the cycles can be extended to overcome bus discharges at the beginning of any transfer.

Similarly, the functional units, e.g. the ALU, are not precharged. Instead, the ALU amortizes its longer operation time as compared to the bus transfer time, by working across clock phases. During phase one, the ALU sources are driven from the register file to the ALU and the ALU begins operation; at the end of phase one, the ALU inputs are detached from the bus. During phase two, the ALU completes its cycle and drives the result out on the bus. Its effective execution period runs from the middle of phase one to the beginning of phase two, including the non-overlap period between the clock phases.

Each instruction passes through five pipeline stages: Instruction Fetch (IF), Instruction Decode (ID), Operand Decode (OD), Operand Store/Execution (SX), and Operand Fetch (OF). The pipeline is fully synchronous and contains three active instructions; thus, only two sets of stages can be active simultaneously: IF with OD and OF, or ID with SX. In the absence of unexpected external events, e.g., cache misses and page faults, the processor will simply toggle between two states. Each stage requires one full clock cycle, so each instruction completes in five cycles (really three pairs of stages). In Figure 2, the activities occurring for a sequence of three instructions are shown; we assume that all instructions are base-offset loads combined with ALU instructions.

3.2 Data path

The relatively complex and long data path requires metal data buses to achieve any reasonable level of performance. Polysilicon is relegated to the control lines crossing the width of the data path. Register transfers thus become dominated by two effects. First, high

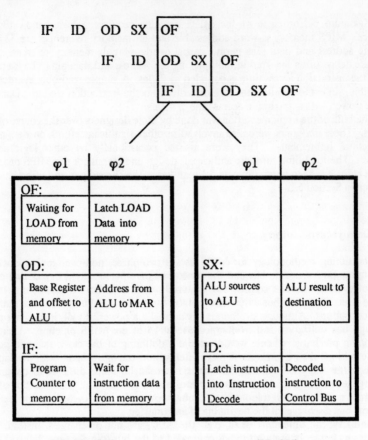

Figure 2: Activity in pipeline stages

polysilicon sheet resistance makes the control signal delay almost completely diffusion-limited; a narrow cell pitch is critical. We have achieved a pitch of 33λ with few performance compromises. Second, without precharging, the transfer of a selected value can be limited by the rise time of the bus. This potential delay is combatted by placing the driving pullups on the bus, rather than in the source data path cell. With the data and source select pulldowns in the cell, the bus then forms a distributed AND-OR-INVERT gate. When the bus becomes deselected between cycles, it rises to an intermediate or high level.

We recognized early in the design that the ALU lay in the critical path for many operations. We decided to devote considerable design time, area and power to maximizing the performance of this unit. The arithmetic and logical function blocks are separated to minimize loading on the adder. We implemented a carry-lookahead tree, with propagate and generate signals computed in pairs. This allows a total ALU add time of under 100ns. The ALU is also responsible for the detection of fourteen different conditions needed for conditional branch and set operations.

Much of the data path design was dictated by the need to optimize the add time. Since the ALU needed a fast carry lookahead and we decided to use a narrow pitch, the layout forced the ALU to be at the end of the data path so that the buses did not need to pass through the PG-tree in metal. This forced the barrel shifter into the middle of the data path.

Adequate support for multiplication and division is an important MIPS objective. However, full multiply and divide instructions are not consistent with our objective of single–cycle execution on all instructions, and the limited silicon area available. We implemented a modified Booth's algorithm for multiplication and division by way of a pair of special registers integrated into the ALU (H and L). These permit multiplication at a rate of two bits per ALU operation (four bits per full instruction) and division at a rate of one bit per ALU operation. A two-bit Booth multiply step requires both a shift operation and an add operation. To complete these within the same time slot as a standard register-register add requires that the H/L registers be closely integrated with the ALU and that a special nonbus communication path be used. The ALU has been extended to 34 bits to support this two–bit shift and add operation.

The barrel shifter is used for the full complement of rotates and logical and arithmetic shifts, plus character insertion and extraction. This variety of functions is controlled by an input multiplexer, which selects the data to go to each word of a two-word combined rotator. The shift amount determines which 32–bit section from the 64-bit concatenation of the two buses goes to the output. The combined rotator is implemented as a pair of cascaded shifters – the first by the *shift amount* div 4, the second by the *shift amount* mod 4. We chose this implementation in the face of severe cell–pitch constraints and the need to drive one bidirectional bus through the shifter. The barrel shifter lies between the registers and the ALU. During ALU operations, two operands must get past the shifter on phase one with the result returning on phase two. This was achieved in a pitch of 33 λ by including one normal bidirectional path (A Bus), and using an internal one-directional bus through the shifter for the other path to the ALU.

Our principal goals for the register array were small size and fast register transfer speed for all operations. Any two cells may be read onto either bus on either phase, or two cells may be written from either bus on phase two. The bus methodology described earlier allows the register pullups to be of very modest strength, so that the power consumption is kept to a minimum. The register array contains both the sixteen general purpose registers and that portion of the Surprise Register that holds the trap code.

The variety of program counter operations and the requirement for instruction restart made the program counter a major challenge. The program counter must hold the current value, three previous values for pipeline restart and one possible future value for branching. On every cycle, one of six possible sources must be selected for the new value: increment, self-refresh, zero (for interrupt service), the branch value, or values from either of the data buses. Simultaneously, the old value must be shifted into the buffer of old values. When an exception occurs, these values are saved by the service routine and restored on return. The complexity of the program counter structure and its central role in the processor put it directly in the critical path. Because of these speed requirements and full utilization of the ALU by user instructions, a separate PC-incrementer with a simple carry-lookahead scheme was used. The pitch constraint of 33λ was extremely severe in this unit, especially where the program counter and address masking functions overlap.

The address masking primitives are integrated into the program counter structure along with the Memory Address Register. This masking allows a machine address to be converted to a process virtual address. Thus, the first level of memory mapping (segmentation) is provided by the chip; the external memory map need only supply a one–level page map to implement a full segmented and paged system. The use of VLSI made the on-chip segmentation relatively straightforward and decreased the off-chip hardware requirement with no significant performance impact.

The size of the process virtual address space is defined by a bit mask in a special register. When masking is enabled for normal operations, the high order bits of the machine address are substituted with a process identifier from another special register. These two special registers are accessible only to processes running in supervisor state. The masking unit also detects attempts to access outside the legal segment and raises an exception to the master pipeline control. Though the virtual address is a full 32-bits, the package constraint only permits 24 address pins. With word addressing, however, this gives an address space of 64 megabytes.

3.3 Control Bus

The control bus encodes both register sources and destination for each type of operation. It includes a set of true/complement pairs for the ALU sources or memory reference sources, and for the ALU destination or memory load destination. The control bus also includes a collection of special signals including the pipestage, the branch condition results, and various exceptional conditions.

Each data path control driver taps this bus with one or more NOR decoders and latches the decoded value into a driver in the phase immediately preceding its active use in the data path. These signals are driven via dynamic bootstrap drivers by the appropriate clock. The bootstrap drivers put a considerable load on the clocks, but great care has been taken to minimize skew by conservative routing of clocks in metal. In a few cases, bootstrap drivers cannot be used because the control signal may need to be active on both clock phases, e.g., in register reads. These drivers are implemented as large static superbuffers. We have recently discovered that both the dynamic and static drivers may be heftier than necessary because the propagation delay time through the polysilicon control lines appears to be limited solely by diffusion delay.

3.4 Instruction Decode Unit

The Instruction Decode Unit (IDU) latches each instruction and translates the instruction pieces into appropriate control signals for the data path. The instruction word is latched during phase one of the Instruction Decode cycle, and the outputs from the IDU are latched on phase two and distributed to the control bus. The design and encoding of the instruction set is carefully tailored to allow a natural decomposition of the instruction word. Each type of instruction field may appear in one and only one position within the word. Thus, decoding may be done in parallel by two independent PLA's, providing a faster total decode time.

The Load-Store-Branch PLA decodes those fields of the instruction that may contain opcode fields for memory operations or program counter operations. It generates encoded register transfer sources and destinations for the appropriate execution cycles of this instruction. Those signals may be placed immediately on the control bus for the Operand Decode cycle or delayed up to one and a half clock periods until they are needed in following cycles.

The second PLA decodes ALU instruction pieces for the Operand Decode (OD) cycle and for the Operand Store/Execute (SX) cycle. Rather than decoding both pieces simultaneously and delaying the SX output for a cycle, the ALU PLA decodes these two pieces in sequence. The instruction fields for the second ALU piece are held for a cycle and then applied to the same PLA. These latter results appear just before the SX cycle, instead of before the OD cycle. This technique doubles the effective throughput of the PLA.

Both PLA's also contain a compact mechanism for the definition of register fields. Some instruction pieces have implied operands, e.g., multiply step uses the special H-L register, but most instructions require explicit general–purpose registers. Substitution of explicit fields for implied ones is controlled by an extra set of OR lines for each register field in the ALU and Load-Store-Branch PLAs. If the explicit register specifier is indicated, a simple multiplexer on the output of the PLA selects these extra OR lines, rather than the implied operand lines. This field substitution mechanism significantly reduces the size of the IDU.

A small instruction class PLA, operating in parallel with the decode PLA's, selects between the outputs of the two decode PLA's. This selection is based only on the few bits which determine one of four possible instruction piece combinations:

1. Load (long offset)
2. Load (short offset) + ALU2
3. ALU3 + ALU2
4. All conditional instructions

This approach allows the complete instruction decode to occur in one cycle, except for the SX field, which is not needed until one cycle later.

The outputs of the PLA's are linked by a Level Sensitive Scan Design shifter. This LSSD chain allows the function of the PLA's and the data path to be confirmed independently. The heavily utilization of all processor resources mandates this kind of fine control of individual clock cycles for testing purposes. The inputs to the control bus processor can be read or written by the LSSD chain. To test the entire data path, only the bus connections need be working, since complete control of the data path can be obtained via the LSSD chain. This limited use of LSSD represents a good compromise. Little area is lost in the LSSD shift register, compared to a scheme that shifts through all the storage in the processor. However, this limited use of LSSD seems to have most of the advantages of a more complete scheme.

3.5 Master Pipeline Control

The difficult design of the Master Pipeline Control (MPC) reflects the complex control problem of interfacing a pipelined processor to an unpredictable environment. Under normal circumstances, the MPC must control the pipeline execution with cache misses and memory wait states that cause the pipeline to be suspended. It must also deal with a host of potential "surprises" that may occur, including arithmetic overflow, privilege violation, internal masking error, illegal instruction, page fault, interrupt, bus error and reset. The processor must not only respond, but also save the processor state accurately for instruction restart. Finally, the MPC is responsible for generating and sensing all control signals to the pins.

The core of the MPC is the fundamental control engine of the processor: a 16-state finite state machine implemented in a PLA with 46 min-terms. Its inputs include indicators of cache hit, memory ready, and encoded exception information. Its outputs indicate the current machine state in two forms: a standard encoding of the 16 states, and special decoded versions of the same information. These decoded versions control the disabling of register writes and the modification of the program counter on exceptions; these latter signals are needed to control the execution of the next pipestage.

This PLA is surrounded by the surprise state register, and a special PLA for sensing and encoding all the different exception cases, four blocks of random logic. The first block forms the link to the IDU. It creates qualified clocks for instruction decoding and masks illegal instruction indicators coming from the IDU. The second block senses masking errors and sends special signals to the program counter for interrupt service. The third block detects ALU overflow conditions and comparison results. The last logic block forms the interface to the chip's control pins; this interface block qualifies the signals and governs the timing. Our LSSD chain extends through the MPC to pick up all the critical PLA outputs.

The Master Pipeline Control turned out to be unexpectedly difficult to implement. It is both slower and larger than originally planned. The complexity of its logical design has also left little time to fully optimize its physical implementation.

3.6 The Test Chips

During Autumn 1981, we decided to submit the pieces of the MIPS processor as test chips. The six major pieces of the design for which we submitted test chips are the ALU, the barrel shifter, the PC unit, the register file, the IDU, and the MPC. We created a test frame consisting of the control bus and a data path bus interface to allow the frame to fit in a 64-pin package. The test frame was general enough to accommodate all the test pieces. The major aims of the test chips were:

1. To get an early check on the individual pieces of the chip before assembling a complete processor. We were (and still are) afraid of getting back a large chip that is fabricated correctly but does not work and cannot be diagnosed.
2. To obtain some early performance measurements. Our goal was calibration of performance estimates and identification of unforseen critical paths.

In addition, we believed that the test chips could be layed out and debugged in parallel with other design activities.

In reality, what we learned from the test chips was very different. We did uncover several bugs in our individual designs, but we were never able to do performance testing. In addition, we discovered the following startling facts:

1. Yield could be a potentially serious problem, at least for the MOSIS runs. For our first fabrication of the register file, we received no working chips from a batch of ten.
2. Complete integration and simulation of parts takes longer than expected.
3. We had to address testing fairly early on. The ICTEST system [16] provided an unified simulation and testing environment that simplified our task and encouraged us to do more complete simulation.
4. Our static checks were not sufficiently comprehensive and hence some designs had ratio errors. We have worked on fixing the static checker so that it does not overlook any potential problems.
5. Many of our tools broke on the test chips, which was a blessing in disguise.
6. The power consumption of the test chips exceeded our estimate by 50%. We have revised the power budget for the full chip accordingly.

Thus, while the test chips took longer than we expected and did not produce the results we desired, they helped us face up to some of the most demanding parts of the project early on. The success of this approach is discussed in more detail in Section 5.2.

3.7 The Package Constraint

We have chosen an eighty-four pin chip carrier for the MIPS chip. This package imposes two constraints on the VLSI implementation: a careful allocation of pins in the external interface and a power budget of roughly two watts. Separate address and data pins are crucial to achieving a high memory bandwidth. A full assortment of control and status pins are necessary to support the full range of faults, interrupts, and status information. The pins may be summarized as follows:

Function	# Pins		Function	# Pins
Address	24		Data I/O	32
Vdd	3		Gnd	3
Substrate	1		Clocks	2
LSSD	4		System Status In	7
MIPS Status Out	8		Total	84

Since the eighty-four pin chip carrier also imposes a power limit of about two watts, we have had to allocate our power budget carefully. Using a (supposedly) conservative 0.1 mA per inverse square of pullup, we arrived at a total of 311 mA for the power budget in Figure 3.

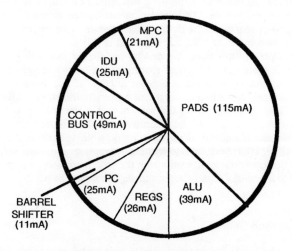

Figure 3: MIPS power budget

Unfortunately, this supposedly conservative estimate is not consistent with our test chip experience. Two different designs, the Program Counter and the Register File test chips from different fab runs (M25KE1,M25HH1,M25KD1), consistently dissipated 1.5 times the expected power. This suggests the pullup current is typically closer to 0.15 mA per inverse square. In this case, we would expect operating current of 470 mA and power of 2.3 W. All these figures represent typical values under nominal temperature and voltage conditions. MIPS appears to use all the power budget available.

4 Limits to Performance

Performance constraints ideally come from the implementation technology and the software environment. Practically, manpower and design tool availability are also important in limiting the performance obtainable in a VLSI design. The first part of this section describes two limitations imposed by the available design tools. The rest of this section describes the interplay of architectural and implementation technology constraints in MIPS. This interaction is demonstrated by examining three performance-enhancing changes that we considered in the the design of the chip. All three of these changes could have been adopted into the architecture with alterations only in the reorganizer and machine-level architectural specification.

4.1 Limitations of Available Design Tools

One limitation of our design environment was the difficulty of identifying critical paths in the processor. We knew that the propagation delay through the ALU should be on the critical path, but we didn't know during layout whether it would be, or what the other delays along this path would be. Delays within the data path were easy to estimate via circuit simulations of bit slices. The control structure, however, is irregular in comparison with the data path and cannot be simulated in slices. Circuit simulation of the whole control structure was impractical due to its size, and partitioning was impactical due to its complexity. During layout of the control sections, we relied mainly on our limited intuition for transistor sizing. This guessing led to queasiness in the designers and the development of TV. TV is a timing analyzer begun in the summer of 1982; it is described in [12].

Another example of limitations imposed by our design tools can be seen in the final layout of the MIPS processor. The global routing in the control area takes up significantly more space then we initially estimated. A major fraction of this extra area comes from an initially poor placement of blocks and the use of a (sometimes) space-inefficient but simple-to-implement routing scheme. Given the existence of a reasonable automatic router, we could have more carefully optimized the placement and allowed the router to do the bulk of the work; iterating the placement would also be possible because redoing the placement involves little designer effort compared to redoing the routing. Alternatively, we could spend several weeks of designer time hand–optimizing the placement and routing. For the present, we have decided to live with a slightly larger and slightly slower chip.

4.2 Reduced Branch Delay

In a pipelined processor, the instruction fetch unit fetches the next sequential instruction before the previous instruction has completed execution. If the instruction in execution is a branch and if the branch is taken, then the fetched instruction is not the next instruction to be executed. Furthermore, if the pipe is deeper than two instructions, the instruction following the branch can have already begun execution. In most pipelined architectures, this effect is handled by the hardware: if the branch succeeds, any instructions sequentially following the branch are backed-out (if needed) and thrown away; the instruction fetch unit then gets the instructions from the branch destination and execution continues. Because of the high frequency of branches and their aggravating effects on pipelines, various schemes such as branch prediction are used to mitigate the effect.

A simple alternative to complicating the instruction pipeline or implementing a prediction scheme is to have the instructions that are already fetched executed regardless of the effect of the branch. This scheme is called *delayed branching*; a delayed branch of length n means that the n sequential instructions after the branch are always executed. The compiler is given the task of making those instructions safe to execute, and, whenever possible, using the instructions to shorten the execution time of the program [7]. The data shown later in the section shows that this problem can be successfully solved when the branch delay is short.

In the original design of the architecture, MIPS had a uniform branch delay of two for all branches. The ALU computation currently performed in SX was instead performed in a pipestage following OF called EX. This made it easy to synthesize instructions of the form "add memory to register" by combining a load/store piece and an ALU piece. However, it increases the branch delay for all conditional and unconditional relative and absolute branches by one (i.e., from one to two). These types of branches typically account for 18.9% of all instructions executed ($\sigma = 7.1\%$). Table 1 shows that for a uniform branch delay of two, the average number of branch slots filled would go down to 66.7% (of two instructions) from 90% (of one instruction). A branch delay of length two would increase execution time of programs by an average of 14.3% (0.86 slots empty/branch, 18.9% branches).

Program Name	% Filled Branch Slots for branch delay		
	1	2	3
Fibonacci	90.0	65.0	56.6
Hanoi	85.7	50.0	38.0
Puzzle I	95.0	79.9	72.2
Puzzle II	75.9	56.5	48.9
Puzzle III	97.0	76.2	67.8
Queens	96.5	72.4	59.7
Average	90.0	66.7	57.2

Table 1: Utilization of branch delay lengths 1, 2, and 3

The costs of moving to a branch delay of one were nontrivial. First, it placed much tighter requirements on page–fault detection. In both cases, the memory address leaves the chip at the end of the SX pipestage. In the original design, detection of a page fault is not required until one and a half pipestages later, when the destination of a load instruction is written. In the reduced branch delay design, a page fault must be detected before the ALU result is written only half a pipestage later. Faster virtual memory mapping hardware was investigated and found to be implementable with current technology. Also, two instruction pieces synthesizing a memory–to–register instruction could no longer be packed into the same instruction, but now needed to be reordered into separate instructions. This sometimes increases code size if other instructions cannot be packed with the first piece and the second pieces, but it is not significant in practice. The suggestion for moving up the EX cycle to allow branch delay of one was originally suggested by Jud Leonard in the early fall of 1981 and adopted shortly thereafter.

4.3 Bypassing

The original architectural specification left undefined whether the result of the ALU computation in the SX pipestage could be stored to main memory in the same pipestage by using the same register for the ALU destination and the store source. Spice simulations showed the time required to write the ALU computation through a register to the memory interface would add an extra delay of about 40ns; this delay occurs because the write-through might need to recharge a bus discharged by the original register value. This extra time would be required in every pipestage, whether there was a write-through or not. Thus, allowing write-through would increase the length of each processor cycle by approximately 16%. Bypassing was proposed as an alternative; to bypass, the result of the ALU would be written to both the A and B bus of the data path, and the write-through of the register to the B bus on its way to the storage interface would be disabled. Calculations showed this could be done for an increase in the complexity of the control and a lengthening of each pipestage by about 20ns, or a penalty of about 8% of every cycle.

This left us with three design alternatives: write-through without bypassing, bypassing with disabling of write-through, and redefinition of write-through as illegal in the architecture with enforcement provided by the pipeline reorganizer. If no other pieces were available for packing, this last alternative could decrease code density, and thus increase execution time, by forcing two pieces previously packable into one instruction into two separate instructions. We define the *packing rate* to be the number of instruction pieces per instruction word. To choose among these alternatives, the pipeline reorganizer and MIPS simulator were used to take data on the usefulness of bypassing. The results of these measurements appear in Table 2. Based on this data, we conclude that the improvement in execution speed obtained by adding bypassing to the MIPS architecture is small. We compared the costs of including either write-through or bypassing, and chose to prohibit storing of a register into memory and modifying it by an ALU operation in the same cycle.

Program Name	Packing rate	Improvement with Bypassing		
		Density	Size	Time
Fibonacci	3.5 %	4.6 %	4.5 %	4.5 %
Puzzle I	23.2 %	16.4 %	12.9 %	5.6 %
Puzzle II	4.2 %	0.7 %	0.6 %	0.3 %
Puzzle III	3.0 %	0.8 %	0.8 %	0.5 %
Queens	6.8 %	5.4 %	5.0 %	5.2 %
Avg.	8.1 %	5.6 %	4.8 %	3.2 %

Table 2: Evaluation of bypassing

4.4 Increased Pipelining

Figure 4 shows the basic timing within one pipestage. Although the bus is heavily used, additional bus bandwidth is available. First, little use of the bus is made during an ALU computation, as well as during the set-up time for the next clock phase. During most of this time, the bus pullups are "precharging" the bus in preparation for the next bus transfer. This precharging could be more effectively done by use of a precharging circuit triggered by the rising edge of each clock phase. This would free up the buses for approximately 50% of each pipestage. This bus time could be utilized by overlapping bus transfers with ALU usage. If the 70ns for next state determination of the Master Pipeline Control were unchanged, this could reduce the cycle time per pipestage from 250ns to 170ns. This overlapping would increase the execution speed of the processor by about 50%. Unfortunately, this overlapping would place severe demands on the rest of the processor. If the same instruction cache and instruction decoding hardware were used, another pipestage would need to be added between IF and ID, to provide the same time for instruction fetch and decode. This would increase branch delays by one, or two for most branches and by three for indirect branches (used on procedure return). This would take away 15 to 20 percent of the added performance (according to calculations similar to those made in Section 4.2). A similar reduction in time available for data references would best be handled by the addition of a data cache.

Other problems with this approach arise in its VLSI implementation. The Master Pipeline Control hardware would be significantly increased in complexity. This hardware is currently a critical path in the machine, so increasing its complexity and requiring it to operate 50% faster could prove extremely difficult. Much of the delays in the control section are caused by large capacitive loads of global control signals, so it might be possible to attain a speed up by increasing the size of control line drivers, although this could cause problems with the overall power budget. Also, this organization would require a three bus design, because the previous ALU or Barrel Shifter result might have to be substituted for either

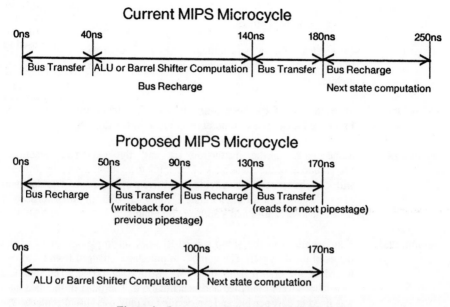

Figure 4: Increased microengine pipelining

operand.   Not allowing computations to be available in the next pipestage would be unacceptable due to severe degradation of performance and code density, especially for heavily arithmetic computations, e.g. full-word multiplies and divides. Using a three bus data path would increase pitch by approximately 21%, the area of the chip by about 7%, and would also increase bus transfer times by about 10ns due to increased diffusion delay in lengthened polysilicon control lines crossing the data path.   This is a penalty of 13% if we assume a transfer time of 40ns and 25ns for precharging.

Without better cache technology and lower polysilicon resistance, this improved design immediately loses over half of its potential performance increase. The impact on the Master Pipeline Control is hard to estimate; increased demands on it could remove any remaining performance advantage, as well as further increase the size of the chip, causing additional delays due to critical propagation times in the master control lines. The increased complexity of this design would also require a larger design team and more design time due to its substantially more complex logical structure. Lastly, this idea was only proposed after much of the grueling layout for the two bus design had been completed and we had long passed the point of making such a major design change.

5 Lessons, Successes, and Failures

In this section we discuss some of our experiences in designing a large and architecturally novel intergated circuit within a university environment. While some of our experiences may be peculiar to the strange manner in which large projects are attempted at universities, we believe that many of our experiences will be common to any large, complex, and evolving IC design project.

5.1 Chronology

The MIPS project began in earnest two years ago. The following timetable shows some of the milestones of the development process:

Winter 1981   First ideas for a streamlined processor, discussions of a VLSI implementation. Initial thoughts about a pipelined, compiler-supported architecture.

Spring 1981   Development of the basic foundation of the architecture by the VLSI Processor Design class. Proposal of an initial instruction set.

Summer 1981   Stabilization of the instruction set and the data path resources. Development of the framework of the implementation: a two bus structure with each bus potentially carrying data twice per pipe stage. Development of a code generator for the portable C compiler. Start of initial versions of the Code Reorganizer and a Software Simulator.

Autumn 1981   Four members of the group started to work on implementations of the pieces of the data path. The number of pipestages changed from six to five to reduce the branch delay from two instructions to one. The final few instructions were examined and benchmarked. The pitch of the data path was fixed at 33λ per bit, and we decided to only use buried contacts. We investigated code reorganization algorithms. An ISP description of the processor was completed in December 1981.

Winter 1982   The first of the test chips was sent for fabrication. It included the register file and its control bus decoders. Serious work began on specifying the control portion of the processor. Bypassing of ALU results was rejected. The instruction set stopped changing.

Spring 1982   Completion of the program counter test chip. Design of the instruction decode unit. First circulation of the Master Pipeline Control document. A U-Code to MIPS assembly language code generator was started to provide an optimizing Pascal compiler for MIPS. The Reorganizer was rewritten to include all desired optimizations.

Summer 1982   Submission of the instruction decode test chip and the barrel shifter test chip; implementation of the Master Pipeline Control (MPC) unit.

Autumn 1982   Submission of the MPC test chip, completion of the ALU test chip. Assembly of the 6 test chips into a processor began. The Timing Verifier was developed and the Pascal compiler was producing high quality MIPS assembly language. By Christmas 1982, chip assembly was completed and simulation of the entire processor began.

Figure 5 shows how the available manpower was divided over the various tasks during the entire design process by showing the tenths of person-years for the first and second years of the project as well as the total. The total amount of effort that has been expended is about

6.1 person–years. The figures show that the project manpower grew from the first to second years as the implementation became the dominant concern. What does *not* show in these figures is that a total of 15 people have at one time or another been working on the main design.

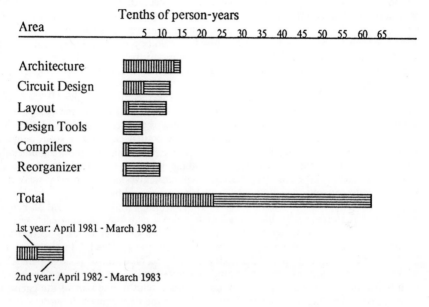

Figure 5:   Effort distribution in the MIPS design

5.2 Successes

We strongly believe that instruction set selection on the basis of hard data from compiled code was an important and successful component of our methodology. The fact that we had a C compiler targetted to a MIPS simulator very early in the design allowed us to make design decisions based on data that was directly relevant to the architecture. Many of the last instructions and features considered were borderline in terms of benefits versus hardware cost. Being able to quantify the benefits made the decisions much more straightforward and provided the impetus to resist constant additions to the architecture.

The specification of an assembly language interface that is distinct from the machine language was also a great advantage. Not only were our early compiler efforts isolated to a certain extent from the turmoil of a changing architecture, but the hardware designers were freer to change things with the knowledge that most software changes would be limited to the reorganizer.   Another advantage is that different versions or implementations of the architecture need not behave exactly alike. Only the reorganizer need be changed from one implementation to the next. This approach saved us considerable work when we realized that the PC-saving instructions, used for procedure call, could not save the value of the PC that we had initially intended because the PC was incremented at that point in the pipe. Because this instruction was encapsulated by a macro-level call instruction, the required change to the architecture and software was trivial.

One goal throughout the MIPS design was to simplify the hardware implementation whenever possible. This meant keeping the VLSI implementation regular at the cost of sometimes complicating the software. Also, the streamlined instruction set philosophy and our goal of simplicity in the implementation drove us to eliminate all state not visible from the instruction set. Eliminating internal state meant that complex state savings sequences and information flows to save internal state were not needed. Despite our best efforts, one nonorthogonal feature crept into the design. To perform Booth multiplication two bits at a time, one extra bit, which extends HL to 65 bits, is required. We call this bit the D-bit, for obvious reasons. This single bit of state caused the addition of several very irregular one-bit data paths as well as additional critical timing paths and many headaches in the layout and documentation. The D-bit serves to remind us how valuable simplicity and regularity are in VLSI implementations.

The experience of the RISC chip [14] and the Geometry Engine [2] shows that if performance is not carefully and comprehensively considered during the design, then the results are bound to be disappointing. Our attack on this problem was two-fold. First, we did extensive Spice simulations of a few obvious potential bottlenecks early on in the design cycle. This was to verify the attainable performance range. Second, and more significantly, we undertook to build a timing analyzer, TV (Timing Verifier), that would locate the critical paths in the completed circuit [12]. Several large capacitive loads driven by weak pullups, which might have easily gone unnoticed, have been uncovered by TV.

The test chips were a definite success. Despite disappointing yields and an inability to do thorough performance testing, the effort to place a pad frame around each section was quite worthwhile. Completing the test chip forced the designer of each section to fully simulate his section of the processor before it was incorporated into the final design. Furthermore, this simulation and debugging of each section could be done in parallel. If the pieces had been assembled without thorough separate simulation, this debugging would have been much more sequential. As in any large system, bugs in one part of the circuit might have caused misleading behavior in another, making the debugging still harder. If the test chips had not been assembled, the processor might have been ready for simulation perhaps four to six weeks earlier. In our estimation, this time is shorter than the length of the added debugging effort that would have been required if the pieces were assembled directly. Additionally, we uncovered several bugs in our design aids early. Had the design aids broken on the entire chip, we would have had a more difficult time tracking down these bugs and delayed our final design still further.

Another good decision was to use a layout language called SILT [4] as our chip assembly tool. Each of the six basic blocks of the processor was laid out with the graphical layout editor ICARUS [5]. After layout, blocks were placed and interconnected with SILT. This approach saved considerable effort in the assembly of the test chips because they were all quite similar and shared a lot of SILT code; without this approach, the test chips would have been prohibitively expensive to construct. Graphical editors are very convenient when a designer is laying out the leaf cells of a design; these cells are typically very random and tightly packed. However, modifying a large design can be very painful with a graphical editor, as are more global tasks such as block placement and routing. If a wire has been forgotten, a great many rectangles might have to be manually changed to separate two major functional blocks by 7λ.

In contrast, a procedural layout tool like SILT is very awkward for designing leaf cells because of an extended edit-compile-plot cycle, and because it is often very difficult to visualize what a long sequence of *place box* commands produces. SILT is very good at laying out geometry that can be algorithmically expressed. An example of this phase of layout is the requirement to "place the barrel shifter above the ALU". If the ALU should happen to change height and the SILT description was written with some care, the barrel shifter and all its connections should move appropriately. SILT also has a simple, built-in river router that provides a useful tool for connecting arrays of locations, such as buses, groups of control lines and even single wires in some cases.

SLIM [8] also proved to be a great labor saving tool in the design of the ten PLA's on the chip. It allowed each PLA to be abstracted from the level of Boolean equations to a higher functional description that could be written, read, and maintained with significantly less effort.

5.3 Failures

In any large project with many decisions, some of them will be wrong. In our case, many of the wrong decisions really represent a lack of decision. Specifically, a number of things continued to change after the layout had begun. Among the worst of these was the pitch of the data path. Due to changes in the support of buried and butting contacts, we decided late in Fall 1981 to banish butting contacts and use a data path pitch of 33λ. This forced a significant amount of layout effort to be repeated. While talented designers may be very effective in doing the layout of a tight data path cell, having such designers redo a layout is a waste of talent. The squeeze from 35λ to 33λ done during the redesign with buried contacts made the layout very difficult; the small performance improvements gained were probably not worth the effort expended in the design with such a narrow pitch.

The instruction set also was in a state of flux until quite late in the design. Minor changes were being made well into 1982, as the pieces of the data path were being designed. The best example is the *count leading zeros* instruction which was added late in Fall 1981, dropped around May 1982, then temporarily added again for about a week in June. It is useful to point out that this instruction was added and dropped based solely on its ability to be implemented by the hardware. Despite that fact that most of the late changes resulted in definite improvements (the change from 6 to 5 pipe stages is a good example), they inevitably caused repetition of effort.

Another area in which a lack of forethought, or rather lack of parallel thought, caused problems was in the external interface. Because all the designers were fairly busy, the details of the Master Pipeline Control had to be worked out after the architecture was firmly entrenched. That meant solving all the hard problems relating to privileges, page faults, and exceptions within a fixed framework. The result is a few minor quirks in the exception handling system that *may* have been avoidable if there were more flexibility in the structure of the mechanism. A prime example is the difficulty in returning from an interrupt with the lock-step, fixed length, pipeline. The problem arises because one needs to restart the three instructions that were in the pipe at the time of the exception using three successive indirect jumps to those three instructions. However, complications arise because the last indirect jump has to do its data fetch from system space after the first target instruction is being fetched from user space. If redesigning the existing program counter block had been an option at the time this problem was realized, a conceptually simpler mechanism might have been found.

As in many projects without rigorous enforcement of documentation standards, the current state of the design rested only in the heads of those most intimately involved with the project. This meant that our communication problems were larger than they should have been. In several instances, the same problem had to be thought out a second and third time because the answers were not written down the first time. The timing of the control buses was derived on at least three separate occasions before the Master Pipeline Control document was written.

The documentation problem was amplified by our manpower situation. A lot of the people that were working on the project were only peripherally involved or only involved for a quarter or two. They spent a great deal of time learning the basics of the processor. Furthermore, the core project team spent valuable time teaching new contributors the basic ins and outs of the design.

Worse still was a misunderstanding that arose on several occasions. Someone on the edge of the project would undertake to do a small piece of the processor without fully understanding the interfaces to the surrounding blocks or the timing requirements. Unable to find any hard documentation or some person who knew all the answers, the designer would, knowingly or otherwise, make assumptions, some of which inevitably turned out to be false. When the erroneous assumption was finally caught at the supposed completion of the task, significant redesign was often needed.

5.4 Lessons

Many of the lessons that we think we have learned are the same ones learned by VLSI designers, and engineers in general, time and time again. One problem that can be particularly serious in VLSI is synchronizing the design of various subsystems; one aspect of the design should not get way ahead the others. If manpower and time permit, it appears best to have all the portions enter circuit design and subsequent layout together. A more coordinated design might have saved much aggravation in the design of the control portions of the chip.

Many engineering projects are sparsely documented; this is especially troublesome when the work force is likely to change and the design is evolutionary. Documentation is a difficult and unexciting task, but it is seldom a waste of time.

The effort required to design a VLSI part makes it important to decide on as many things as possible before the drawing of transistors begins in earnest. Particularly when a number of people will be working on interacting parts, all aspects of these interactions, be they physical, electrical or temporal, should be at least considered, if not set in concrete, before either party proceeds. This is not to say that small sections should not be laid out on a trial basis during earlier stages of the design to get size, power and speed estimates, but rather that everybody should be aware that these initial efforts are likely to be made obsolete by future developments. If the high-level design can also be partially flexible to implementation-driven changes, the design will be dramtically eased.

The design tools being used on an ambitious design *will* break. Of course, the tools always break just before a deadline and just after the maintainer has gone on a long backpacking trip into the wilderness. We found some design aids that broke because the design was too big, but more often our design tools could not handle the less constrained design methodology that we used. This often produced unexpected results in tools believed to be very stable. In other cases, we found that our design could not be recognized as "standard" by the tools and was rejected outright (our bus-pullups caused this difficulty).

There comes a time in the design process when one can no longer afford to go back and change things. Changing things early is far less expensive than changing them later. A great deal of engineering time and aggravation can be saved by fixing the freeze date on the project, continually moving that date can cause constant redesign.

MIPS also taught us some things about designing processors in VLSI. First, high performance and Mead-Conway design techniques are not mutually exclusive for large designs. Given the right tools one can make large chips go as fast as similar small chips. However, to break into the 100ns clock period range requires circuit and analysis techniques that are beyond most Mead-Conway designs and designers. Self-timed precharging, dynamic logic, sense amps, and similar approaches are needed to make the most of the technology.

Adopting a streamlined instruction set has allowed us to experiment with pipelining as a media to attain high performance. Without the simplifications of a streamlined instruction set and the two-part definition of the architecture, an implementation would have been out the question. Both the practical size and the manageable complexity of our integrated circuit stem directly from these two ideas.

VLSI is a technology that is much more bottom-up driven than other implementation media. The advantages and disadvantages of alternative implementations are difficult to assess without actually carrying out the designs. Low-level layout problems encountered late in the design cycle sometimes force changes in the specification. To whatever extent possible, the noncritical design specifications should be alterable as the design progresses and unforseen problems demand change to the specification.

If at all possible, every aspect of the processor should be analyzed from the context of hardware costs and software benefits. Hardware resources are far from free, particularly within the context of a single–chip implementation. The costs in manpower, area, power and complexity must be compared with the performance improvement resulting from the feature.

Conclusions

Our biggest success was completing the design of a microprocessor that combines high performance with the correct handling of systems issues and external interrupts. We were able to realize this project within the bounds of our environment because we started with few preconceived notions about processor design and then considered design decisions without bias. Instruction set design can be turned into a scientific process where experimental evidence backs design decisions. This approach allowed us to increase preformance by implementing some unique hardware-software tradeoffs: functions with low utility but a high cost of realization in hardware are relegated to software where the problems are more efficiently handled.

As this paper goes to press, we are completing the final stages of simulation of the entire MIPS chip and expect to submit the processor for fabrication shortly, provided our tools and energy hold up for the next few weeks.

Acknowledgments

Many people have contributed to the MIPS project; especially worthy of note are John Gill, Forest Baskett, and Jud Leonard.

The MIPS project has been supported by the Defense Advanced Research Projects Agency under contract # MDA903-79-C-0680.

References

1. Chaitin, G.J., Auslander,M.A., Chandra,A.K., Cocke,J., Hopkins,M.E., Markstein,P.W. Register Allocation by Coloring. Research Report 8395, IBM Watson Research Center, 1981.

2. Clark, J. The Geometry Engine: A VLSI Geometry System for Graphics. Proc. SIGGRAPH '82, ACM, 1982.

3. Davidson, S., Landskov, D., Shriver, B.D., and Mallett, P.W. "Some Experiments in Local Microcode Compaction for Horizontal Machines." *IEEE Trans. on Computers C-30*, 7 (July 1981), 460 - 477.

4. Davis, T. and Clark, J. SILT: A VLSI Design Language. Technical Report 226, Computer Systems Laboratory, Stanford University, October, 1982.

5. Fairbairn, D. and Rowson, J. An Interactive Layout System. In *Introduction to VLSI Systems*, Mead, C. and Conway, L., Eds., Addison-Wesley, 1980, ch. 4.4, pp. 109-115.

6. Fisher,J.A. "Trace Scheduling: A Technique for Global Microcode Compaction." *IEEE Trans. on Computers C-30*, 7 (July 1981), 478-490.

7. Gross, T.R. and Hennessy, J.L. Optimizing Delayed Branches. Proceedings of Micro-15, IEEE, October, 1982, pp. 114-120.

8. Hennessy, J.L. A Language for Microcode Description and Simulation in VLSI. Proc. of the Second CalTech Conference on VLSI, California Institute of Technology, January, 1981, pp. 253-268.

9. Hennessy, J.L. and Gross, T.R. "Postpass Code Optimization of Pipeline Constraints." *ACM Trans. on Programming Languages and Systems* (1983). Accepted for publication.

10. Hennessy, J.L., Jouppi, N., Baskett, F., and Gill,J. MIPS: A VLSI Processor Architecture. Proc. CMU Conference on VLSI Systems and Computations, October, 1981, pp. 337-346.

11. Hennessy, J.L., Jouppi, N., Baskett, F., Gross, T.R., and Gill, J. Hardware/Software Tradeoffs for Increased Performance. Proc. SIGARCH/SIGPLAN Symposium on Architectural Support for Programming Languages and Operating Systems, ACM, Palo Alto, March, 1982, pp. 2 - 11.

12. Jouppi, N. TV: An nMOS Timing Analyzer. Proceedings 3rd CalTech Conference on VLSI, California Institute of Technology, March, 1983.

13. Patterson, D.A. and Sequin C.H. RISC-I: A Reduced Instruction Set VLSI Computer. Proc. of the Eighth Annual Symposium on Computer Architecture, Minneapolis, Minn., May, 1981, pp. 443 - 457.

14. Patterson, D.A. and Sequin, C.H. "A VLSI RISC." *Computer 15*, 9 (September 1982), 8-22.

15. Radin, G. The 801 Minicomputer. Proc. SIGARCH/SIGPLAN Symposium on Architectural Support for Programming Languages and Operating Systems, ACM, Palo Alto, March, 1982, pp. 39 - 47.

16. Watson, I., Newkirk, J., Mathews, R. and Boyle, D. ICTEST: A Unified System for Functional Testing and Simulation of Digital ICs. Proc. Cherry Hill International Test Conference, Philadelphia, 1982.

Fundamental Issues in the Electrical Design of VLSI Circuits

Daniel W. Dobberpuhl
Digital Equipment Corporation
Hudson, Massachusetts

The Mead-Conway approach to VLSI provides a path for non-specialists to enter the world of VLSI design. The approach condenses most of the more esoteric aspects of circuits and devices into simpler rules of thumb, which, when applied in the context of the prescribed methodology, should result in functional designs. In addition, the simplifications have caused some experienced designers to reshape their own techniques and methodology towards a more clearly focused kernel.

However, as new designers gain experience in VLSI, many become curious to know more about the underlying technology. They suspect that there may be more efficient approaches to certain configurations or they would like to know more about the accuracy and applicability of the rules of thumb in different circumstances.

This presentation will provide an introduction to the regions under the Mead-Conway surface. To provide an example of this subterranian culture we will explore the electrical design of a simple ratioed stage in some depth. We will also look at some potential rewards that may be realized from fine tuning of circuitry.

Circuit Timing

Crystal: A Timing Analyzer for nMOS VLSI Circuits
J.K. Ousterhout

TV: An nMOS Timing Verifier
N.P. Jouppi

Optimizing Synchronous Circuitry by Retiming
C.E. Leiserson, F.M. Rose and J.D. Saxe

Crystal: A Timing Analyzer for nMOS VLSI Circuits

John K. Ousterhout
Electrical Engineering and Computer Sciences
University of California
Berkeley, CA 94720

Abstract

Crystal is a timing analyzer for nMOS circuits designed in the Mead-Conway style. Based on the circuit extracted from a mask set, Crystal determines the length of each clock phase and pinpoints the longest paths. The analysis is independent of specific data values and uses critical path techniques along with a simple RC model of delays. Several additional techniques are used to improve the speed and accuracy of the program, including separate up/down timing, static level assignment, flow control for pass transistors, and precharging.

1. Introduction

As the density of integrated circuits increases into the hundreds of thousands of transistors, the intuitions of circuit designers become more and more fallible. The large number of components and their complex interrelationships make it almost impossible for designers to foresee all the consequences of each design decision. To decrease the likelihood of costly errors, designers depend on assistance from computer tools. Programs ranging from layout rule checkers to simulators are used to validate designers' intuitions and root out errors before they are cast in silicon.

This paper is concerned with one particular aspect of design validation: performance. Even if a circuit design is logically correct and free of layout errors, it will be of little or no use unless it can operate at an acceptable speed. For small designs, circuit simulators such as SPICE [6] can be used to verify performance. For large circuits, however, designers are left largely on their own, since circuit simulators become intolerably slow for even a few thousand transistors. Designers can extract pieces of a large design and simulate the pieces, but this depends on the designers' ability to select the right pieces. Experience with VLSI chips at U.C. Berkeley suggests that this is an error-prone approach [2,8]. The initial speed limitations were due not to major elements like ALU carry chains, but to seemingly trivial logic where small transistors were accidentally used to drive large loads. In one case [2] the errors were not discovered until after fabrication; in the other case [8] several such errors were discovered before fabrication by an early version of Crystal.

Crystal is a data-independent timing analysis program. Its input is a description of an nMOS circuit. The description is extracted from the mask layout and includes interconnect resistance and capacitance as well as transistor sizes and types. Crystal determines the length of each clock phase and identifies the slowest paths. The analysis is fast: one example circuit of 45000 transistors can be processed in about 20 minutes of CPU time on a VAX-11/780. However, since Crystal uses a very simple model of timing, it produces less accurate results than circuit simulators. Our goal is to achieve overall timing estimates within ±20% of the simulation results of SPICE.

Section 2 introduces the basic mechanism used by Crystal. In its simplest form, this approach produces grossly pessimistic timing estimates. Furthermore, the basic approach suffers from severe computational inefficiencies that make it impractical for any real circuits. Section 3 discusses these problems and describes several additional mechanisms used in Crystal to increase the efficiency of the program and the accuracy of its results. Section 4 presents the limits of the timing model, and Section 5 describes our experiences using the program.

2. The Basic Mechanism

The notion of timing analysis has existed for some time, and several timing analyzers have been described in the literature [1,4,5]. A timing analyzer can be thought of as a program that simulates a single clock cycle with all possible combinations of data values at the same time. In this way it computes worst case delays and verifies that the timing requirements of the circuit's various memory elements will be satisfied.

Most existing analyzers were designed for bipolar technologies such as ECL and TTL. MOS circuits, particularly those designed in the Mead-Conway style, differ in several important ways from bipolar circuits, and thus suggest a different approach to timing analysis. Whereas bipolar circuits tend to use complicated asynchronous clocking schemes, MOS circuits in the Mead-Conway style use simple, synchronous clocking mechanisms based on non-overlapping clock phases. Thus Crystal does not deal with minimum delay times or set-up and hold-times, which account for much of the complexity of bipolar timing anaylzers. Crystal's approach is simpler: for each clock phase, it merely determines how long it takes after the clock changes for all the effects of that change to propagate throughout the circuit. This results in a worst-case estimate for the length of the clock phase.

Crystal also differs from bipolar timing analyzers in its computation of the basic delays. In bipolar technologies the delay of a piece of logic is almost independent of the way the logic is used, and wire delays can be combined with logic delays in a simple fashion. The delay for each logic type is provided by the user. In MOS, the delay calculation is a complex function of the logic type, its geometry, and the way it is used in the circuit. The delay for one piece of logic may even be impacted by the logic that provides its inputs. This makes the basic delay calculations much more complex for MOS analyzers than for bipolar analyzers.

To compute how long it takes for a clock change to propagate, Crystal makes a recursive, depth-first pass over the circuit starting with the clock. At any given time, Crystal considers a change in value at a particular transistor gate, called the *driver* (see Figure 1). When the driver changes value, it may cause certain other transistor gates, called *targets*, to change value at later times. Crystal finds each target of the driver and

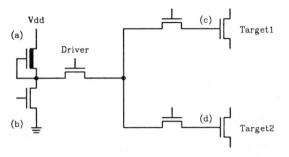

Figure 1. When it is discovered that the driver changes value at a particular time, Crystal firsts looks on one side of the driving transistor to find possible paths to Vdd and Ground at (a) and (b). Then it searches on the other side of the driving transistor for possible paths to other gates at (c) and (d). Delays are computed for each separate path between a target gate and Vdd or Ground (4 paths in this case).

computes the delay between driver and target. This information is then propagated recursively by treating the target as driver and figuring out which other gates it can affect. If a target is reached from several different drivers, then it isn't necessary to perform the recursive step unless the target's new delay time is the latest one seen for it.

Crystal finds the targets of a particular driver in two steps, as illustrated in Figure 1. First, starting at the source terminal of the driving transistor, Crystal searches through sources and drains of other transistors to find all paths from the driver source to Vdd or Ground. Crystal assumes that all transistors could conceivably be on. For each such path, it then searches from the drain of the driving transistor through sources and drains to all possible gates. Each of these gates is considered a target, and the path from the target through the driver to Vdd or Ground is used to compute the delay from driver to target.

To compute the delay, Crystal approximates each transistor with a resistance value. The resistance values were chosen based on SPICE simulations with expected processing parameters (see Table 1). Different resistance values are used depending on whether the target is being pulled to Vdd or Ground, and depending on whether the gate of the transistor is known to have a particular value (see Section 3). All the resistances and capacitances of transistors and nodes are summed along the path from target to Vdd or Ground, and the RC product is used as the delay from driver to target (load devices require special processing; see Section 3.1

Transistor Type	Ohms/square (pulling to 1)	Ohms/square (pulling to 0)
Enhancement	30000	15000
Depletion Load	22000	–
Super-Buffer (depletion, gate 1)	5500	5500
Depletion (gate 0 or unknown)	50000	7000

Table 1. Resistance values used for transistors.

for details). Only the resistance and capacitance directly along the path from target to supply rail is considered. If there are side paths separated from the main path by pass transistors, the capacitance from those side paths is ignored.

For each path from the driver source to Vdd or Ground, all possible targets are found on the drain side of the driver and delays are calculated. Then the drain side of the driver is examined for paths to Vdd or Ground, and for each of these the source side is examined for targets.

The algorithm described above is just a depth-first critical path analysis. A depth-first search is used rather than a breadth-first one because circuits such as static memory elements contain feedback loops; by marking the pending nodes during the search, infinite loops can be avoided. In a depth-first search, the processing time could in the worst possible case be exponential in the size of the circuit, whereas breadth-first search is linear. Fortunately, the graphs describing integrated circuits tend to be shallow (only a few gate delays per clock period), and the loops tend to be short (usually just two gates). This results in running times that are almost linear in the size of the circuit.

Note that this analysis does not consider any specific data values at various nodes: all possible paths are assumed to be valid, and it is assumed that all transistors could conceivably be turned on. A data-independent analysis is more powerful than a simulation based on particular data values because it is certain to find the worst-case time; simulations will be effective only to the extent that the test cases are complete.

3. Improvements in Speed and Accuracy

In actual chips, not all transistors will always be turned on, and not all data values will be possible. A completely data-independent analysis, like that of Section 2, will chase many paths that can never be exercised in the real circuit, so it is likely to produce ridiculously pessimistic time estimates. Furthermore, the time required to chase all the paths will make the program too slow to be practical. Thus, although Crystal uses the simple mechanism as the core of its analyzer, it also employs several additional techniques that incorporate a few data dependencies in order to make its estimates more exact and its running times more reasonable.

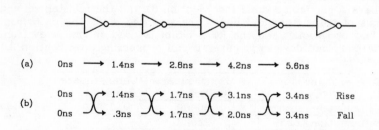

Figure 2. Crystal keeps separate rise and fall delay times for each node. If only the worst case times were used (the rising delay), then the pessimistic delay estimates in (a) would result. Instead, Crystal recognizes the inverters and combines rise delays for one stage with fall delays for the next, as shown in (b).

3.1. Up/Down Times and Loads

In a purely data-independent analysis, Crystal would have to use the worst-case time for each driver-target delay. This will almost always be the low-to-high transition, since nMOS rise times are usually several times longer than fall times. However, real circuits consist almost exclusively of inverting logic. For example, in Figure 2 it is unnecessarily pessimistic to assume that each of the several inverter stages is rising. Instead, Crystal keeps separate times for high-to-low and low-to-high transitions at each node and computes delays to reflect the level inversions that occur. Figure 2 shows how delay estimates are reduced by keeping separate up- and down-times.

There are three distinct cases that can occur in nMOS: pulldown, enhancement pullup, and depletion pullup. These are illustrated in Figure 3. Figure 3(a) shows the pulldown case. In this case a level inversion occurs: when the driver transistor turns on it pulls the target to ground. Only the high-to-low time of the target is affected, and it is determined by adding the low-to-high time of the driver to the delay through the path.

The enhancement pullup case is illustrated in Figure 3(b). No level inversion occurs here: when the pass transistor turns on, it pulls the target to a high voltage. For this path, only the low-to-high time of the target is affected. It is determined by adding the low-to-high time of the driver to the delay through the path. Because the target is being pulled high,

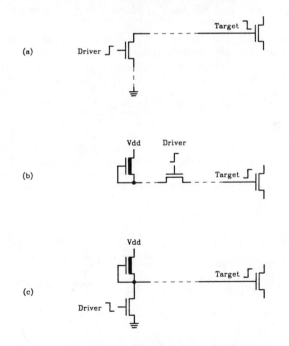

Figure 3. The three different ways that targets are driven in nMOS: a) a pulldown transistor turns on, causing the target to be pulled to 0; b) a pass transistor turns on, causing the target to be pulled to 1; c) a pulldown turns off, causing the target to be pulled to 1 through a depletion load.

different resistances are used for the transistors along the path than in the pulldown case. In the case of Figure 3(b), it is likely that there is also a path through the pass transistor to ground; this path will be analyzed separately using the pulldown rules to compute a new high-to-low time for the target.

The third case, depletion pullup, is illustrated in Figure 3(c). After the driver transistor turns off, the load will pull the target to a high level. When chasing out the paths, Crystal watches for load devices along the path, and remembers the closest load to each target. The low-to-high time for the target is computed by adding the high-to-low time for the driver to the delay to Vdd through the load. Only the closest load to the target is considered. Furthermore, if a depletion load is seen along a particular path, then enhancement pullup times are ignored for the path: Crystal assumes that the depletion load will be responsible for all low-to-high transitions at the target.

3.2. Fixed Values

In real circuits, certain nodes will have fixed values during each clock phase. These fixed values will prevent some delay paths from occurring when the chip runs. The most obvious examples are the clock signals themselves, as illustrated in the dynamic shift register example of Figure 4. Without any knowledge that Phase2 is zero when Phase1 is on, Crystal will attempt to propagate the input signal through all the shift register stages during Phase1, and will thus produce a very long worst-case time for Phase1.

Before invoking the delay analysis for a particular clock phase, users can indicate that certain signals (most notably the other clock phases) have fixed values. Crystal performs a static logic simulation to fix as many other node values as possible. If one input of a NAND structure is fixed at 0, Crystal will infer that the output must be fixed at 1. If one input of a NOR structure is fixed at 1, Crystal will infer that the output must be fixed at 0, and so on.

When searching for delay paths, Crystal assumes that there is a zero delay to all nodes with fixed values. It also refuses to propagate delays through enhancement transistors whose gates are fixed at zero. In the case of Figure 4, this means that only a few short paths will be considered during Phase1, and only a (different) few short paths during Phase2.

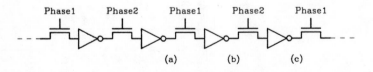

(a)   (b)   (c)

Figure 4. In analyzing the delays for Phase1 it is important to realize that Phase2 is zero. Otherwise, delays will be computed under the assumption that data could flow from end to end in a single clock phase. For the Phase1 analysis, the user indicates that Phase2 is fixed at 0; Crystal will evaluate the path from (a) to (b) but not from (b) to (c).

3.3. Flow Control

Pass transistors cause special problems, especially when the pass transistors are bi-directional. An example is in Figure 5. Without any extra information, Crystal will try a path where Input1 drives Output2 by passing forwards and backwards through the multiplexor structure. In practice, only one of the mux pass transistors will be turned on at a time, so the path is impossible. A more severe problem arises for barrel shifters and other structures where arrays of pass transistors are used to re-arrange data, as in Figure 6. In searching for all possible paths between a driver on the left and a target at the bottom, Crystal will analyze tortuous long paths through the pass transistor array. This has two consequences. First, a large number of pass transistors will be examined in series, leading to unrealistic long delay estimates (in reality only a few pass transistors are enabled at any one time). Second, the number of possible paths through such a structure grows exponentially with its size. For even a 16-bit barrel shifter the analysis takes too long to be practical.

One approach to the problem is to use fixed values to limit the possible paths. For example, in the case of the 2-input mux, 2 separate timing analyses could be performed, one with *Select* fixed to 0 and $\overline{Select}$ fixed to 1, and one with the values reversed. This is the approach taken by the SCALD system. However, this approach has two disadvantages. First, it is tedious and expensive to make separate analyses with different fixed values, especially for more complex structures where there are many valid combinations of control lines. The second problem is that no delays are calculated through fixed nodes. Thus, if the critical path involves the $\overline{Select}$ signal, Crystal will never detect that fact.

Crystal's solution to the problem is to allow designers to indicate the direction of signal flow through transistors. This feature is used in two ways. For the mux case, where information always flows in one direction, designers indicate which side of each pass transistor is the source of the

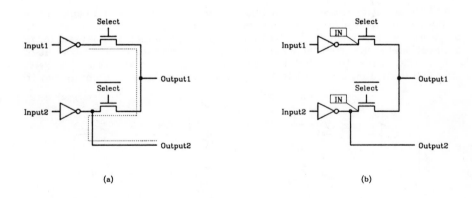

(a)   (b)

Figure 5. Without any information about how information flows through pass transistors, Crystal will examine the dotted path in (a), which can never occur in practice. If the user indicates that 0/1 signals always flow into the two mux pass transistors at their left sides, as in (b), then Crystal will not examine the bogus path.

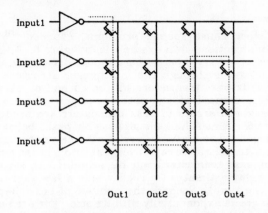

Figure 6. Another example where bogus paths, like the dotted one, will be examined unless information is provided about how the structure is used.

zero or one signal, as shown in Figure 5(b). The indication is made directly in the mask layout and passed through to Crystal by the circuit extractor. Crystal will then ignore any paths through the pass transistor with the zero or one source on the wrong side.

A similar but more powerful technique is used for bidirectional structures, such as the one in Figure 7. For bidirectional structures, designers tag one side of each transistor in the structure with a particular name other than "In" or "Out". When a potential path contains transistors tagged this way, Crystal will consider the path only if information flow in the path is always unidirectional with respect to the tags. This permits separate paths passing in opposite directions across the structure, but ignores paths that go back and forth. Each different structure in the circuit uses a separate tag for its pass transistors, so Crystal handles them independently.

Although it might seem that this kind of flow control would require the designer to spend a large amount of time tagging his design, in practice the work for this is small. In the RISC II Cache [8], which has several large bidirectional structures, approximately 12000 transistors have tags out of 46000 total transistors. But because the design uses arrays extensively, only 103 individual tags had to be entered by the designers, compared to

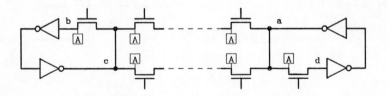

Figure 7. To handle bidirectional structures, the designer indicates a direction of flow by tagging one side of each transistor in the structure. Crystal permits paths in either direction, e.g. from *a* to *b* or from *c* to *d* or from *c* to *b*, but not paths that pass back and forth with respect to the tags, as from *a* to *c* to *d*.

about 900 other distinct node labels. If all pass transistors are unidirectional, it may be possible for the timing analyzer to infer the directionality and eliminate flow tagging. This is the approach taken by the TV program [3].

3.4. Busses and Precharging

Consider the structure of Figure 8, where several memory cells connect to a single bus and each cell can be written from or read out onto the bus. When analyzing this circuit, Crystal will consider paths from each memory cell out onto the bus and into each other memory cell. Thus, for N cells on the bus, Crystal will consider N^2 paths. For large memory arrays this is both expensive and unnecessary.

As long as the capacitance of the bus is much greater than the internal capacitance of any of the memory cells, the delay from one cell to another can be approximated by two separate delays: one from a cell onto the bus, and a second "independent" delay from the bus to each other cell. Thus, instead of N^2 paths, Crystal need only examine 2N paths: N paths from cells onto the bus, and N paths from the bus back into individual cells. There are two ways that Crystal finds out about busses. First, users may indicate this explicitly. Second, there is a user-settable capacitance threshold above which Crystal automatically considers a node to be a bus. The default threshold is 1pf.

Crystal also allows users to specify that certain nodes are precharged before certain clock phases. When this happens, Crystal ignores all low-to-high transitions for the precharged nodes.

3.5. Cross-Phase Signals

It is quite common for a particular piece of combinational logic to stabilize across several clock phases. For example, the input to an ALU might be loaded at the beginning of Phase 1 and the output latched at the end of Phase 2. In this case, Crystal will "bill" the entire delay of the ALU to Phase 1: it assumes that anything that could change in a clock phase must change during that clock phase. Work is currently underway to relax this restriction. Instead of requiring all nodes to settle during each clock

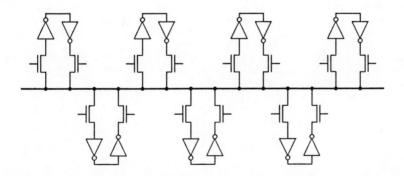

Figure 8. To avoid examining separate paths from each memory cell onto the bus then out to each other memory cell, Crystal computes separate delays from cells to the bus and from the bus to cells.

phase, only static and dynamic memory nodes loaded by that clock must settle. Other nodes can continue settling during the next clock phase. This mechanism has only recently been implemented and is still undergoing testing. It appears to be important, though: all of the designs that have used Crystal so far contain such cross-phase signals.

4. Weakness of the Delay Model

The simple resistive approximation for transistors can occasionally produce large errors in delay estimates. Consider the situation of Figure 9. In (b) the inverter will take much longer to drive its output to zero than in (a) because the gate voltage is much lower than Vdd when the output is being driven in (b). In general, the effective drive power of a transistor is a function both of the input waveform and of the load being driven, whereas Crystal considers only the load being driven. If the characteristic resistance for a transistor is chosen based on fast inputs, the delay for a given device may be underestimated by as much as an order of magnitude. In the case of Figure 9(b), the delay of the input signal is much greater than the delay of the inverter itself, so the underestimate for the inverter's delay will cause only a small percentage error in the overall delay estimate for the circuit. However, there are many situations where overall errors of as much as 50% or more will occur if a single resistance value is used for each transistor type (see Section 5).

The solution to this problem is to include the input waveform in the delay calculation. An approach that we are currently exploring is to include slopes in delay calculations. In the slope model, the low-to-high and high-to-low times for each node will be accompanied by slopes. The resistance values of transistors will be specified as functions of the transistor type, input slope, and load being driven.

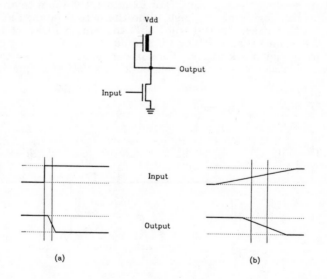

(a)   (b)

Figure 9. If the input to an inverter rises quickly to 1, as in (a), the output is driven quickly to 0. If the input rises very slowly as in (b), the pulldown transistor is driven by a lower gate voltage, so the output falls more slowly.

5. Experiences Using Crystal

It was fortunate, and not entirely coincidental, that the RISC II Cache chip was undergoing final pre-fabrication validation at the same time that Crystal was developed. The cache has been used throughout the development of Crystal as a large test case. It contains about 46000 transistors, over half of which are in a large memory array. It uses a four-phase non-overlapping clocking scheme. Crystal uncovered a half-dozen performance bugs that would have forced the clock to run 40% slower than planned. The most important bug found was a small transistor accidentally left to drive very long line, resulting in a 60ns delay for that one wire. In virtually all cases, the long delays were due to single isolated transistors; in each case there was plenty of space to increase the driving size. Thus, the performance tuning was easy to do, once the slow transistors were identified.

Approximately 20 minutes of CPU time are required to process the four clock phases of the cache on a VAX-11/780. Of this time, only 6 minutes was spent in delay analysis (only one of the four clock phases required more than 1 minute of analysis time); 7 minutes were required just to read in the 2.5 megabyte circuit description file, 4 minutes were used to propagate fixed node values, and the remaining 3 minutes were used in scanning the database to print out results. Although Crystal is relatively fast, it is not small: the program requires nearly 5 megabytes of virtual address space to process the cache chip, or about 100 bytes per transistor.

The RISC II microprocessor has also been analyzed using Crystal. RISC II contains over 40000 transistors, and the design has been carried out with relatively ambitious performance goals. The designers did not initially expect Crystal to be ready in time for their use, so they made extensive timing checks using SPICE and hand calculations. The Crystal analysis of this circuit pinpointed only one performance bug, which would have slowed the clock time by 40%. Crystal also uncovered a functional error that was not detected during simulation because none of the test cases triggered it. The timing analysis for the RISC II microprocessor requires over an hour of CPU time.

The third Crystal experience to date is a small circuit (2100 transistors) provided by the VLSI Systems Area at Xerox PARC. This circuit had been fabricated and tested before running it through Crystal. The initial version of the circuit had required 200ns for Phase 1, instead of the hoped-for time of 50ns. The designers found and fixed a performance bug, and the second fabrication of the circuit ran at 50ns. We ran both versions of the circuit through Crystal. In the first version, Crystal pinpointed the performance bug and estimated a clock time of around 190ns. In the second version of the circuit, Crystal estimated a clock time of about 55ns. The entire Crystal analysis of both versions of the circuit, including modifying the mask layout to produce the second version and extracting the circuit, took less than one day of real time. Crystal's analysis required less than a minute of CPU time.

Crystal provides several kinds of output. It identifies the longest path for each clock phase, and can also print out the longest paths. If desired, Crystal will identify single-stage delays that are longer than a given value, so designers can identify all the slow drivers. Crystal will also pinpoint nodes with large capacitance or resistance values. Output is provided both textually and graphically. Graphical output is provided by generating a command file for the Caesar layout editor [7]: when Caesar processes the command file it identifies the critical path with patches of a special layer along with text giving the delay to each point. Caesar commands can then be used to view each transistor in the path in detail, from beginning to end.

Crystal also generates a SPICE deck for the critical path that it finds. The SPICE deck includes all the transistors along the path as well as parasitic capacitance and resistance for the interconnect. No information is output for side paths. Table 2 compares Crystal's and SPICE's estimates for several sample circuits. For small circuits Crystal's estimates are very close to SPICE's more detailed calculations. For the larger cache circuits, Crystal's error relative to SPICE is 40-60%. The deviation between SPICE and Crystal is almost entirely due to problems of the sort discussed in Section 4.

6. Summary

The following list contains the approximations made by Crystal in computing delays:

[1] Each transistor type is characterized by an effective resistance when pulling to Vdd and another effective resistance when pulling to Ground.

[2] Capacitance from side paths is not included: Crystal considers only the direct path from a particular gate to Vdd or Ground.

[3] When examining a path, the closest depletion load device to the target gate is assumed to be responsible for pulling the gate high.

[4] For nodes with large capacitance, delays through the node are broken down into separate delays to the node and from the node.

In spite of these approximations, I believe Crystal will be a useful tool for designers, particularly if the model can be upgraded to include slope information. In any case, even the simplest model has enabled designers to find performance bugs and to pinpoint potential trouble spots for more detailed analysis with SPICE.

7. Acknowledgements

Berkeley's VLSI community provides a fertile environment for experimentation with new design tools. The ambitious design projects led by

Circuit	Crystal	Spice	Error
Test	20ns	21ns	5%
PLA	17ns	18ns	6%
Cache Phase 1	79ns	111ns	41%
Cache Phase 2	176ns	270ns	53%
Cache Phase 3	61ns	83ns	37%
Cache Phase 4	104ns	170ns	64%

Table 2. Comparison of Crystal and SPICE delay calculations for several sample circuits. Test consists of 4 AND gates and inverters in series. PLA is a small programmed logic array with 4 inputs, 6 minterms, 4 outputs. The last four circuits are the critical paths (as determined by Crystal) for the four clock phases of the RISC II Cache chip.

Prof. Dave Patterson motivate much of the tools work and validate the results. Lynn Conway, Alan Bell, and Gaetano Borriello of the VLSI Design Area at Xerox PARC provided early advice to us as well as the test case discussed above. Dimitris Lioupis, Bob Sherburne, and Gaetano Borriello were early users and critics of Crystal and have provided important feedback. Dan Fitzpatrick modified our circuit extractor to provide flow information for transistors. Dave Patterson and Carlo Séquin have provided helpful comments on this paper.

The work described here was supported in part by the Defense Advanced Research Projects Agency (DoD), ARPA Order No. 3803, monitored by the Naval Electronic System Command under Contract No. N00039-81-K-0251.

8. References

[1] Bening, L.C., et al. "Developments in Logic Network Path Delay Analysis." *Proc. 19th Design Automation Conference*, 1982, pp. 605-615.

[2] Foderaro, J.K., Van Dyke, K.S., and Patterson, D.A. "Running RISCs." *VLSI Design*, Vol. III, No. 5, Sept./Oct. 1982.

[3] Jouppi, N.P. "TV: An NMOS Timing Verifier." *3rd Caltech Conference on VLSI*, 1983.

[4] Kirkpatrick, T.I. and Clark, N.R. "PERT as an Aid to Logic Design." *IBM Journal of Research and Development*, Vol. 10, March 1966, pp. 135-141.

[5] McWilliams, T.M. "Verification of Timing Constraints on Large Digital Systems." *Proc. 17th Design Automation Conference*, 1980, pp. 139-147.

[6] Nagel, L.W. "SPICE2: A Computer Program to Simulate Semiconductor Circuits." ERL Memo ERL-M520, Univ. of California, Berkeley, May 1975.

[7] Ousterhout, J.K. "Caesar: An Interactive Editor for VLSI." *VLSI Design*, Vol. II, No. 4, Fourth Quarter 1981, pp. 34-38.

[8] Patterson, D.A., et al. "Architecture of a VLSI Instruction Cache." To appear, *10th Int'l Symposium on Computer Architecture*, 1983.

TV: An nMOS Timing Analyzer

Norman P. Jouppi
Computer Systems Laboratory
Stanford University
Stanford, California 94305

Abstract

TV is a timing analyzer for nMOS designs. Based on the circuit obtained from existing circuit extractors, TV determines the minimum clock duty and cycle times and verifies that the circuit obeys the MIPS clocking methodology. The delay analysis is an event driven simulation that only uses the values stable, rise, fall, as well as information about clock qualification. TV stresses fast running time, small user input requirements, and the ability to offer the user valuable advice. It calculates as much as possible statically, including the direction of signal flow, use, and clock qualification of all transistors.

1 Introduction

Critical path identification in integrated circuits is becoming increasingly difficult. Circuits have become so large that the designer rarely knows which paths are the most critical. Moreover, circuit simulators such as SPICE [7] can only be applied to a small area of the total chip at any time, because of their computation requirements are large and grow worse than linearly in the number of gates and connections. Designers have resorted to circuit simulation of small functional units, making assumptions about the interaction of these units. The speed of paths that cross several functional units needs to be known, since they often determine chip performance. Our design experience with the MIPS microprocessor [4, 3] underscored the importance of this need. The circuit designers had flexibility to trade off power for speed by choosing the size of transistors. However, no one knew what paths would be critical in practice, so designers relied on hunches that often proved wrong.

Many timing analysis programs exist for bipolar technologies. In these technologies, gate delays are known from manufacturer's specifications. Wire delays can be approximated or computed exactly depending on the availability of placement and wiring information. The goal for this previous work [6, 5] has been the identification of critical paths and verification of a clocking methodology in these "mainframe" technologies. As designs become large, i.e., greater than 10K gates, such verification tools are indispensable.

TV differs from previous timing analyzers in that it must calculate its own delays from the output of circuit extractors. This in turn requires an understanding of the function of nMOS circuits. Moreover, delay models for nMOS pose a major complications over ECL analyzers; in this respect, TV builds on work done by other researchers. Another important difference is that TV must determine the direction of signal flow through the transistors of the circuit.

TV is similar to bipolar timing analyzers in that it is data independent. The only simulation values used are *stable*, *rise*, and *fall*. If a changing signal reaches an inverter, its sense is inverted; if it enables a pass transistor, both rising and falling changes must be processed. Since this analysis does not account for specific signal values, paths that are not used in practice will participate in critical path analysis. For example, paths that are anded with a signal and its complement, or paths that are possible in a data path of a machine but are never used by the microcode, could be reported. Selection of these paths as the most critical does not occur often in practice, and as in McWilliams' ECL timing verifier [6], these paths may be disabled with user input. Despite this problem, the data-independent approach offers the advantage of complete coverage and greatly reduced running time.

To permit fast running time and ease of use, TV statically calculates as much as possible with minimal user input. This includes the direction of signal flow, function, and clock qualification of all transistors. Efficient delay analysis is performed by a breadth-first algorithm. To provide the designer with quick feedback on techniques to increase circuit performance, TV has an "interactive advisor".

2 Understanding nMOS Circuits

Before delay analysis, TV determines the direction of signal flow through all transistors, the use made of each transistor, clock qualification of circuit nodes and gates, and propagates any user-specified values for case analysis. Statically determined flow control is in contrast to combinations of designer-assisted and runtime determination of flow, as in Crystal [9]. The algorithms used are robust and work for a wide range of design methodologies. They perform well on both the Geometry Engine [2], which makes extensive use of precharging and pass transistors, and MIPS, which contains no precharging and unconventional restoring-logic circuits.

2.1 Determining Signal Flow

TV makes a first pass over the circuit to determine the direction of signal propagation through every transistor. This is a very important precursor to delay analysis; without it delay analysis could be rendered worthless by finding impossible or useless paths. TV's principal input is from Terman's circuit extractor [1]. TV also uses information about ratio rules for inverters (e.g., 4:1 for restored logic vs. 8:1 for pass logic) as well as other information about the types of circuits used in the design methodology. This approach requires that the circuit be free of static errors before reliable TV operation can proceed.

Transistors whose signal flow direction has been determined are called *set*. This direction-finding pass uses nine rules to set transistors; eight of these are *safe*, and one is *unsafe*. Safe rules are guaranteed not to choose direction incorrectly, given the design methodology. Unsafe rules usually work, but are not guaranteed. Transistors set via unsafe rules can be checked later by the user if desired. A set transistor's *drain* is a sink of signal flow; its *source* is a source of signal flow. Non-gate terminals to an unset transistor are called *channels*.

The first rule, *constant propagation*, knows any transistor channel connected to Vdd, ground, or a clock must be a sink of signal flow; the other terminal must be a source. This typically accounts for 50 to 65 percent of the transistors in circuits analyzed so far.

The second rule sets transistor directions according to *user input specifications*. This is not required for the analysis of either the MIPS microprocessor or the Geometry Engine. The user can specify directions for an array of transistors by specifying the gate, source and drain of one transistor in the corner of the array, plus iteration counts and spacings in lambda to define the positions of the other transistors in the array. Our current design methodology does not use bidirectional pass transistors; if these exist they need to be manually input.

structure common in MIPS but uncommon in Mead-Conway designs are bus pull-downs (see Figure 1) where there is a pull-up on each side. Sizes of the internal and bus pull-ups and pull-downs are calculated with the aid of Thevenin's equivalents. Because of these electrical requirements, the bus pull-down must always transmit signal flow towards the stronger pull-up from the weaker pull-up. The third rule uses *bus pull-down detection* to identify and safely set the direction of bus pull-downs.

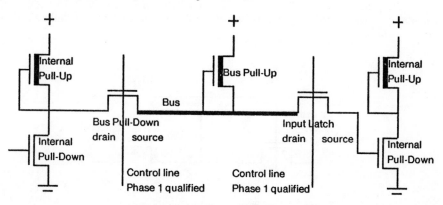

Figure 1: MIPS Bus Structure

The fourth rule is based on *Kirchoff's current law*. Figure 2 illustrates how it applies to signal flow. If all but one of the transistors connected to a node have known direction, and the known transistors all sink or all source the node, then the unknown transistor must transmit signal flow in the opposite direction with respect to the node. Moreover, if all but two of the transistors connected to a node have known and equal direction, and these two unknown transistors' gates are complements of each other, then both transistors must transmit signals in the opposite direction with respect to the others. Only simple inverters (where there is only one pull-down and pull-up) are detected. Despite this restriction it is very useful for determining signal flow in the presence of 2-way multiplexors. The basic Kirchoff rule and its extension, *complement gate detection* (the fifth rule), are safe.

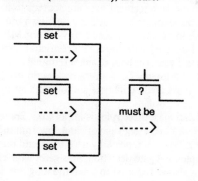

Figure 2: "Kirchoff" Direction Setting Rule

TV uses the fact that *k-ratio rules* for inverters must be obeyed in the sixth rule. By finding the minimum resistance to ground through each unset transistor connected to a pull-up, we can determine whether it can act as a pull-down (with signal flow towards the pull-up), or as a pass transistor (with signal flow away from the pull-up). It is safe to set transistors that cannot satisfy ratio rules to be pass transistors; however, nothing is certain about unset transistors that can satisfy ratio rules.

If two pass transistors connected to a node *n* have as their gates the same node, then in our design methodology the direction of signal flow for both these gates relative to node *n* must be the same (see Figure 3). This case does not occur often relative to the number of transistors, but is necessary to solve many circuits. It is implemented by the seventh rule, *pass transistors with same gates.*

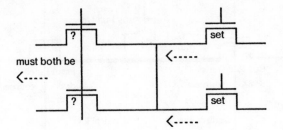

Figure 3:   Two Pass Transistors with Same Gate

The eight rule, *analogy*, identifies replicated cells in which one cell has a transistor set while in another cell the corresponding transistor is unset. This rule is useful for solving structures like Manchester carry chains. Looking only at the center of such a chain, it is not possible to determine the direction of signal flow, but once the edge and restoring bit slices are located, the task becomes simple. Analogy is currently limited to cases where the set and unset transistors are connected to each other via channel terminals. To determine that two transistors come from different instances of a cell, it tests that size, either x or y coordinate, the number of gates connected to these two transistors, and these gates' x or y positions are the same. Analogy would be much easier to implement within a hierarchical circuit extraction system, where information on replicated cells is readily available.

The ninth rule, *forced storage nodes*, attempts to identify storage nodes. If all the transistors connected to an enhancement gate are sources to it except for *n* unknowns, then this node is assumed to be a storage node and the unknown transistors must source to the gate as well. This is not safe in general but seems to work for MIPS and the Geometry Engine. First, the rule is applied for $n=1$ with the added restriction that no transistors are set that are connected to enhancement gates on both channel terminals. Then the rule is applied without the restriction for $n=1$ and for $n=2$. If all transistors do not have set directions after input forcing with $n=2$, TV aborts further analysis and outputs information on all remaining unset transistors.

The order of application of these rules is important. Some of them, like inverter k-ratio and constant propagation, have a high payoff and low running time and are executed first. Since the Kirchoff rule is fast, it is repeatedly used to extend the solutions opened up by other rules. Others, like complement gate detection, pass gates with same gates, analogy, and forced storage node are almost always required to solve the toughest cases but require more time to execute. The order of application and the percentage of transistors set by each rule appear in Figure 4 for the Geometry Engine. As in MIPS, over 90% of the transistors are set via constant propagation, inverter k-ratio, and the Kirchoff rules. The other rules, although they do not contribute much in number, make complete solution of the circuit possible without any user input. Only rarely do two or more rules not contribute at all.

Transistors resolved	Rule used
63.33%	constant propagation
0.00%	user input specifications (none given)
0.00%	bus pull-up detection
3.82%	kirchoff pass 1
1.03%	complement gate detection pass 1
20.60%	inverter k-ratio
11.93%	kirchoff pass 2
0.01%	complement gate detection pass 2
0.03%	pass gates with same gate
0.00%	kirchoff pass 3
0.28%	forced storage node: n=1, restricted
0.23%	analogy
0.00%	kirchoff pass 4
0.02%	forced storage node: n=1, no restriction
100.00% (39877 transistors)	

Figure 4: Application Order of Direction-Finding Rules

2.2 Determining Transistor Usage

After the direction of signal flow through all transistors has been determined, TV classifies all transistors into one of eight possible types. This classification is based on their usage in the circuit and is used to guide the remainder of the analysis. All enhancement transistors whose drain is Vdd and whose gate is connected to a clock are labeled as precharging transistors. Those with a drain connected to a clock are assumed to be bootstrap drivers. Discharging transistors (which precharge a node low) are found during clock qualification propagation. Pull-ups are depletion transistors with their drain connected to Vdd. If their gate is not connected to their source, they are assumed to be superbuffers. Pull-downs are found recursively: any enhancement source connected to a pull-up is a pull-down, and any enhancement source connected to a pull-down drain is also a pull-down. Any transistors not labeled by the previous rules are assumed to be pass transistors.

2.3 Propagating Clock Qualification

Next, TV propagates clock qualification information through restoring logic until it reaches gates of pass transistors. Starting at each clock phase, it walks up through pull-downs and marks restored nodes as qualified by that clock phase. Restored nodes that have more than one pull-down chain without any transistors gated by clock phases are not marked as qualified (Figure 5). Instead, the pull-down transistor gated by a clock is marked as a precharging (low) transistor. The product terms in the main Geometry Engine PLA provide an example of this structure. Without recognizing this case, clock qualification would propagate through the entire PLA as if it were a complicated control driver. The results of clock qualification, pass transistors qualified by clocks, are used to start and end paths in the remainder of the analysis.

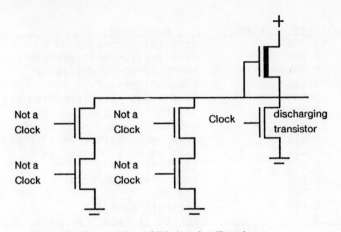

Figure 5:   Recognition of Discharging Transistors

2.4 Propagating User-Input Node Values: Case Analysis

Since the analysis is data-independent, it may make worst-case assumptions about values of nodes; these assumptions may not be possible in practice. Figure 6 shows an example where case analysis is required in MIPS. ID and OD are true in the ID and OD pipestages, respectively, and are mutually exclusive. Without user input, TV will assume the path from point A to point C is possible. Eliminating this problem will require performing the analysis twice: once with ID true and OD false, and once with ID false and OD true. TV will propagate any user supplied constants as far as possible through the circuit. (Zeroes propagate through NANDs to become ones, etc.) TV also needs to be fast because of case analysis: if many cases need to be analyzed we can only afford to spend a small amount of time on each (currently a complete analysis is performed for each case). The number of cases required in the MIPS analysis has been small (around 4), so a method of partial re-analysis has not been pursued.

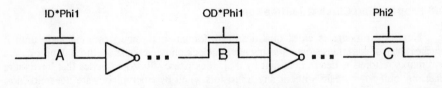

Figure 6:   Case Analysis in MIPS

3 Delay Analysis

Delay analysis finds the maximum clock frequency for a circuit and the duty cycle for the clocks. This clock timing is determined by the critical paths of the circuit. Paths can start at either the rising or falling edge of a clock, and can end at either the falling edge of the same clock or the rising edge of the next clock, giving the following three types of paths (see Figure 7):

1. Paths that start at a gate qualified by a clock and are latched by the same clock determine the minimum time that clock must be high. These paths are not allowed in a strict two phase clocking approach, but appear in MIPS.
2. Paths that start at a gate qualified by one clock and end with the rise of a second clock determine the minimum time between the rise of the first clock and the rise of the second clock.
3. Paths that start at the completion of precharging on one phase and end with the rise of a second phase determine the minimum time between the fall of the first clock and the rise of the second clock.

In TV, paths are terminated at the rising edge of a second clock. This is because the second phase can permit propagation of the signal into areas of the circuit precharged by the first phase. In this case, the data must be stable at the rising edge of the second clock, or the nodes may be erroneously discharged. Multiple clock-cycle paths are not currently permitted as they would add significant amounts of complexity or running time. Paths crossing multiple cycles are not present in either MIPS or the Geometry Engine.

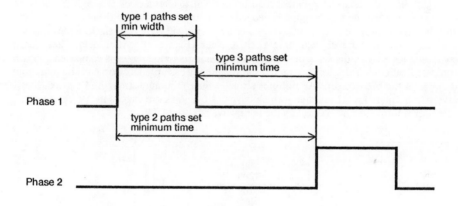

Figure 7: Constraints on Clock Duty and Cycle Times

3.1 Breadth-First Delay Analysis

To perform the delay analysis efficiently, TV implements it with a breadth-first application of the delay model. With a breadth-first algorithm, the delay modeling routine is called only once for each node active during the given clock phase. In contrast, a depth-first algorithm may apply the modeling routine to a node many times.

To perform a breadth-first analysis, the number of predecessors for each node must be computed. This computation is complicated by a multiple-phase clocking discipline. The number of predecessors to a node cannot be determined simply by looking at the connections to that node, since some of these connections may only be active on phase 1, others on both clock phases, etc. To determine the number of predecessors to a node on a given clock phase, a depth-first search over the circuit starting at the clock node and inputs, is made. The running time for each clock phase of the predecessor finding pass is $O(n+e)$, where n is the total number of nodes and e is the total number of edges in the circuit. In order to correctly handle loops in the circuit, the depth-first algorithm must maintain a linked list of nodes for which the predecessor routine is currently called. Before incrementing the predecessor count of a node, this list is searched. If the node is found in the list, its predecessor count is not incremented. The list could be as long as the longest possible path for each phase of the clock. In practice this search is very fast and accounts for less than 1% of the running time of the program. This loop prevention works for dynamic latches as used in our design methodology, but does not work for static latches (such as a pair of cross coupled NAND gates).

After the number of predecessors to each node has been computed, the delay modeling routine is applied breadth-first. Originally a depth-first application of the delay model was used in TV, and it typically applied the delay modeling routine to each node an average of between three and four times per clock phase. The combination of a depth-first predecessor-finding plus breadth-first application of the delay model is about three times faster overall. This difference in performance will become more significant as delay models of greater complexity are developed.

3.2 User Output

The *delay* of a node is the time it requires to stabilize after a clock is raised. *Slack* is the difference between the time a signal is needed so as not to increase the delay of its successor and the time it is produced. Nodes on the critical path will have zero slack; nodes that are overtaken by another path will have positive slack (Figure 8). After delays have been computed for a node, the slack between that node and its predecessors is computed. This is compared with any previously computed slack value for each predecessor, and the minimum is stored at the predecessor. All pull-ups stronger than minimum strength (4 square for 4:1 inverters, 8 square for 8:1 inverters) which have positive slack are flagged as "power wasters". The slack information is also used by the interactive advisor of TV.

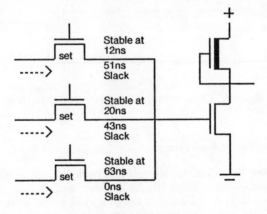

Figure 8: Slack Calculation

TV should to give users intelligent advice. One way to fail in this would be to report only one worst path, or worse yet, to report many paths that are either *equivalent* or *relatives*. Reporting one worst path would force the designer to work on only one problem at a time, or if this path did not exist in practice, force reanalysis of the circuit. TV allows a user-selectable number of paths (currently from one to twenty) to be reported. *Equivalent* paths are those considered equal in function and delay characteristics: examples of these are parallel paths in a data path. *Relative* paths have a common prefix responsible for the majority of their delay.

When TV reports the n worst paths, it reports n non-equivalent, non-relative worst paths, along with the last node of any that are equivalent or relatives to the n worst. TV determines equivalence as follows:

1. Equivalent paths must attain stability on their final node within a predefined percentage of each other, called the equivalence range.
2. Equivalent paths must have gate and stray capacitance within equivalence range of each other on their last node.
3. Equivalent paths must have the same number of steps.

The first qualification gives a necessary condition for delay equivalence, while the second and third give necessary conditions for functional equivalence. These qualifications have the advantage that they can be verified in a short amount of running time. Relative detection also eliminates reporting any subpaths of a critical path. Otherwise, if one path were considerably worse than the others, TV might report n different subpaths of the worst path (e.g., if the worst path had m nodes, it might report n paths: $1..m,1..m-1,..,1..m-n$). Currently, relatives must have a common prefix accounting for at least 70% of each path's delay.

TV keeps gate capacitances and parasitics separate during the analysis. After the delay analysis is complete, it calculates for all nodes on critical paths what percentage of the delay incurred was due to gates and what percentage was due to strays. Nodes that added to the delay more than twice the average are considered "worst offenders". Detailed information is only output for the worst offenders to highlight them and to prevent the designer from being overwhelmed with data. Where most of the delay is from parasitics, it can be overcome by using more power in the form of stronger pull-ups. When most of the delay is from other gates, such as in bus structures with many writers, it is more difficult to improve performance. By looking at these figures, the designer can see at a glance whether it is easy to speed up the circuit.

One of the major motivations for TV was that SPICE analysis of large circuits as a whole is impractical due to large computer requirements, and that partitioning the design into sections to allow the use of SPICE is highly subject to human error. However, as a check of TV's accuracy SPICE analysis of the critical paths is very useful. TV outputs a SPICE input deck for as many of the critical paths as the user desires. Two types of decks are possible. All transistors attached to nodes in the critical paths are generated in a detailed version. A version that can be simulated in about one-fourth the running time lumps all transistors attached to a node in the critical path but not in the path themselves into a single capacitance. These SPICE results give the users of TV greater confidence in its results and provides a calibration source for TV.

TV outputs a plot of the critical paths to scale with our standard node plots, as well as tabular information. This allows the designer to quickly identify which areas of the circuit and which types of paths are most critical.

Not only is it important to provide the designer with information on critical paths quickly, but the designer needs fast feedback on the performance implications of possible changes. TV meets these needs with an interactive advisor (IA). After the previously described analysis is complete, TV writes out its database. With the IA, the designer can query the database on any node or transistor. Slack calculations are available so that the designer can concentrate his efforts on nodes where the delay comes mainly from one predecessor.

4 Delay Models

Since TV generates SPICE decks for critical paths it finds, it can afford to place less stringent demands on its delay models. Whereas circuit analysis programs like SPICE strive for absolute accuracy, TV requires primarily relative accuracy. Relative accuracy means that the delay through each stage be modeled accurately in ratio to the delay calculated for other types of stages. Unfortunately, relative accuracy is not easier to implement than absolute accuracy, because accurate modeling of pass transistors, inverters, bootstrap drivers and other structures relative to each other is the most difficult part of constructing a good delay model. Both relative and absolute accuracy need be no more accurate than the specifications provided by the fabrication process.

TV calculates an RC time constant for each node relative to the time its *driver's* input becomes stable. A driver is either a pull-up or a pull-down; the signal from a driver may propagate through a tree of pass transistors to storage nodes or may directly drive another pull-down. In cases where propagation activates the gate of a pass transistor, the difference in RC time constants between the source and drain of the pass transistor is used as the incurred delay. Only one path is assumed open through a tree of pass transistors. This simplifies the RC tree analysis into RC line analysis. From our experience, this seems to be a valid assumption except in the case of red-green function blocks. In this case, several paths of varying length will exist from the output of the function block back into it, forming a tree. TV calculates delays for every driven node from Penfield and Rubenstein's [10] approximation of RC trees by RC time constants. They derived worst case min and max bounds; TV averages them to obtain a "typical" value. All transistors are assigned a rising and falling resistance from tables based on their use in the circuit.

TV has two modeling routines that calculate RC time constants. The first computes worst-case constants statically (i.e., before the start nodes are propagated). Static delay calculations were originally chosen because they were easier to implement and required less running time. Static RC analysis has the disadvantage that when a node is driven from many sources with different impedances, TV choses the worst-case RC time constant for the node. If the highest impedance source is not the node's zero-slack predecessor, this choice is overly pessimistic. Most often, alternate drivers of a node are of similar size, but exceptions to this can cause large errors. Our results indicate worst-case static analysis may typically increase delay estimates over a path by 70%. Moreover, this delay is lumped on a few specialized nodes (typically buses), and violates our goal of relative accuracy.

The second modeling routine (dynamic delay modeling) hunts back along the path it has traveled to find the nearest driver. From this it calculates RC time constants from the driver to driven nodes along the actual path taken. This modeling routine was used for the results described in Section 6. The dynamic delay modeling routine currently uses about eight times the running time of static analysis.

5 Clocking Verification

Verification in VLSI design can be simplified by use of a strict two-phase clocking discipline [8]. This methodology requires that data transfers are set up on one clock phase and latched on another. During the MIPS design this methodology was found to be too confining and posed an undesired limit to performance. Instead, we chose to write and read nodes on the same clock phase. This requires that the writer's output remain valid within some interval around the time the reader latches data. This methodology, unlike the strict two phase, requires timing information to verify that the methodology is obeyed. TV does clocking verification in a separate optional pass from critical path analysis. Unlike the critical path calculations, verification requires both minimum and maximum delays.

The canonical case that TV clocking verification analyzes is shown in Figure 1. TV needs to be verify that the signal on the bus remains valid for a set up time and a hold time around the time the reader latches the data. Set up times are provided by the delay analysis when it outputs the minimum length of each clock phase. Hold times guarantee a safety factor. The hold time is the minimum delay along the path from the clock through the writer to the reader minus the maximum time from the clock to the reader's latch. This verification problem occurs most often on busses, and care must be taken so that the verification does not run in $O(n^2)$ time; this could occur if there are n readers and n writers per bus. Only the time from each reader and writer to the bus need to be calculated; then the earliest writer instability and latest reader latching for the bus are compared. This is an example of path equivalence performed before the analysis to cut down on analysis effort. This reduction in required analysis could also result if the input to TV was hierarchical in nature. Clocking verification is currently being implemented.

6 Results: A Case Study

The first version of TV became available while we were completing the last test chips for MIPS. TV was run on the MIPS ALU test chip with interesting results. A plot of the critical path appears in Figure 9. The propagation delay in each functional block along the critical path of the ALU according to TV and SPICE are compared below:

ALU section	Time required to propagate through section (ns)	
	SPICE	TV
Read Booth register	21	21
Bit slice (latch, etc.)	25	23
Propagate generate tree	22	19
Carry tree	29	38
Bit slice (sum, etc.)	22	30
Total	118	131

TV predicted times from 86% to 136% of that predicted by SPICE. From these results it was seen that the time spent in the bit slice portion of the ALU was much larger than expected (48% of the total add time, not including the register read). This prompted a redesign of the bit slice to improve its performance.

In December 1982 the complete chip became available for analysis. Initial results from TV were very disappointing, but we discovered case analysis was necessary to split paths that crossed two pipestages. These split paths were still much longer than expected, however. Our design goal was a 250ns clock cycle, but TV and SPICE pointed out paths totalling about 950ns between phase one and two. Most of this delay was from inadequately driven nodes within the control structure. By including more superbuffers and beefing up existing drivers, the cycle time should be able to be cut in half.

TV's estimates are within SPICE simulations by about 20% for the critical path during phase 1 and phase 2. Most of this delay error is in very slowly rising stages. So far it has been assumed in delay modeling that the input to a driver is about the same slope as its output. An example of this is a string of equally sized inverters; the delay from one stage to the next is not simply 1 unit or k units, but rather is almost identical for both the rising and falling cases. This assumption works well for carefully designed circuits, like the ALU. For global paths, (especially in the first iterations of a design), the delay is mainly associated with a few nodes with significantly slower slopes. For stages with slopes significantly different than average, the resulting delay calculations are inaccurate. Work is currently in progress to include slope in delay calculations.

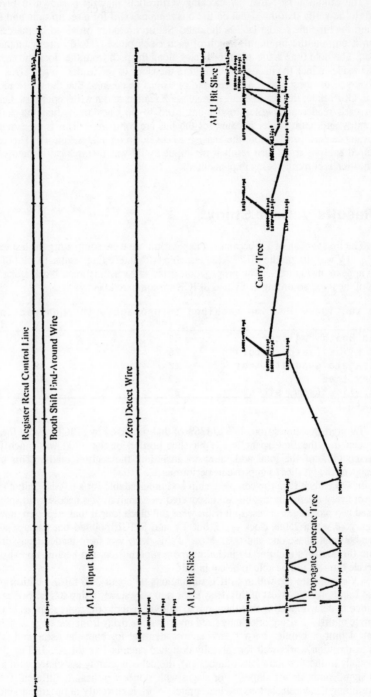

Figure 9: ALU critical path

TV's running statistics on a DEC VAX 11/780 for several chips follow:

Chip	# tran-sistors	TV time for entire chip	SPICE time for critical path	Transistors on critical path
MIPS ALU	3362	59sec	287sec	64
MIPS	24020	375sec	360sec	69
Geometry Engine	39877	620sec	545sec	121

Currently TV must have the circuit being analyzed resident in memory, otherwise thrashing will result. This requires about 84. bytes of storage per node and 46 bytes per transistor (using unpacked records), for a total of about 5 Megabytes in the case of the Geometry Engine. TV is implemented in Pascal.

7 Future Work

7.1 Enhancements to Improve Accuracy of Delay Models

TV receives information about transistors and parasitic capacitance of nodes from Chris Terman's circuit extractor, but not resistance of interconnect. Thus, diffusion delay in long polysilicon wires is ignored. As work on TV progresses, polysilicon diffusion delay will be estimated by breaking up nodes with large amounts of parasitic poly capacitance connected to many gates (identifying attributes of polysilicon busses or control lines) into a series of nodes connected by resistors.

7.2 Interactive Advisor

In the future the interactive advisor of TV will be expanded. The goal is that the designer will be able to specify changes to the circuit (e.g., all transistors in a replicated bit slice should change size) and that reanalysis of the circuit may proceed quickly from these changes. These changes affect the circuit characteristics of only the nodes connected to these transistors, and the delay models need to be reapplied only to those nodes. The new times calculated can be propagated to their successors, checking slack times to see whether other paths become critical. Since slack comparisons are made at each stage instead of delay calculations, this reanalysis can be made very fast. Also, the reanalysis is likely to affect only a small part of the total circuit. As TV can perform a complete analysis of chips with 40,000 transistors in less than 15 minutes of VAX 11/780 CPU time, the designer should receive interactive response on how his ideas impact chip performance on a global level. This will give the designer the ability to quickly experiment with circuit performance enhancements in cases where such information is not even currently available.

7.3 Application to CMOS

Many of the techniques used in TV should be applicable to CMOS. Resistance values based on circuit simulations could be used for nMOS as well as pMOS transistors. Direction-finding routines, similar to those for our nMOS design methodology could be used. Direction-finding based on ratio rules would not be applicable, however. This is unfortunate as TV typically finds the direction of over 80% of those transistors not directly connected to a constant by the inverter k-ratio rule and its following Kirchoff pass. Replacing this rule and keeping reasonable running time could be challenging.

8 Conclusions

TV has already proved its worth as timing analyzer for nMOS designs. It is fast and its running time grows linearly with the number of transistors. Small demands are placed on the user, yet TV has the ability to offer the user valuable advice. A significant contribution is the ability to calculate much information statically including the direction of signal flow, usage, and clock qualification of all transistors.

Much future work remains, however. The interactive advisor can be expanded so that the user can experiment with ways to increase circuit performance. Work still needs to be done on timing verification. TV's delay models also leave room for further research.

Acknowledgments

John Hennessy and Jim Clark have given me very helpful advice. The MIPS and Geometry Engine teams have provided much inspiration for this project. The MIPS team has provided helpful reviews of this paper. This work was supported in part by a National Science Foundation Fellowship and the Defense Advanced Research Projects Agency under contract # MDA903-79-C-0680.

References

1. Baker, C., and Terman, C. "Tools for Verifying Intergrated Circuit Designs." *Lambda* (Fourth Quarter 1980), 22-30.

2. Clark, J. The Geometry Engine: A VLSI Geometry System for Graphics. Proc. SIGGRAPH '82, ACM, 1982.

3. Hennessy, J.L., Jouppi, N., Baskett, F., and Gill,J. MIPS: A VLSI Processor Architecture. Proc. CMU Conference on VLSI Systems and Computations, October, 1981.

4. Hennessy, J., Jouppi N., Przybylski, S., Rowen, C., and Gross, T. Design of a High Performance VLSI Processor. Proceedings Third Caltech VLSI Conference, 1983.

5. Hitchcock, Robert B. Sr. Timing Verification and the Timing Analysis Program. 19th Design Automation Conference, IEEE, June, 1982, pp. 594-604.

6. McWilliams, T.M. Verification of Timing Constraints on Large Digital Systems. Proc. of the 17th Design Automation Conference, IEEE, Minneapolis, , 1980, pp. 139-147.

7. Nagel, L. SPICE2: A Computer Program to Simulate Semiconductor Circuits. Tech. Rept. UCB ERL-M250, University of California, Berkeley, May, 1975.

8. Noice, D., Mathews, R., and Newkirk, J. A Clocking Discipline fo Two-Phase Digital Systems. Proc. ICCC 82, IEEE, September, 1982.

9. Ousterhout, John K. Crystal: A Timing Analyzer for nMOS VLSI Circuits. Proceedings of the third Caltech VLSI Conference, 1983.

10. Penfield, Paul Jr., and Rubenstein, Jorge. Signal Delay in RC Tree Networks. 18th Design Automation Conference, IEEE, June, 1981, pp. 613-17.

Optimizing Synchronous Circuitry by Retiming
(Preliminary Version)

Charles E. Leiserson
Flavio M. Rose
Laboratory for Computer Science
Massachusetts Institute of Technology
Cambridge, Massachusetts 02139

James B. Saxe
Department of Computer Science
Carnegie-Mellon University
Pittsburgh, Pennsylvania 15213

Abstract—This paper explores circuit optimization within a graph-theoretic framework. The vertices of the graph are combinational logic elements with assigned numerical propagation delays. The edges of the graph are interconnections between combinational logic elements. Each edge is given a weight equal to the number of clocked registers through which the interconnection passes. A distinguished vertex, called the host, represents the interface between the circuit and the external world.

This paper shows how the technique of *retiming* can be used to transform a given synchronous circuit into a more efficient circuit under a variety of different cost criteria. We give an easily programmed $O(|V|^3 \lg|V|)$ algorithm for determining an equivalent circuit with the smallest possible clock period. We show how to improve the asymptotic time complexity by reducing this problem to an efficiently solvable mixed-integer linear programming problem. We also show that the problem of determining an equivalent circuit with minimum state (total number of registers) is the linear-programming dual of a minimum-cost flow problem, and hence can also be solved efficiently. The techniques are general in that many other constraints can be handled within the graph-theoretic framework.

This research was supported in part by the Defense Advanced Research Projects Agency under Contract N00014-80-C-0622 and by the Office of Naval Research under Contract N00014-76-C-0370. James B. Saxe was supported in part by an IBM graduate fellowship. Flavio Rose's current address is Mailstop HLO1-1/R07, Digital Equipment Corporation, 75 Reed Road, Hudson, Massachusetts 01749.

1. Introduction

The goal of VLSI design automation is to speed the design of a system without sacrificing the quality of implementation. A common means of achieving this goal is through the use of optimization tools, which improve the quality of a quickly designed circuit. Optimization tools include, for example, sticks compaction [2], boolean simplification [6, p. 90], and gate-array optimization [20]. In this paper we show how to optimize clocked circuits by relocating registers so as to reduce combinational rippling. Unlike pipelining, *retiming* does not increase circuit latency.

In order to illustrate the optimization method used in this paper, let us consider the problem of designing a digital correlator. The correlator takes a stream of bits $x_0, x_1, x_2, \ldots$ as input and compares it with a fixed-length pattern $a_0, a_1, \ldots, a_k$. After receiving each input x_i ($i \geq k$) the correlator produces as output the number of matches

$$y_i = \sum_{j=0}^{k} \delta(x_{i-j}, a_k), \tag{1}$$

where δ is the comparison function

$$\delta(x, y) = \begin{cases} 1 & \text{if } x = y, \\ 0 & \text{if } x \neq y. \end{cases}$$

Figure 1 shows a design of a simple correlator for the case when $k = 3$. Correlator 1 consists of two kinds of functional elements, adders and comparators, whose I/O characteristics are shown in the figure. The boxes between the comparators are registers which act to shift the x_i's to the right down the length of the correlator. On each tick of the global clock, each x_i is compared with a character of the pattern, and the adders sum up the number of matches.

This design, though easy to understand, has poor performance. Between ticks of the clock, the partial sums of the matches ripple up the length of the correlator. Suppose, for instance, that each adder has a propagation delay of 7 esec,[1] and each comparator has a propagation delay of 3 esec. Then the clock period must be at least 24 esec—the time for a signal to propagate from the register labeled A through one comparator and three adders.

A design that gives better performance is shown in Figure 2. The clock period for Correlator 2 may be as little as 13 esec—the time for a signal to propagate from the register labeled B through two comparators and one adder. Remarkably, the I/O characteristics of this correlator are *identical* to those of Correlator 1. Notice that the the two designs use the same functional elements connected in the same manner, except for the locations of registers. That Correlator 2 indeed has the I/O characteristic specified by Equation (1) may be verified by the reader, but it should be apparent that this verification requires considerably more effort than the verification of Correlator 1.

In this paper we study the problem of optimizing a circuit by relocating the registers so as to minimize the clock period while preserving the I/O characteristics. The methods given in this paper are powerful enough to generate automatically the

---

[1]Recall that one eptosecond (esec) equals one one-zillionth of a second.

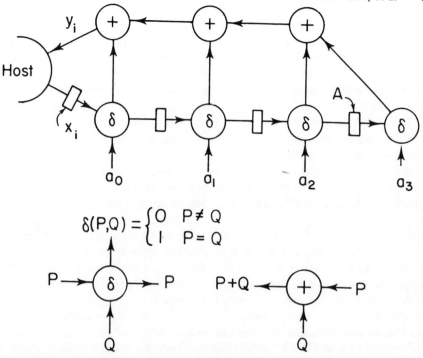

Figure 1. Correlator 1: A simple circuit made of two kinds of functional elements. Each comparator δ has a propagation delay of 3 esec, and each adder $+$ has a propagation delay of 7 esec. The longest path of combinational rippling starts at the register labeled A, and thus the clock period of the circuit is 24 esec.

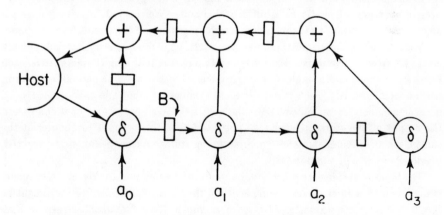

Figure 2. Correlator 2: A circuit functionally equivalent to Correlator 1 but which is more efficient. The longest path of combinational rippling begins at register B, and thus the clock period of this circuit is 13 esec.

optimized Correlator 2 given a description of Correlator 1. The algorithms are efficient, and they produce the best circuit that can be obtained by relocating registers.

Section 2 presents the graph-theoretic model of synchronous circuits used in this paper. In Section 3 we describe the operation of *retiming* [12], whereby the locations of registers in a circuit can be changed in such a way that the I/O characteristics are preserved. Section 4 gives a polynomial-time algorithm for determining a retiming of a circuit that minimizes clock period. In Section 5 we show how to reduce the problem of finding the optimal retiming of a circuit to an efficiently solvable mixed-integer programming problem, and thus improve the asymptotic efficiency with which we can solve the problem.

Sections 6, 7, and 8 discuss extensions of the main result. Section 6 considers the special case where all functional elements have identical propagation delays and shows that the optimal retiming can be found more efficiently in this case. It also discusses the relationship of this work to *systolic* computation and shows how to improve the performance of many systolic circuits in the literature. Section 7 examines the problem of retiming a circuit so as to minimize the total amount of state (number of registers). In particular, we show that the problem can be reduced to the linear-programming dual of a minimum-cost flow problem, and hence can be solved optimally in polynomial time. Section 8 extends our methods to a more general circuit model in which individual functional elements may have nonuniform propagation delays—*e.g.*, the low-order output bit of an adder may be available earlier than the high-order bit.

We briefly mention some further extensions and offer conclusions in Section 9. An appendix is provided which contains background on the combinatorial techniques used in the paper.

2. Preliminaries

In this section we define the notations and terminology needed in the paper and present our graph-theoretic model of digital circuits. We conclude by giving a simple algorithm for determining the minimum feasible clock period of a circuit from its graph.

We can view a circuit abstractly as a network of *functional elements* and globally clocked *registers*. The registers are assumed to have the following characteristics: each has a single input and a single output; all are clocked by the same periodic waveform; and at each clock tick, each storage element samples its input and the sampled value is made available at the output until the next tick. We also assume that changes in the output of one storage element do not interfere with the input to another at the same clock tick. An example of a noninterfering storage element is an edge-triggered, master-slave, D-type flip-flop [18].

The functional elements provide the computational power of the circuit. Our model is unconcerned with the level of complexity of the functional elements—they might be NAND gates, multiplexors, or ALU's, for example. Each functional element has an associated *propagation delay*. The outputs of each functional element at any time are defined as a specified function of its inputs, provided that all the inputs have been stable for at least a time equal to its propagation delay. We make the conservative assumption that when an input to a functional element changes, the outputs may behave arbitrarily until they settle to their final values.

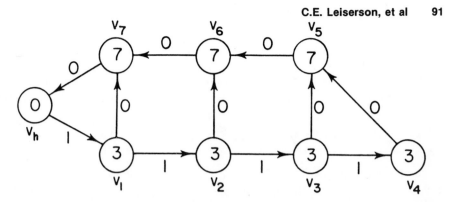

Figure 3. The graph model of Correlator 1 from Figure 1.

To be precise, we model a circuit as a finite, rooted, vertex-weighted, edge-weighted, directed multigraph $G = \langle V, E, v_h, d, w \rangle$ (henceforth, we shall simply say "graph" or, more frequently, "circuit"). Figure 3 shows the graph of Correlator 1 from Figure 1. The vertices V of the graph model the functional elements of the circuit. Each vertex $v \in V$ is weighted with its numerical propagation delay $d(v)$. A root vertex v_h, called the *host*, is included to represent the interface with the external world, and it is given zero propagation delay. (Multiple, independent external interfaces can be handled without difficulty, but their inclusion unnecessarily complicates the model.)

The directed edges E of the graph model interconnections between functional elements. Each edge $e \in E$ connects an output of some functional element to an input of some functional element and is weighted with a *register count* $w(e)$. The register count is the number of registers along the connection.[2] Between two vertices, there may be multiple edges with different register counts.

We shall use the following terminology extensively. To avoid confusion between vertex-weight functions such as d and edge-weight functions such as w, we shall use the term *weight* for edge-weight functions only. In fact, the only vertex-weight functions we use are the propagation delays $d(v)$, and in general we shall refer to the particular edge weights $w(e)$ of a circuit as register counts. If e is an edge in a graph that goes from vertex u to vertex v, we shall use the notation $u \xrightarrow{e} v$. In the event that the identity of either the head or the tail of an edge is unimportant, we shall use the symbol ?, as in $u \xrightarrow{e} ?$.

For a graph G, we shall view a path p in G as a sequence of vertices and edges. If a path p starts at a vertex u and ends at a vertex v, we use the notation $u \xrightarrow{p} v$. A *simple path* contains no vertex twice, and therefore the number of vertices exceeds the number of edges by exactly one.

We extend the register count function w in a natural way from single edges to arbitrary paths. For any path $p = v_0 \xrightarrow{e_0} v_1 \xrightarrow{e_1} \cdots \xrightarrow{e_{k-1}} v_k$, we define the *path weight* as the sum of the weights of the edges of the path:

---

[2] If an output of a functional element fans out to more than one other functional element, the single interconnection can be treated, without loss of generality, as several edges, each with an appropriate weight. Any optimization can be translated from the model back to a circuit with fanout. Section 7 examines fanout more closely.

$$w(p) = \sum_{i=0}^{k-1} w(e_i).$$

Similarly, we extend the propagation delay function d to simple paths. For any simple path $p = v_0 \xrightarrow{e_0} v_1 \xrightarrow{e_1} \cdots \xrightarrow{e_{k-1}} v_k$, we define the *path delay* as the sum of the delays of the vertices of the path:

$$d(p) = \sum_{i=0}^{k} d(v_i).$$

In order that a graph $G = \langle V, E, v_h, d, w \rangle$ have well-defined physical meaning as a circuit, we place nonnegativity restrictions on the propagation delays $d(v)$ and the register counts $w(e)$:

D1. *The propagation delay $d(v)$ is nonnegative for each vertex $v \in V$.*

W1. *The register count $w(e)$ is a nonnegative integer for each edge $e \in E$.*

We also impose the restriction that there be no directed cycles of zero weight:

W2. *In any directed cycle of G, there is some edge with (strictly) positive register count.*

We define a *synchronous circuit* as a circuit that satisfies Conditions D1, W1, and W2. The reason for including Condition W2 is that whenever an edge e between two vertices u and v has zero weight, a signal entering vertex u can ripple unhindered through vertex u and subsequently through vertex v. If the rippling can feed back upon itself, problems of asynchronous latching, oscillation, and race conditions can arise. By prohibiting zero-weight cycles, Condition W2 prevents these problems from occurring, provided that system clock runs slowly enough to allow the outputs of all the functional elements to settle between each two consecutive ticks.

For any synchronous circuit G, we define the (minimum feasible) *clock period* $\Phi(G)$ as the maximum amount of propagation delay through which any signal must ripple between clock ticks. Condition W2 guarantees that the clock period is well defined by the equation

$$\Phi(G) = \max\{ d(p) \mid w(p) = 0 \}.$$

For the circuit graph in Figure 3 the clock period is 24, which corresponds to the sum of the propagation delays along the path $v_4 \to v_5 \to v_6 \to v_7$.

Determination of the clock period $\Phi(G)$ is relatively simple. The algorithm we present here is similar to an algorithm that forms a part of a design tool developed at American Microsystems, Inc. [16].

Algorithm CP (*Compute the clock period of a circuit*). This algorithm computes the clock period $\Phi(G)$ for a synchronous circuit $G = \langle V, E, v_h, d, w \rangle$.

1. Let G_0 be the subgraph of G that contains precisely those edges e with register count $w(e) = 0$.

2. By Condition W2, G_0 is acyclic. Perform a topological sort on G_0, totally ordering its vertices so that if there is an edge from vertex u to vertex v in G_0, then u precedes v in the total order.

3. Go through the vertices in the order defined by the topological sort. On visiting each vertex v, compute the quantity $\Delta(v)$ as follows:

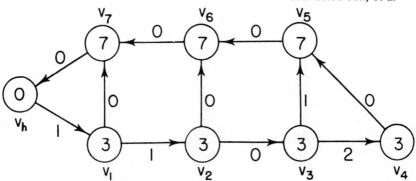

Figure 4. The graph of a correlator equivalent to Correlator 1, but with vertex v_3 having a lead of 1 clock tick.

    a. If there is no incoming edge to v, set $\Delta(v) \leftarrow d(v)$.

    b. Otherwise, set $\Delta(v) \leftarrow d(v) + \max\{\,\Delta(u) \mid u \overset{e}{\to} v$ and $w(e) = 0\,\}$.

4. The clock period $\Phi(G)$ is $\max_{v \in V} \Delta(v)$. ∎

The algorithm works because for each vertex v, the quantity $\Delta(v)$ equals the maximum sum $d(p)$ of vertex delays along any zero-weight directed path p in G such that ? $\overset{p}{\twoheadrightarrow} v$. The running time is $O(|E|)$.

3. Retiming

Retiming transformations can alter the clock period of a circuit by relocating registers. Consider the circuit graph in Figure 4, which differs from the graph in Figure 3 (Correlator 1) only in the placement of registers. In particular, the difference is that vertex v_3 in Figure 4 has one more register on each outgoing edge and one fewer register on its incoming edge. Intuitively, the two circuits have the same input-output behavior as seen from the host, but the computations performed by v_3 in Figure 4 *lead* by one clock tick the same computations performed by v_3 in Figure 3. Alternatively, we say that vertex v_3 in Figure 3 *lags* by one clock tick the corresponding vertex in Figure 4.

Figure 5 shows the result of further *retiming* the circuit from Figure 4. In this figure, vertex v_4 has been assigned a lead of one clock tick, or equivalently a lag of -1. Also, the clock period of the circuit in Figure 5 is shorter—21 instead of 24.

A retiming can be viewed as an assignment of a lag to each vertex in a circuit, and this is how we shall define it formally. A *retiming* of a circuit $G = \langle V, E, v_h, d, w \rangle$ is an integer-valued vertex-labeling $r : V \to \mathbf{Z}$ such that $r(v_h) = 0$. The retiming specifies a transformation of the original circuit in which the registers are added and removed so as to change the graph G into a new graph $G_r = \langle V, E, v_h, d, w_r \rangle$, where the edge-weighting w_r is defined for an edge $u \overset{e}{\to} v$ by the equation

$$w_r(e) = w(e) + r(v) - r(u). \tag{2}$$

In the example of Figure 3, the retiming that assigns -1 to functional elements v_3 and v_4, and 0 to all other vertices, yields the circuit of Figure 5.

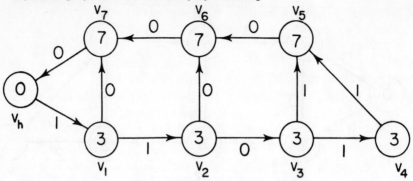

Figure 5. The graph of Correlator 1 modified so that both v_3 and v_4 have leads of 1 clock tick.

Equation (2), which tells how retiming affects the register counts of edges, extends naturally to paths.

Lemma 1. *Let $G = \langle V, E, v_h, d, w \rangle$ be a synchronous circuit, and let r be a retiming. Then for any path $u \xrightarrow{p} v$ in G, we have*

$$w_r(p) = w(p) + r(v) - r(u).$$

Proof. Suppose p is composed of vertices and edges $v_0 \xrightarrow{e_0} v_1 \xrightarrow{e_1} \cdots \xrightarrow{e_{k-1}} v_k$. We have

$$
\begin{aligned}
w_r(p) &= \sum_{i=0}^{k-1} w_r(e_i) \\
&= \sum_{i=0}^{k-1} (w(e_i) + r(v_{i+1}) - r(v_i)) \\
&= \sum_{i=0}^{k-1} w(e_i) + \sum_{i=0}^{k-1} (r(v_{i+1}) - r(v_i)) \\
&= w(p) + r(v_k) - r(v_0)
\end{aligned}
$$

because the sum on the right telescopes. ∎

Corollary 2. *Let $G = \langle V, E, v_h, d, w \rangle$ be a synchronous circuit, and let r be a retiming on the vertices of G. Then for any cycle p in G, we have $w_r(p) = w(p)$.*

Proof. Immediate from Lemma 1. ∎

A retiming of r of a circuit G is *legal* if the retimed graph G_r satisfies Conditions W1 and W2. An arbitrary assignment of lags to the vertices of a circuit G may cause the retimed circuit G_r to violate Condition W1, which says that no edge may have a negative register count. This condition must be checked explicitly in order to ensure that a retiming is legal. Interestingly enough, Condition W2 need not also be checked because of the following consequence of Corollary 2.

Corollary 3. *Let* $G = \langle V, E, v_h, d, w \rangle$ *be a synchronous circuit, and let* r *be a retiming on the vertices of* G. *Then the retimed graph* $G_r = \langle V, E, v_h, d, w_r \rangle$ *satisfies Condition W2.*

Proof. Let p be any cycle in G. We must show that p includes at least one edge e such that $w_r(e) > 0$. Since graph G satisfies Conditions W1 and W2, the register count $w(p)$ of the cycle in G must be postive. But the register count $w_r(p)$ of the cycle in G_r is by Corollary 2 equal to $w(p)$ and is therefore positive. Hence there must be an edge on the cycle in G_r that has positive register count. ∎

To conclude this section, we comment that it is necessary to prove that when a circuit G is retimed to produce a new graph G_r, the new circuit is functionally *equivalent,* as seen by the external world, to the original—provided, of course, that G_r satisfies Conditions W1 and W2. Such a proof can be found in [12], which also contains a technical definition of the term "equivalent."

Moreover, retiming is, in a sense, the most general possible method for changing the register counts within a circuit without disturbing the circuit's functionality. Although we do not formally prove it here, we outline the thread of reasoning. Without loss of generality, assume that any circuit $G = \langle V, E, v_h, d, w \rangle$ under discussion has the following two properties. 1. Every vertex $v \in V$ is connected by a path to the host vertex v_h. (Otherwise, computations performed at v can never affect the external behavior of the circuit.) 2. Every vertex $v \in V$ has at least one input. (Otherwise, v computes a constant function.[3]) Given the graph of such a circuit, but no knowledge of what functions are computed by the functional elements, it is impossible, other than by retiming, to alter the register counts on the edges and be assured that the external behavior is unchanged. For any relabeling of the edge weights that is not a retiming, an adversary can specify the functional elements in such a way that the new circuit behaves differently from the original circuit. We omit the details of this argument. (For an example of a similar adversary argument used to prove a slightly weaker result than the one we are claiming here, see the proof of Theorem 3 of [12]).

4. Determining an optimal retiming

This section presents a polynomial-time algorithm for relocating registers within a circuit so as to maximize the performance of the circuit. Specifically, we will solve the following problem: *Given a circuit graph* $G = \langle V, E, v_h, d, w \rangle$, *find a legal retiming* r *of* G *such that the clock period* $\Phi(G_r)$ *of the retimed circuit* G_r *is as small as possible.* The solution of this problem depends on some basic results from combinatorial optimization and graph theory. In particular, we rely on the fact that the following linear programming problem can be solved efficiently.

Problem LP. *Let* S *be a set of* m *linear inequalities of the form*

$$x_j - x_i \leq a_{ij} \tag{3}$$

on the unknowns $x_1, x_2, \ldots, x_n$, *where the* a_{ij} *are given real constants. Determine feasible values for the unknowns* x_i, *or determine that no such values exist.*

---

[3]In the graph (Figure 1) of Correlator 1, for example, we do not use functional elements to input the constants a_i, but have instead incorporated them into the comparators.

W	v_h	v_1	v_2	v_3	v_4	v_5	v_6	v_7
v_h	0	1	2	3	4	3	2	1
v_1	0	0	1	2	3	2	1	0
v_2	0	1	0	1	2	1	0	0
v_3	0	1	2	0	1	0	0	0
v_4	0	1	2	3	0	0	0	0
v_5	0	1	2	3	4	0	0	0
v_6	0	1	2	3	4	3	0	0
v_7	0	1	2	3	4	3	2	0

D	v_h	v_1	v_2	v_3	v_4	v_5	v_6	v_7
v_h	0	3	6	9	12	16	13	10
v_1	10	3	6	9	12	(16)	13	10
v_2	17	20	3	6	9	13	10	17
v_3	24	27	30	3	6	10	17	24
v_4	24	27	30	33	3	10	17	24
v_5	21	24	27	30	33	7	(14)	21
v_6	14	17	20	23	26	30	7	(14)
v_7	7	10	13	(16)	19	23	(20)	7

Figure 6. Tables showing the values of the functions W and D for Correlator 1. The quantity $W(u,v)$ is the number of registers on a minimum-weight path from u to v, and $D(u,v)$ is the maximum propagation delay along any such *critical path*. The distinct entries in the table for D include all possible clock periods for any retiming of Correlator 1. A legal retiming r has a clock period less than c if and only if $W_r(u,v) > 0$ wherever $D(u,v) > c$. Circled entries in the table for D are explained in the last paragraph of Section 4.

Constraint systems in which each equation has the form of Inequality (3) have been studied extensively. Any such system of linear inequalities can be satisfied—or determined to be inconsistent—in $O(mn)$ time by the Bellman-Ford algorithm [9, p. 74]. (The appendix contains a brief review of this algorithm.) VLSI applications include compaction [2, 14] and placement for river routing [11].

The algorithm for optimizing the clock period of a circuit is based on an alternate characterization of clock period in terms of two quantities which we now define:

$$
\begin{aligned}
W(u,v) &= \min\{\, w(p) \mid u \xrightarrow{p} v \,\}, \\
D(u,v) &= \max\{\, d(p) \mid u \xrightarrow{p} v \text{ and } w(p) = W(u,v) \,\}.
\end{aligned}
\tag{4}
$$

The quantity $W(u,v)$ is the minimum number of registers on any path from vertex u to vertex v. We call a path $u \xrightarrow{p} v$ such that $w(p) = W(u,v)$ a *critical path* from u to v. The quantity $D(u,v)$ is the maximum total propagation delay on any critical path from u to v. Both quantities are undefined if there is no path from u to v. Figure 6 shows the values for Correlator 1.

Lemma 4. Let $G = \langle V, E, v_h, d, w \rangle$ be a synchronous circuit, and let c be any positive real number. The following are equivalent:

4.1. $\Phi(G) \leq c$.

4.2. For all vertices u and v in V, if $D(u,v) > c$, then $W(u,v) \geq 1$.

Proof. (4.1 $\Rightarrow$ 4.2): Suppose $\Phi(G) \leq c$, and let u and v be vertices in V such that $D(u,v) > c$. If $W(u,v) = 0$, then there exists a path p from u to v with propagation delay $d(p) = D(u,v)$, which is greater than c, and register count $w(p) = W(u,v) = 0$. Contradiction.

(4.2 $\Rightarrow$ 4.1): Suppose 4.2 holds, and let $u \xrightarrow{p} v$ be any zero-weight path in G. Then we have $W(u,v) = w(p) = 0$, which implies $d(p) \leq D(u,v) \leq c$. ∎

It is not difficult to compute W by solving the all-pairs shortest-paths problem in G. Common ways of solving this problem are the Floyd-Warshall method [9, p. 86], which runs in $O(|V|^3)$ time, and Johnson's algorithm [4], which runs in $O(|V||E| \lg|V|)$ time. (The Floyd-Warshall algorithm is included in the appendix.) The basic operations on weights used by these algorithms are addition and comparison. The following algorithm shows that with a suitably chosen weight function, an all-pairs shortest-paths algorithm can be used to compute both W and D.

Algorithm WD (*Compute W and D*). Given a synchronous circuit $G = \langle V, E, v_h, d, w \rangle$, this algorithm computes $W(u, v)$ and $D(u, v)$ for all $u, v \in V$ such that u is connected to v in G.

1. Weight each edge $u \xrightarrow{e} ?$ in E with the ordered pair $(w(e), -d(u))$.
2. Using the weighting from Step 1, compute the weight of the shortest path joining each connected pair of vertices by solving an all-pairs shortest-paths algorithm. (In the all-pairs algorithm, add two weights by performing componentwise addition. Compare weights using lexicographic ordering.)
3. For each shortest-path weight (x, y) between two vertices u and v, set $W(u, v) \leftarrow x$ and $D(u, v) \leftarrow d(v) - y$. ∎

The reason that W and D are important is that they behave nicely under retiming.

Lemma 5. *Let $G = \langle V, E, v_h, d, w \rangle$ be a sychronous circuit, and let W and D be defined on G by the equations (4). Let r be a legal retiming of G, and let W_r and D_r be defined analogously on G_r. Then*

　　5.1. *a path p is a critical path of G_r if and only if it is a critical path of G,*
　　5.2. $W_r(u, v) = W(u, v) + r(v) - r(u)$ *for all connected vertices $u, v \in V$, and*
　　5.3. $D_r(u, v) = D(u, v)$ *for all connected vertices $u, v \in V$.*

Proof. Assertion 5.1 follows from Lemma 1 because retiming changes the weights of all paths from u to v by the same amount, and then 5.2 follows immediately. Assertion 5.3 is a consequence of 5.2 together with the fact that retiming does not alter propagation delays. ∎

The next result is a corollary to Lemma 5 which shows that the range of D contains the clock periods of all circuits obtainable by retiming G. In Figure 6 the 20 distinct values in the table for D include all possible clock periods for any retiming of Correlator 1.

Corollary 6. *Let $G = \langle V, E, v_h, d, w \rangle$ be a synchronous circuit, and let r be a retiming of G. Then the clock period $\Phi(G_r)$ is equal to $D(u, v)$ for some $u, v \in V$.*

Proof. By the definition of clock period, the circuit G_r contains some zero-weight path $u \xrightarrow{p} v$ such that $d(p) = \Phi(G_r)$. Thus $W_r(u, v) = w_r(p) = 0$. Moreover, no zero-weight path in G_r has greater propagation delay than p. Hence, $D_r(u, v) = d(p)$, and by Lemma 5 we have $\Phi(G_r) = D_r(u, v) = D(u, v)$. ∎

Lemma 4 and Lemma 5 also allow us to characterize the conditions under which a retiming produces a circuit whose clock period is no greater than a given constant.

Theorem 7. *Let $G = \langle V, E, v_h, d, w \rangle$ be a synchronous circuit, let c be an arbitrary positive real number, and let r be a function from V to the integers. Then r is a legal retiming of G such that $\Phi(G_r) \leq c$ if and only if*

　　7.1. $r(v_h) = 0$,
　　7.2. $r(u) - r(v) \leq w(e)$ *for every edge $u \xrightarrow{e} v$ of G, and*
　　7.3. $r(u) - r(v) \leq W(u, v) - 1$ *for all vertices $u, v \in V$ such that $D(u, v) > c$.*

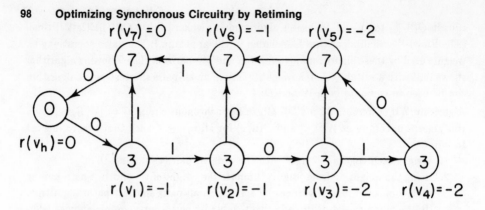

Figure 7. The graph of Correlator 2 from Figure 2. For each vertex v, the value $r(v)$ is the lag of v with respect to the corresponding vertex in Correlator 1. The retimed weight of an edge e from u to v is given by $w_r(e) = w(e) + r(v) - r(u)$.

Proof. By Corollary 3, the retiming r is legal if and only if Assertions 7.1 and 7.2 hold. If r is indeed a legal retiming of G, then by Lemma 4 the retimed circuit G_r has clock period $\Phi(G_r) \leq c$ precisely under the condition that $W_r(u, v) \geq 1$ for all vertices $u, v \in V$ such that $D_r(u, v) > c$. Since by Lemma 5 we have $W_r(u, v) = W(u, v) + r(v) - r(u)$ and $D_r(u, v) = D(u, v)$, this condition is equivalent to Assertion 7.3. ∎

Theorem 7 provides the basic tool needed to solve the optimal retiming problem. Notice that the constraints on the unknowns $r(v)$ in the theorem are linear inequalities involving only differences of unknowns, and thus we have an instance of Problem LP.[4] Therefore, using the Bellman-Ford algorithm to test whether a retimed circuit exists with clock period less than some constant c takes $O(|V|^3)$ time since there can be only $O(|V|^2)$ inequalities.

We now present an algorithm to determine a retiming for a circuit G such that the clock period of the retimed circuit is minimized.

Algorithm OPT1 (*Optimally retime a circuit*). Given a synchronous circuit $G = \langle V, E, v_h, d, w \rangle$, this algorithm determines a retiming r such that $\Phi(G_r)$ is as small as possible.

1. Compute W and D using Algorithm WD.
2. Sort the elements in the range of D.
3. Binary search among the elements $D(u, v)$ for the minimum achievable clock period. To test whether each potential clock period c is feasible, apply the Bellman-Ford algorithm to determine whether the conditions in Theorem 7 can be satisfied.
4. For the minimum achievable clock period found in Step 3, use the values for the $r(v)$ found by the Bellman-Ford algorithm as the optimal retiming. ∎

---

[4]Actually, we have the *integer linear programming* version of the problem (see Problem ILP in the appendix) because the unknowns $r(v)$ are required to be integer. Since the value on the right hand side of each equation is integer, however, the Bellman-Ford algorithm will produce an integer optimal solution if one exists.

Algorithm OPT1 runs in $O(|V|^3 \lg|V|)$ time. An algorithm with better asymptotic performance is given in the next section, but for some circuits, we may improve the performance of Algorithm OPT1 by using a smaller set of inequalities. The key observation is that we may eliminate any inequality $r(u) - r(v) \leq W(u,v) - 1$ from Item 7.3 if either $D(u,v) - d(v) > c$ or $D(u,v) - d(u) > c$. The intuition behind this optimization is that there is no need to explicitly require a path p to have positive weight $w_r(p)$ if we already require some subpath of p to have positive weight.

As an example, Figure 7 shows the circuit graph of Correlator 2, which was obtained from Correlator 1 by applying the Bellman-Ford algorithm to the inequalities from Theorem 7 with clock period $c = 13$. There are 11 inequalities (one for each edge) that must be satisfied to ensure a legal retiming—Item 7.2 in Theorem 7. Of the potential 34 inequalities arising from cases where $D(u,v) > 13$—Item 7.3 in the theorem—only five need be included if we eliminate those for which either $D(u,v) - d(v) > 13$ or $D(u,v) - d(u) > 13$. In the table for D in Figure 6, those entries corresponding to the five relevant inequalities are circled.

5. A more efficient algorithm for optimal retiming

In this section we describe an asymptotically more efficient algorithm for optimal retiming. Specifically, we will show that the *feasible clock period test* in Step 3 of Algorithm OPT1, which determines whether there exists a retiming of G with clock period at most c, can be performed in $O(|V||E| \lg|V|)$ time, a significant improvement over $O(|V|^3)$ for sparse graphs. The algorithm is based on a reduction of the feasible clock period test to the following mixed-integer programming problem.

Problem MILP. *Let S be a set of m linear inequalities of the form $x_j - x_i \leq a_{ij}$ on the unknowns $x_1, x_2, \ldots, x_n$, where the a_{ij} are given real constants, and let k be given. Determine feasible values for the unknowns x_i subject to the constraint that x_i is integer for $i = 1, 2, \ldots, k$ and real for $i = k + 1, k + 2, \ldots, n$, or determine that no such values exist.*

Although mixed-integer programming is in general NP-complete (because integer programming is [1, p. 245]), this special case can be solved in $O(mn + km \lg n)$ time [13]. (The appendix gives a less efficient, but easily programmed, algorithm to solve Problem MILP.) The reduction of the feasible clock period test to Problem MILP makes use of the following lemma.

Lemma 8. *Let $G = \langle V, E, v_h, d, w \rangle$ be a synchronous circuit, and let c be a positive real number. Then the clock period $\Phi(G)$ is less than or equal to c if and only if there exists a function $s : V \rightarrow [0, c]$ such that $s(v) \geq d(v)$ for every vertex v and such that $s(v) \geq s(u) + d(v)$ for every zero-weight edge $u \xrightarrow{e} v$.*

Proof. For each vertex v, let $\Delta(v)$ be the maximal sum of the combinational delays along any path that both ends at v and contains only zero-weight edges. (This is the same Δ as in Algorithm CP.) By definition, $\Phi(G) \leq c$ if and only if $\Delta(v) \leq c$ for all v. If $\Phi(G) \leq c$, the function Δ satisfies the desired properties for s. Conversely, if a function s exists having the desired properties, then we have $\Delta(v) \leq s(v) \leq c$ for every vertex v. ∎

Lemma 8 and Corollary 3 together give a characterization of when it is possible to retime a circuit so that the retimed circuit has a clock period of c or less.

Lemma 9. *Let $G = \langle V, E, v_h, d, w \rangle$ be a synchronous circuit, and let c be a positive real number. Then there exists a retiming r of G such that $\Phi(G_r) \leq c$ if and only if there exists an assignment of a real value $s(v)$ and an integer value $r(v)$ to each vertex $v \in V$ such that the following conditions are satisfied:*

9.1. $-s(v) \leq -d(v)$ *for every vertex $v \in V$,*

9.2. $s(v) \leq c$ *for every vertex $v \in V$,*

9.3. $r(u) - r(v) \leq w(e)$ *wherever $u \xrightarrow{e} v$,*

9.4. $s(u) - s(v) \leq -d(v)$ *wherever $u \xrightarrow{e} v$ such that $r(u) - r(v) = w(e)$, and*

9.5. $r(v_h) = 0$.

Proof. Conditions 9.3 and 9.5 capture the requirements for r to be legal retiming, as given in Corollary 3, namely that r must be a mapping from V to $\mathbf{Z}$ such that $r(v_h) = 0$ and G_r satisfies Condition W1. Conditions 9.1, 9.2, and 9.4 capture the requirement for G_r to have a clock period $\Phi(G_r) \leq c$ as given in Lemma 8. (Recall that G_r is defined to have $w_r(e) = w(e) + r(v) - r(u)$ for each edge $u \xrightarrow{e} v$.) ∎

Unfortunately, this result does not quite allow us to recast a feasible clock period recognition problem as an instance of Problem MILP because of the qualifying clause "such that $r(u) - r(v) = w(e)$" in Condition 9.4. The next theorem shows that the conditions can be expressed without such a clause.

Theorem 10. *Let $G = \langle V, E, v_h, d, w \rangle$ be a synchronous circuit, and let c be a positive real number. Then there is a retiming r of G such that $\Phi(G_r) \leq c$ if and only if there exists an assignment of a real value $R(v)$ and an integer value $r(v)$ to each vertex $v \in V$ such that the following conditions are satisfied:*

10.1. $r(v) - R(v) \leq -d(v)/c$ *for every vertex $v \in V$,*

10.2. $R(v) - r(v) \leq c$ *for every vertex $v \in V$,*

10.3. $r(u) - r(v) \leq w(e)$ *wherever $u \xrightarrow{e} v$, and*

10.4. $R(u) - R(v) \leq w(e) - d(v)/c$ *wherever $u \xrightarrow{e} v$.*

Proof. Any solution to the conditions in Lemma 9 can be converted to a solution to the above conditions by using the same values for the $r(v)$ and taking $R(v) = r(v) + s(v)/c$ for each vertex v. Conversely, any solution to the conditions above is easily converted to a solution in which $r(v_h) = 0$. Such a solution yields a solution to the conditions in Lemma 9 using the substitution $s(v) = c(R(v) - r(v))$. ∎

Theorem 10 is the basis for the following improvement on Algorithm OPT1.

Algorithm OPT2 (*Optimally retime a circuit*). Given a synchronous circuit $G = \langle V, E, v_h, d, w \rangle$, this algorithm determines a retiming r such that $\Phi(G_r)$ is as small as possible.

1. Compute W and D using Algorithm WD.
2. Sort the elements in the range of D.
3. Binary search among the elements $D(u, v)$ for the minimum achievable clock period. To test whether each potential clock period c is feasible, solve Problem MILP to determine whether the conditions in Theorem 10 can be satisfied.

4. For the minimum achievable clock period found in Step 3, use the values for r found by the algorithm that solves Problem MILP as the optimal retiming. ∎

This algorithm can be made to run in $O(|V||E|\lg^2|V|)$ time by choosing efficient algorithms for each of the steps. If Johnson's all-pairs shortest-paths algorithm [4] is used in Algorithm WD, Step 1 runs in $O(|V||E|\lg|V|)$ time. Since there are only $O(|V|^2)$ elements in the range of D, Step 2 runs in $O(|V|^2\lg|V|)$ time. Each iteration of the binary search in Step 3 requires solving an instance of Problem MILP with $|V|$ integer variables, $|V|$ real variables, and $2|V|+2|E| = O(E)$ inequalities. Thus the total time for Step 3 is $O(|V||E|\lg^2|V|)$ if the algorithm from [13] is used. (The appendix contains a less efficient algorithm.) The optimal retiming from Step 4 is produced as a side effect of Step 3.

In fact, the bound is somewhat better than $O(|V||E|\lg^2|V|)$ for graphs with many more edges than vertices. Both Johnson's all-pairs shortest-paths algorithm and the algorithm [13] for solving Problem MILP actually run in time $O(|V||E| + h|V|^{2+\frac{1}{h}})$, where h is an arbitrary positive integer that may be chosen after an instance of the problem is given. The choice $h = \lceil\lg|V|\rceil$ gives the bound stated in the previous paragraph. For families of circuits in which $|E| = \Omega(|V|^{1+\epsilon})$ for some constant $\epsilon > 0$, the choice $h = \lceil 1/\epsilon\rceil$ gives an $O(|V||E|)$ bound on the time required for the feasible clock period test, thereby allowing optimal retimings to be found in $O(|V||E|\lg|V|)$ time.

An asymptotic improvement of almost a logarithmic factor is possible even for very sparse graphs. The result is based on using van Emde Boas's priority queue scheme [19] (with modifications suggested by Johnson [5]), which requires $O(\lg\lg n)$ time per operation, to implement one of the steps in the algorithm [13] for Problem MILP. The improvement is based on the idea that the set of values that ever need to be placed in the priority queue is small, and that an "addition table" can be constructed to aid in maintaining a correspondence between these values and small integers. The total running time can thereby be reduced to $O(|V||E|\lg|V|\lg\lg|V|)$.

6. Unit propagation delay, systolic circuits, and slowdown

This section examines circuits in which the propagation delays of all functional elements are equal. Not surprisingly, the optimal retiming problem can be solved more efficiently for such circuits than for arbitrary circuits. Of more general interest, however, is the relation of this class of circuits to *systolic computation* [7], [8], [10], [12].

We define a circuit $G = \langle V, E, v_h, d, w\rangle$ to be a *unit-delay* circuit if each vertex $v \in V$ has propagation delay $d(v) = 1$. The next theorem gives a characterization of when a unit-delay circuit has clock period less than or equal to c. The theorem is phrased in terms of the graph $G - 1/c$, which is defined as $G - 1/c = \langle V, E, v_h, d, w'\rangle$ where $w'(e) = w(e) - 1/c$ for every edge $e \in E$. Thus $G - 1/c$ is the graph obtained from G by subtracting $1/c$ from the weight of each edge in G.

Theorem 11. *Let $G = \langle V, E, v_h, d, w\rangle$ be unit-delay synchronous circuit, and let c be any positive integer. Then there is a retiming r of G such that $\Phi(G_r) \leq c$ if and only if $G - 1/c$ contains no cycles having negative edge weight.*

Proof. Consider first the case where $G - 1/c$ has no negative-weight cycles. We shall produce a retiming r of G such that $\Phi(G_r) \leq c$. Assume without loss of generality that there is a path from each vertex v of G to v_h (if not, add edges of the form $v \to v_h$ with sufficiently large weight so that no negative-weight cycles are introduced), and let $g(v)$ be the weight of the shortest path from v to v_h in $G - 1/c$.

Arbitrarily choose an integer $k \in \{1, 2, \ldots, c\}$. For each vertex v, let $R(v) = k/c + g(v)$ and $r(v) = \lceil R(v) \rceil - 1$. The values of $r(v)$ and $R(v)$ so defined satisfy the conditions listed in Theorem 10—the key observation is that Condition 10.4 of the theorem follows directly from the definition of the $R(v)$ in terms of the shortest path weights $g(v)$.

On the other hand, suppose $G - 1/c$ contains some cycle p with negative weight. We must prove that G cannot be retimed to have a clock period of c or less. Let n be the number of edges in the cycle p. By the definition of $G - 1/c$, we have $w(p) - n/c = w'(p)$, where w' is the edge-weight function for $G - 1/c$. But by supposition, $w'(p)$ is negative which means that $w(p) - n/c < 0$, that is, p contains fewer than n/c registers in G. But retiming leaves the number of of registers on any cycle unchanged (Corollary 2). Thus, no matter how the fewer than n/c registers are distributed on the cycle of n vertices, there will be no registers on some string of more than c vertices. Consequently, G cannot be retimed to have a clock period of c or less. ∎

To test whether there is a retiming r of a unit-delay circuit G such that $\Phi(G_r) \leq c$, we can use the Bellman-Ford algorithm to find the weight $g(v)$ of the shortest path in $G - 1/c$ from each vertex v to the host vertex v_h.[5] If the shortest-path weights are not well defined, the Bellman-Ford algorithm will detect a negative-weight cycle, which means that no retiming exists. Thus the feasible clock period test can be performed in $O(|V||E|)$ time for unit-delay circuits.

A *systolic* circuit is a unit-delay circuit in which there is at least one register along every interconnection between two functional elements. Thus the clock period of a systolic circuit is the minimum possible—the propagation delay through a single functional element. Systolic circuits have been studied extensively ([7], [8], [10], [12]), and they have many applications including signal processing, matrix manipulation, machine vision, and raster graphics.

Interpreted in the context of systolic circuits, Theorem 11 is a generalization of the Systolic Conversion Theorem from [12], which says that G can be retimed to be systolic if if the constraint graph $G - 1$ has no cycles of negative weight. (Simply restrict Theorem 11 to the case where $c = 1$.) The Systolic Conversion Theorem is generalized in a different way in [12], however, through the idea of *slowdown*.

For any circuit $G = \langle V, E, v_h, d, w \rangle$ and any positive integer c, the *c-slow* circuit cG is the circuit obtained by multiplying all the register counts in G by c. That is, the circuit cG is defined as $cG = \langle V, E, v_h, d, w' \rangle$ where $w'(e) = cw(e)$ for every edge $e \in E$. All the data flow in cG is slowed down by a factor of c, so that cG performs the same computations as G, but takes c times as many clock ticks and communicates with the host only on every cth clock tick. In fact, cG acts as a set of c independent

---

[5]The appendix describes how the single-source shortest-paths problem can be solved by the Bellman-Ford algorithm. The single-destination shortest-paths problem can be solved similarly.

versions of G, communicating with the host in round-robin fashion.

What is the importance of c-slow circuits? Their main advantage is that they can be retimed to have shorter clock periods than any retimed version of the original. For many applications, throughput is the issue, and multiple, interleaved streams of computation can be effectively utilized. A c-slow circuit that is systolic offers maximum throughput. (We not only call cG a c-slow version of G, but also any circuit obtained by retiming cG.)

The following corollary to Theorem 11 tells when a circuit has a c-slow version which is systolic.

Corollary 12. *Let $G = \langle V, E, v_h, d, w \rangle$ be a unit-delay synchronous circuit, and let c be an arbitrary positive integer. Then the following are equivalent:*

12.1. *The graph $G - 1/c$ has no negative-weight cycles.*

12.2. *The circuit G can be retimed to have clock period less than or equal to c.*

12.3. *The c-slow circuit cG can be retimed to be systolic.*

Proof. That 12.1 and 12.2 are equivalent is exactly Theorem 11. The equivalence of 12.1 and 12.3 follows by applying Theorem 11 to cG with clock period 1, and observing that $cG - 1$ has a negative-weight cycle if and only if $G - 1/c$ has a negative-weight cycle. ∎

The registers of a c-slow circuit cG are naturally divided into c *equivalence classes*. Given any two registers A and B in cG, the number of registers on any two paths from A to B will be congruent modulo c.[6] Consequently, the registers are naturally divided into equivalence classes according to their distances (modulo c) from the host.

At any given time step, any two registers in different equivalence classes contain data from independent streams of computation—data that can never arrive at inputs of the same functional element at the same time. Although retiming destroys the individual identities of the registers, Lemma 1 guarantees that the registers of any retimed c-slow circuit can still be partitioned into c such equivalence classes.

Using the notion of equivalence classes of registers, the following scenario illustrates the relationships given in Corollary 12. Let $G = \langle V, E, v_h, d, w \rangle$ be a unit-delay synchronous circuit. Find an integer c such that $G - 1/c$ has no negative-weight cycles, and consider the circuit cG. There is a retiming r of cG such that $(cG)_r$ is systolic. If we remove all the registers in the c-slow circuit $(cG)_r$ except for those in one equivalence class, the resulting circuit will be a retimed version of the original circuit G whose clock period is less than or equal to c. The choice of the equivalence class to retain corresponds precisely to the arbitrary choice of the constant k in the proof of Theorem 11.

Many systolic circuits appearing in the literature are fundamentally 2-slow or 3-slow—even if the ideas of slowdown and retiming were not explicitly used in their design. For example, the systolic algorithms for band-matrix multiplication and LU-decomposition from [8] are 3-slow and can support three independent, interleaved streams of computation. If all independent streams of computation cannot be utilized in a c-slow circuit, it may be desirable to remove all registers except for those in one

---

[6]We define a "path" between two registers, as well as the register count of such a path, in the obvious way.

equivalence class. The following algorithm determines if a circuit is actually a c-slow version of another, and if so, produces the *reduced* circuit.

Algorithm R (*Remove all but one equivalence class of registers in a circuit*). Given a synchronous circuit $G = \langle V, E, v_h, d, w \rangle$ this algorithm determines the largest c and produces a reduced circuit G' such that G is a c-slow version of G'. (We assume without loss of generality that there is a path from each vertex to the host and that each vertex has at least one input.)

1. For each vertex $v \in V$, set $dist(v)$ to the weight of some path from v to the host vertex v_h.

2. Compute $c = \gcd\{w(e) + dist(v) - dist(u) \mid u \xrightarrow{e} v\}$.

3. For each vertex $v \in V$, set $r(v) = dist(v) \bmod c$.

4. Produce $G' = \langle V, E, v_h, d, w' \rangle$, where $w'(e) = (w(e) + r(v) - r(u))/c$ for each edge $u \xrightarrow{e} v$.

Proof of correctness. We first show that for each edge $u \xrightarrow{e} v$, the value $w'(e)$ is a legal register count. By construction, c evenly divides $w(e) + r(v) - r(u)$ because c divides $w(e) + dist(v) - dist(u)$. (If $c = 0$, the graph G is a retiming of a graph with all zero edge weights, which is not possible.) Thus for any edge, the register count $w'(e)$ produced in Step 4 is guaranteed to be an integer. In addition, $w'(e)$ is guaranteed to be strictly greater than -1 because $r(u)$ must be less than c and $w(e) + r(v)$ is at least 0. Since we have just shown that $w'(e)$ is integer, it must be nonnegative.

The construction in Step 4 directly provides the identity $G' = G_r/c$, and thus G is a c-slow version of G'. We now show that the c computed in Step 2 is the largest possible. Suppose there is a c' such that G is a c'-slow version of another circuit G'. We wish to show that c' divides $w(e) + dist(v) - dist(u)$ for each edge $u \xrightarrow{e} v$, and thus that c' divides c. For every vertex v, the weights of all paths in $c'G'$ from v to v_h are congruent modulo c'. Since retiming changes all path weights between two vertices by the same amount (Lemma 1), it must be the case that in G, the weights of all paths from v to v_h are congruent modulo c'. In particular, the weight of the path $u \twoheadrightarrow v_h$ that determines $dist(u)$ and the weight $w(e) + dist(v)$ of the path $u \xrightarrow{e} v \twoheadrightarrow v_h$ must be congruent modulo c'. Hence c' divides $w(e) + dist(v) - dist(u)$. ∎

Step 1 of Algorithm R can be performed in time $O(|V| + |E|) = O(|E|)$ by depth-first search. Step 3 runs in $O(|V|)$ time, and Step 4 takes $O(|E|)$ time. Step 2 takes more work, but not that much more. The computation of the greatest common divisor of $|E|$ integers can be performed in $O(|E| + \lg x)$ time, where x is the least nonzero absolute value of any of the numbers. Just start with this value x—which can be found in $O(|E|)$ time—as a tentative gcd, and gcd in each of the other numbers in any order. Each (mod) operation in Euclid's algorithm either uses up one of the $|E|$ numbers, or else divides the current tentative gcd by the golden ratio $(1 + \sqrt{5})/2$.[7] (As a practical matter, starting with any of the $|E|$ numbers as the initial tentative gcd would give reasonable performance, since the number of registers in a typical circuit is much less than exponential in the number of edges.) Thus the total running time of Algorithm R is $O(|E| + \lg x)$.

---

[7]That is, if n mod operations are performed to compute a new tentative gcd, then it will be smaller than the old tentative gcd by at least a factor of $((1 + \sqrt{5})/2)^{n-1}$.

Observe that Algorithm R works not only for unit-delay circuits, but for any synchronous circuit. Furthermore, when removing the extra equivalence classes of registers from a c-slow circuit, the clock period of the reduced circuit is not lengthened unduly. The definition of w' in Step 4 guarantees that for any path p of weight $w(p) \geq c$ in G, we have $w'(p) > C$, which implies that $\Phi(G') \leq c\Phi(G)$. To achieve the optimal clock period, however, it will usually be necessary to retime the reduced circuit.

A systolic circuit that is naturally c-slow can be converted using Algorithm R into a circuit that performs an operation on every clock tick and whose clock period is bounded by c. This conversion can result in a performance advantage because in practice, there are time penalties associated with the loading of registers. Because of this overhead, c clock ticks of a circuit with nominal period 1 typically use more time than one clock tick of a circuit with nominal period c. Also, a reduction in registers may save chip area, which can lead to further performance improvements since the wires will in general be shorter. A possible disadvantage of reducing the number of equivalence classes of registers is that throughput is also reduced in cases where the independent streams of computation might be effectively utilized.

7. Register minimization and fanout

Thus far, we have concentrated on clock period as the objective function for determining a retiming. In Section 6, however, we showed that the number of registers in a circuit could sometimes be reduced by a method other than retiming. This section shows the problem of retiming a circuit to minimize the total *state* of a circuit is polynomial-time solvable by reduction to a minimum-cost flow problem. We also show that the total state of a circuit can be minimized subject to a bound on the clock period. These results can be extended to reflect the widths of the interconnections and ways by which *fanout* is modeled.

For a given circuit $G = \langle V, E, v_h, d, w \rangle$, the problem is to determine a retiming r such that the total state $S(G_r) = \sum_{e \in E} w_r(e)$ of the retimed circuit is minimized. By the definition of w_r, we have

$$S(G_r) = \sum_{e \in E} w_r(e)$$
$$= \sum_{u \overset{e}{\to} v} (w(e) + r(v) - r(u))$$
$$= S(G) + \sum_{v \in V} r(v)(\text{indegree}(v) - \text{outdegree}(v)).$$

Since $S(G)$ is constant, minimizing $S_r(G)$ is equivalent to minimizing the quantity

$$\sum_{v \in V} r(v)(\text{indegree}(v) - \text{outdegree}(v)), \tag{5}$$

which is a linear combination of the $r(v)$ since $(\text{indegree}(v) - \text{outdegree}(v))$ is constant for each v. The minimization is subject to the constraint that for each edge $u \overset{e}{\to} v$, the register count $w_r(e)$ is nonnegative—that is,

$$r(u) - r(v) \leq w(e).$$

The optimization problem, cast in this manner, is the linear-programming dual of a minimum-cost flow problem and is solvable in polynomial time [9, p. 131]. (As a practical matter, this problem is efficiently solved by the primal simplex technique [3, p. 186].) Furthermore, since the constants $w(e)$ in the problem are all integers, there is an integer optimal solution to the $r(v)$.

More complicated problems can be solved within the same framework. For example, the total state of a circuit can be minimized subject to a bound on the clock period. Given a maximum allowable clock period c, we wish to find a retiming r that minimizes the state $S(G_r)$ of the retimed circuit subject to the condition that $\Phi(G_r) \leq c$. In this case, we must minimize the quantity (5) subject to the constraints from Theorem 7, which require that $r(u) - r(v) \leq w(e)$ wherever $u \overset{e}{\to} v$, and $r(u) - r(v) \leq W(u, v) - 1$ wherever $D(u, v) > c$. The problem is still the dual of a minimum-cost flow problem, but the minimization is subject to more constraints.

Another consideration in the state minimization problem is that the cost of adding a register is not the same on all edges. For example, it may be cheaper to add a register along a one-bit wide control path than along a 32-bit wide data path. We may model this phenomenon by assigning to each edge e a *breadth* $\beta(e)$ proportional to the cost of adding a register along e. The objective function which we must minimize is then given by

$$\sum_{v \in V} r(v) \left(\sum_{\substack{e \\ ? \to v}} \beta(e) - \sum_{\substack{e \\ v \to ?}} \beta(e) \right), \tag{6}$$

and the constraints on the $r(v)$ are unchanged. The problem is still the dual of a minimum-cost flow problem since the quantity in the large parentheses is a constant for each v. Although the $\beta(e)$ need not be integers, if there is a solution to this problem, there is an integer optimal solution because the linear programming tableau for the problem is unimodular and the right-hand side is an integer vector.

In a physical circuit, a signal from a register or functional element may *fan out* to several functional elements. As was described in Section 2, we model this situation with several different edges in the circuit graph. For the clock period optimization problem, there was no harm in modeling fanout in this manner, but for the register minimization problem, there can be. The difficulty that arises in the register minimization problem is that registers can be shared along the physical interconnection. The objective functions (5) and (6) do not take sharing into account.

Fanout can be incorporated into the model in several ways that allow the sharing of registers to be accounted for exactly. We begin by looking at the situation in Figure 8(a) where one vertex u has an output that fans out to two vertices v_1 and v_2. To properly deal with this situation in the register minimization problem, it is sufficient to introduce a dummy vertex $\hat{u}$ with zero propagation delay which models the fork of the interconnection, as is shown in Figure 8(b). When the circuit is retimed to minimize the number of registers, either the edge from $\hat{u}$ to v_1 or the edge from $\hat{u}$ to v_2 will have zero register count, and the edge from u to $\hat{u}$ will have the shared registers. In Figure 8(c) the edge to v_1 ends up with zero weight after retiming so as to minimize the total number of registers.

Large multiway forks present some modeling alternatives not encountered in the two-way case. If a physical interconnection is to be modeled, the fork can be decom-

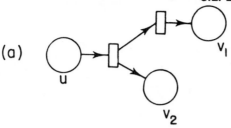

(a)

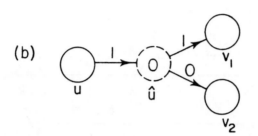

(b)

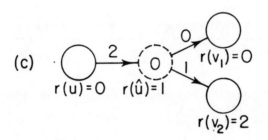

(c)

Figure 8. Modeling two-way fanout with an extra vertex having delay zero.

posed into several two-way forks. (In fact, our concern for modeling the physical interconnection prompted us to design Correlator 1 with the x_i running through the comparators rather than with multiway fanout directly from the host.)

For logical design, however, it may be undesirable to model the physical interconnection. In the case of a three-way fork, for instance, we might wish to share the largest possible number of registers between the two edges with greatest register counts, regardless of which two edges these end up being. Modeling a k-way fork for $k \geq 3$ by decomposing the interconnection into two-way forks will not work.

A solution to this problem of modeling k-way fanout with maximum register sharing is depicted in Figure 9. An output of vertex u, having breadth β, fans out to $v_1, \ldots, v_k$ along edges $u \xrightarrow{e_1} v_1, u \xrightarrow{e_2} v_2, \ldots, u \xrightarrow{e_k} v_k$. In the retimed circuit, the cost of this fanout should be β times the maximum of retimed edge weights $w_r(e_i)$. So that the register count cost function $S(G_r)$ will properly model the register sharing, we first add a dummy vertex $\hat{u}$ with zero propagation delay. Letting $w_{\max} = \max_{1 \leq i \leq k} w(e_i)$, we add edges $v_i \xrightarrow{\hat{e}_i} \hat{u}$ with weights $w(\hat{e}_i) = w_{\max} - w(e_i)$. Finally, we give all edges e_i and

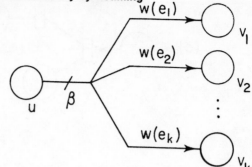

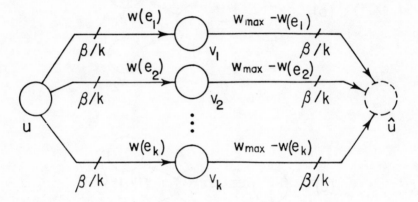

Figure 9. A gadget for modeling the cost of multiway fanout with maximal sharing of registers.

$\hat{e}_i$ breadths of β/k.

The modified circuit graph accurately models the sharing of registers among the edges e_i involved in the fanout when the state is minimized. For any retiming r, Lemma 1 dictates that the weights $w_r(p_i)$ of all paths $p_i = u \overset{e_i}{\rightarrow} v_i \overset{\hat{e}_i}{\rightarrow} \hat{u}$ will be identical since they are identical in the unretimed circuit. The retimed register counts $w_r(e_i)$ are constrained by the rest of the circuit, but the weights $w_r(\hat{e}_i)$ will be as small as possible because $\hat{u}$ is a sink in the graph. Thus the register count of one of the $\hat{e}_i$ will be zero, and therefore the weight of each path p_i will be $\max_{1 \le i \le k} w_r(e_i)$. Since there are k paths, each with breadth β/k, the total cost of the paths will be $\beta \cdot \max_{1 \le i \le k} w_r(e_i)$ as desired.

8. A more general model for propagation delay

In this section we extend the methods of Section 5 to deal with functional elements in which the propagation delays through individual functional elements are nonuniform. In an adder, for example, the propagation delay from a low-order input bit to a high-order output bit may be far greater than the propagation delay from a low-order input bit to a low-order output bit or from a high-order input bit to a high-order output bit. Thus the worst-case propagation delay through two cascaded

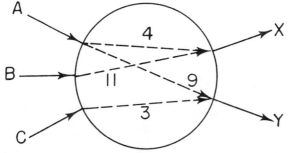

Figure 10. A functional element with nonuniform propagation delays. The time at which output X must settle is either 4 esec after input A settles or 11 esec after input B settles, whichever is later.

adders can be much less than twice the worst-case propagation delay through a single adder.

We may take into account such nonuniform propagation delays by modifying the model for synchronous circuits given in Section 2, so that from each input to each output of a given functional element, an independent propagation delay may be assigned. Figure 10 shows graphically the "insides" of a functional element in this model. The propagation delay function d, rather than being a function from V to the nonnegative reals, is a partial function from $E \times E$ to the nonnegative reals. The value $d(e_a, e_b)$ is the propagation delay from e_a to e_b through some vertex v, where $? \overset{e_a}{\to} v \overset{e_b}{\to} ?$. Not all such $d(e_a, e_b)$ need be specified, since a given output of a functional element need not depend on all the inputs. In the case of an adder, for example, the values of the high-order input bits have no effect on the low-order bits of the output.

Under the extended model, the clock period $\Phi(G)$ of a circuit $G = \langle V, E, v_h, d, w \rangle$ is the maximum delay $d(e_0, e_1) + d(e_1, e_2) + \cdots + d(e_{n-1}, e_n)$ along any path $? \overset{e_0}{\to} v_1 \overset{e_1}{\to} \cdots \overset{e_{n-1}}{\to} v_n \overset{e_n}{\to} ?$ such that $w(e_i) = 0$ for $i = 1, 2, \ldots, n-1$. We define retiming, which affects w but not d, exactly as it is defined in the model of Section 2.[8] The following results show how the optimal retiming problem for a circuit G under the extended model can be reduced to an instance of Problem MILP having one integer variable for each functional element of G and one real variable for each edge of G. The results, which parallel Lemma 8, Lemma 9, and Theorem 10, are presented without proof.

Lemma 13. *Let $G = \langle V, E, v_h, d, w \rangle$ be a synchronous circuit in the extended model, and let c be a positive real number. Then the clock period $\Phi(G)$ is less than or equal to c if and only if there exists a function $s : E \to [0, c]$ such that*

  13.1. $s(e_b) \geq d(e_a, e_b)$ *wherever $d(e_a, e_b)$ is defined, and*

  13.2. $s(e_b) \geq s(e_a) + d(e_a, e_b)$ *wherever $d(e_a, e_b)$ is defined and $w(e_a) > 0$.*

---

[8]The possibility that the internal connections between the inputs and outputs of a functional element may not be a complete bipartite graph gives rise to some technical differences between the extended model and the model of Section 2. First, Condition W2 need only be imposed for those cycles in which consecutive edges are actually connected by the internal data paths in the vertices. Second, retiming may not be the only way to adjust register counts so that function is guaranteed to be preserved—if the (undirected) graph of internal connections in some functional element is not connected, then the element can be broken up into two or more independent components which can be given different lags.

Lemma 14. *Let $G = \langle V, E, v_h, d, w \rangle$ be a synchronous circuit, and let c be a positive real number. Then there is a retiming r of G such that $\Phi(G_r) \leq c$ if and only if there exists an assignment of a real value $s(e)$ to each edge $e \in E$ and an integer value $r(v)$ to each vertex $v \in V$ such that the following conditions are satisfied:*

14.1. $-s(e_b) \leq -d(e_a, e_b)$ *wherever* $d(e_a, e_b)$ *is defined,*

14.2. $s(e) \leq c$ *for every edge* $e \in E$,

14.3. $r(u) - r(v) \leq w(e)$ *wherever* $u \xrightarrow{e} v$,

14.4. $s(e_a) - s(e_b) \leq -d(e_a, e_b)$ *wherever* $u \xrightarrow{e_a} v \xrightarrow{e_b} ?$, $r(u) - r(v) = w(e_a)$, *and* $d(e_a, e_b)$ *is defined, and*

14.5. $r(v_h) = 0$.

Theorem 15. *Let $G = \langle V, E, v_h, d, w \rangle$ be a synchronous circuit in the extended model, and let c be a positive real number. Then there is a retiming r of G such that $\Phi(G_r) \leq c$ if and only if there exists an assignment of a real value $R(e)$ to each edge $e \in E$ and an integer value $r(v)$ to each vertex $v \in V$ such that the following conditions are satisfied:*

15.1. $r(v) - R(e_b) \leq -d(e_a, e_b)/c$ *wherever* $? \xrightarrow{e_a} v \xrightarrow{e_b} ?$ *and* $d(e_a, e_b)$ *is defined,*

15.2. $R(e_b) - r(v) \leq c$ *wherever* $v \xrightarrow{e_b} ?$,

15.3. $r(u) - r(v) \leq w(e)$ *wherever* $u \xrightarrow{e} v$, *and*

15.4. $R(e_a) - R(e_b) \leq w(e_a) - d(e_a, e_b)/c$ *wherever* $d(e_a, e_b)$ *is defined.*

Theorem 15 says that the recognition problem (determining whether a given clock period is feasible for a circuit $G = \langle V, E, v_h, d, w \rangle$ in the extended model) can be efficiently reduced to an instance of Problem MILP having $k = |V|$ integer variables, $n - k = |E|$ real variables, and $m = 2|E| + 2|d|$ inequalities, where $|d|$ is the number of pairs (e_a, e_b) of edges for which $d(e_a, e_b)$ is defined. Thus the recognition problem can be performed in $O(|d|(|E| + |V| \lg|E|))$ time.

By the same argument used in Corollary 6 for the model of Section 2, the optimal clock period must be equal to $D(u, v)$ for some pair of vertices u and v. The values $D(u, v)$ for all connected pairs of vertices u and v can be found in $O(|d||E| \lg|V|)$ time by an algorithm similar to Algorithm WD. The key step is to apply Johnson's all-pairs shortest-paths algorithm [4] to the edge-weighted graph $H = \langle E, P, wd \rangle$, where the vertex set E of H is the edge set of G, the edge set P of H contains an edge (e_1, e_2) wherever $d(e_1, e_2)$ is defined, and the weighting function is defined by $wd((e_1, e_2)) = (w(e_1), -d(e_1, e_2))$. Once the values for $D(u, v)$ are found, an algorithm similar to Algorithm OPT1 can be used to find an optimal retiming in $O(|d|(|E| \lg|V| + |V| \lg^2|V|))$ time by binary search.

As was observed for Algorithm OPT2, asymptotic performance improvements may be obtained for the optimal retiming problem by using the methods of van Emde Boas [19] and Johnson [4, 5].

9. Concluding remarks

Our goal has been to provide a general framework for the precise understanding of circuit timing. Through the use of a simple graph-theoretic model, we have been able to cast a variety of circuit timing problems in purely combinatorial terms. We believe our approach to be robust. Many other circuit models and many other circuit problems can be handled within the basic framework. We take time here to discuss a few.

Timing at the host. The interface represented by the host may sometimes demand that various timing specifications be met. By augmenting the set of inequalities specified in Theorem 10, it is often possible for the optimization algorithms to act subject to such constraints. For example, if data must be available to the host along some edge $v \xrightarrow{e} v_h$ within some time t after each clock tick, we may express this by the inequality

$$R(v) - r(v_h) \leq \frac{t}{c} + w(e),$$

which is equivalent to saying that we must have $\Delta(v) \leq t$ if the register count $w_r(e)$ of edge e is zero in the retimed circuit. Similarly, if data from the host is not available on an edge $v_h \xrightarrow{e} v$ until some time t after each clock tick, this may be expressed by the inequality

$$r(v_h) - R(v) \leq w(e) - \frac{d(v)}{c} - \frac{t}{c}.$$

Geometric considerations. The optimization methods discussed in this paper may be applied largely independently of geometric considerations because although retiming causes the addition and deletion of registers, it otherwise leaves the functional elements and their pattern of interconnection the same. Thus if a given circuit has an area-efficient layout, chances are that a retimed version of the circuit can be laid out efficiently. In some cases, however, the floorplan of a circuit may limit the number of registers on certain interconnections.

The inequalities that constrain the retimed system can be augmented to express these geometric constraints. For example, to specify an upper bound k on the number of registers that can fit along some edge $u \xrightarrow{e} v$, we may impose the constraint

$$r(v) - r(u) \leq k - w(e).$$

We can also model a situation in which the first k registers on an edge $u \xrightarrow{e} v$ are relatively cheap and additional registers are more expensive. Add an auxiliary vertex $\hat{u}$ in the middle of edge e. Then assign a high cost to registers on the edge $u \rightarrow \hat{u}$ and a low cost to registers on the edge $\hat{u} \rightarrow v$, but constrain $\hat{u} \rightarrow v$ to have at most k registers in the retimed system. Solve the system of constraints as in Section 7.

Multiple hosts. Throughout the paper, we have assumed there was only one external interface. The implication of this assumption is that the exact timing at the host is preserved by retiming. For some applications, notably signal processing and maxtrix computations, data enters and leaves the circuit from many different interfaces. In these applications the exact timing between different interfaces need not be maintained. Our models can be extended to handle this situation of multiple, independent hosts by simply treating them as separate vertices in the circuit graph. Some of the optimization algorithms require minor changes, but all can be made to run in the same time bounds.

The phenomenon of *pipelining* can be explained easily within the context of multiple hosts. If a combinational circuit takes input from one host and provides output to another, retiming will generate the classic pipelined version of the circuit.

Slowdown. In Section 6 we showed how a c-slow circuit, which supports c independent streams of computation, can be reduced to support a single stream of computation by removing registers. The notion of c-slow circuitry offers new insight into many circuit designs that are not technically c-slow. Consider, for example, a 2-slow circuit in which only one stream of computation is being used. The registers in the circuit fall into two equivalence classes, one of which is idle during each clock period. Using Algorithm R to remove all the registers in one equivalence class is one way to optimize such a circuit.

Another way to save registers would be to modify the functional elements to perform slightly different actions on even and odd time steps so that each physical register plays the roles of two logical registers, one in each equivalence class. A cursory examination of the resulting circuit would not reveal that it is 2-slow according to the circuit model, but it would nevertheless communicate with the host only on every other clock tick. Although this method for saving registers may sometimes be acceptable, the overhead of register multiplexing and the complexity of control suggest that Algorithm R is a more reasonable alternative. Moreover, when confronted with a circuit that communicates with the host only on every other clock tick or a circuit whose functional elements perform different operations on alternate clock ticks, we may suspect that it is really a 2-slow circuit in disguise, and that penetrating the disguise may lead to improved performance and simplification of the control logic.

Like any attempt to make different stages of design wholly independent, our methodology has its limitations. Changes in circuit geometry made to accomodate registers introduced by retiming may actually change the speeds of functional elements. Our model has no way of taking such effects into account. Not all cost measures can be handled by the methods of Section 7. For example, we cannot handle the situation in which the first register on an edge is relatively expensive and additional registers are relatively cheap (for example, because the expense of routing a clock signal to the interconnection is the same whenever the number of registers is greater than zero). Indeed, the problem of minimizing the total cost of registers becomes strongly NP-complete (in the sense of Garey and Johnson [1]) in this case.

A major deficiency of the circuit model is its inability to represent combinational logic elements with data dependent propagation delays. For example, if multiplier can produce an answer quickly whenever one of its inputs is zero, its propagation delay is data dependent. One would like to take advantage of the shorter delay whenever possible in order to speed a larger computation.

The model has no way of taking advantage of data dependent timings of functional elements, which is unfortunate. The restrictiveness of the circuit model, however, is more a reflection of the design methodology on which the model is based. It is indeed difficult to take advantage of data dependent timings in globally clocked systems. Perhaps the solution is to investigate retiming in the context of *self-timed* systems [15, p. 242].

Appendix

This appendix explains briefly and without proofs the major algorithms and theoretical results assumed as background in the body of the paper. The algorithms were selected for their simplicity and for some problems are not the most efficient algorithms known. For a more complete discussion of some of these algorithms, the reader is referred to the paper [4] by Johnson and the textbook [9] by Lawler.

A1. The Bellman-Ford algorithm

The Bellman-Ford algorithm is a procedure for solving the linear programming problem LP, which we now restate for the reader's convenience.

Problem LP. *Let S be a set of m linear inequalities of the form*

$$x_j - x_i \leq a_{ij} \tag{3}$$

on the unknowns $x_1, x_2, \ldots, x_n$, where the a_{ij} are given real constants. Determine feasible values for the unknowns x_i, or determine that no such values exist.

There is a natural representation of the constraints S as a *constraint graph* $G = \langle V, E, w \rangle$. For each variable x_i, $i = 1, 2, \ldots, n$, there is a vertex v_i in V. For each constraint $x_j - x_i \leq a_{ij}$ in S, there is an edge $v_i \xrightarrow{e_{ij}} v_j$ in E with weight $w(e_{ij}) = a_{ij}$. For technical reasons, we add to V a *source vertex* v_0 and to E a set of edges e_{0i}, $i = 1, 2, \ldots, n$, with weights $w(e_{0i}) = 0$.

The Bellman-Ford algorithm solves the *single-source shortest-paths problem* on the constraint graph G, which is the problem of determining the weight u_i, $i = 1, 2, \ldots, n$, of the least-weight path from the source v_0 to v_i. More formally, u_i is defined as

$$u_i = \min_{\substack{p \\ v_0 \twoheadrightarrow v_i}} w(p) .$$

If the constraint graph G contains a cycle of negative weight, the least-weight paths from v_0 are not well defined. Under these conditions, the inequalities in S cannot be satisfied by any assignment of values to the x_i. If, on the other hand, the constraint graph G contains no negative-weight cycles, then the values u_i are well defined. Under these conditions, the inequalities in S can be solved by assigning $x_i = u_i$. This assignment is not the only one possible. For any feasible assignment to the x_i, uniformly adding any constant k to each value yields another feasible assignment, as can be checked by substitution in Inequality (3).

We now present the Bellman-Ford algorithm which determines the values u_i or gives an indication that a cycle of negative weight exists in the constraint graph G.

```
1.    for i ← 0 to n do u_i ← 0;
2.    for k ← 1 to n do
3.        for e_ij ∈ E do
4.            if u_j > u_i + w(e_ij) then u_j ← u_i + w(e_ij);
5.    unsatisfiable ← FALSE;
6.    for e_ij ∈ E do
7.        if u_j > u_i + w(e_ij) then unsatisfiable ← TRUE;
```

At the end, if *unsatisfiable* = FALSE, then u_i is the weight of the shortest path from v_0 to v_i and the u_i satisfy the constraints in S. If *unsatisfiable* = TRUE, the graph contains a cycle whose total weight is negative, which means that the constraints in S are unsatisfiable.

The running time of this algorithm is $O(|V||E|)$, because the outer **for** loop in Line 2 is executed $O(|V|)$ times, and the inner **for** loop in Line 3, whose body takes a constant amount of time, is executed $O(|E|)$ times. Thus if there are m constraints in S, the total running time is $O(mn)$. As can be seen from the code, the constant factor is small.

The Bellman-Ford algorithm can also be used to solve the *integer linear programming* version of Problem LP where all the x_i are constrained to be integers.

Problem ILP. *Let S be a set of m linear inequalities of the form $x_j - x_i \leq a_{ij}$ on the unknowns $x_1, x_2, \ldots, x_n$, where the a_{ij} are given real constants. Determine feasible values for the unknowns x_i subject to the constraint that the x_i are integer, or determine that no such values exist.*

The key observation in solving Problem ILP is that if the x_i satisfy the constraints in S and are integer, then they must satisfy the stronger inequalities

$$x_j - x_i \leq \lfloor a_{ij} \rfloor .$$

Applying the Bellman-Ford to the set of stronger inequalities will yield an integer solution whenever one exists.

A2. The Floyd-Warshall algorithm

The Floyd-Warshall algorithm is a procedure for solving the all-pairs shortest-paths problem in an edge-weighted graph $G = \langle V, E, w \rangle$. The problem is to compute for each pair of vertices v_i, v_j in V, the weight u_{ij} of the least-weight path from v_i to v_j, where $1 \leq i, j \leq |V|$.

We could solve the all-pairs problem by using the Bellman-Ford algorithm to solve a separate single-source shortest-paths problem for each vertex in V. The running time for this approach would be $O(|V|^2|E|)$. The Floyd-Warshall algorithm, which follows, is simpler and solves the all-pairs shortest-paths problem in $O(|V|^3)$.

```
1.   for i ← 1 to |V| do
2.       for j ← 1 to |V| do
3.           if e_ij ∈ E then u_ij ← w(e_ij) else u_ij ← ∞;
4.   for k ← 1 to |V| do
5.       for i ← 1 to |V| do
6.           for j ← 1 to |V| do
7.               if u_ij > u_ik + u_kj then u_ij ← u_ik + u_kj;
```

The absence of negative-weight cycles can be tested by checking that $u_{ii} \geq 0$ for $i = 1, 2, \ldots, |V|$ when the algorithm terminates.

Because of the nested **for** loops in Lines 4-6, the running time of the algorithm is $\Theta(|V|^3)$. Johnson's algorithm [4] is more efficient for sparse graphs—its running time is $O(|V||E|\lg|V|)$—but the code is more complex.

A3. An algorithm to solve Problem MILP

Problem LP has not only an integer linear programming relative in Problem ILP, but also a *mixed-integer linear programming* variant.

Problem MILP. *Let S be a set of m linear inequalities of the form $x_j - x_i \leq a_{ij}$ on the unknowns $x_1, x_2, \ldots, x_n$, where the a_{ij} are given real constants, and let k be given. Determine feasible values for the unknowns x_i subject to the constraint that x_i is integer for $i = 1, 2, \ldots, k$ and real for $i = k + 1, k + 2, \ldots, n$, or determine that no such values exist.*

Problem MILP can be solved in $O(n^3)$ time by using the techniques of Sections A1 and A2. We now present a sketch of the algorithm. As before, we rely on a graph G to represent the constraints in S.

1. Use the Floyd-Warshall algorithm to compute the weights u_{ij} of the shortest paths from each vertex v_i to each vertex v_j using the a_{ij} as edge weights. If a negative-weight cycle is detected, then stop because no assignment to the x_i can satisfy all the inequalities in S even if all the x_i are allowed to be real.

2. Set up the integer linear programming problem

$$y_j - y_i \leq \lfloor u_{ij} \rfloor$$

   for $i, j = 1, 2, \ldots, k$, and solve it using the Bellman-Ford algorithm. If this problem has no feasible solution, then stop because no assignment to the x_i can satisfy all the inequalities in S.

3. Let

$$x_j = \begin{cases} y_j & \text{for } j = 1, 2, \ldots, k\,; \\ \min_{1 \leq i \leq k} y_i + u_{ij} & \text{for } j = k + 1, k + 2, \ldots, n\,. \end{cases}$$

   These x_j satisfy the constraints in S and the integer conditions. ∎

The algorithm runs in $\Theta(n^3)$ time because both Steps 1 and 2 can require this much time. A more efficient algorithm exists which runs in $O(mn + km \lg n)$ time [13].

Acknowledgments

Thanks to Dan Gusfield of Yale and Jim Orlin of MIT Sloan School for helpful discussions. Thanks also to Ron Rivest and Rich Zippel of MIT for a discussion that led to our investigation of the problem of minimizing the total state of a circuit. The ARPAnet was an invaluable resource in the preparation of this paper.

References

[1] Michael R. Garey and David S. Johnson, *Computers and Intractability*, W. H. Freeman and Co., San Francisco, 1979.

[2] Min-Yu Hsueh, "Symbolic layout and compaction of integrated circuits," Memorandum No. UCB/ERL M79/80, University of California, Berkeley.

[3] Paul A. Jensen and J. Wesley Barnes, *Network Flow Programming*, John Wiley and Sons, Inc., New York, 1980.

[4] Donald B. Johnson, "Efficient algorithms for shortest paths in sparse networks," *Journal of the Association for Computing Machinery*, Vol. 24, No. 1, pp. 1–13, January 1977.

[5]   Donald B. Johnson, private communication, July 1982.

[6]   Zvi Kohavi, *Switching and Finite Automata Theory,* McGraw-Hill, New York, 1978.

[7]   H. T. Kung, "Let's design algorithms for VLSI systems," *Proceedings of the Caltech Conference on Very Large Scale Integration,* Charles L. Seitz, ed., Pasadena, California, January 1979, pp. 55–90.

[8]   H. T. Kung and Charles E. Leiserson, "Systolic arrays (for VLSI)," *Sparse Matrix Proceedings 1978,* I. S. Duff and G. W. Stewart, ed., Society for Industrial and Applied Mathematics, 1979, pp. 256–282. (An earlier version appears in Chapter 8 of [15] under the title, "Algorithms for VLSI processor arrays.")

[9]   Eugene L. Lawler, *Combinatorial Optimization: Networks and Matroids,* Holt, Rinehart and Winston, New York, 1976.

[10] Charles E. Leiserson, *Area-Efficient VLSI Computation,* Ph.D. dissertation, Department of Computer Science, Carnegie-Mellon University, October 1981. Published in book form by the MIT Press, Cambridge, Massachusetts, 1983.

[11] Charles E. Leiserson and Ron Y. Pinter, "Optimal placement for river routing," *Carnegie-Mellon University Conference on VLSI Systems and Computations,* H. T. Kung, Bob Sproull, and Guy Steele, ed., Pittsburgh, Pennsylvania, October 1981, pp. 126–142.

[12] Charles E. Leiserson and James B. Saxe, "Optimizing synchonous systems," *Twenty-Second Annual Symposium on Foundations of Computer Science,* IEEE, October 1981, pp. 23–36.

[13] Charles E. Leiserson and James B. Saxe, unpublished notes, July 1982.

[14] Thomas Lengauer, "The complexity of compacting hierarchically specified layouts of integrated circuits," *Twenty-Third Annual Symposium on Foundations of Computer Science,* November 1982, pp. 358–368.

[15] Carver A. Mead and Lynn A. Conway, *Introduction to VLSI Systems,* Addison-Wesley, Reading, Massachusetts, 1980.

[16] Pauline Ng, Wolfram Glauert, and Robert Kirk, "A timing verification system based on extracted MOS/VLSI circuit parameters," *Eighteenth Design Automation Conference Proceedings,* IEEE, 1981.

[17] Flavio M. Rose, *Models for VLSI Circuits,* Masters Thesis, Department of Electrical Engineering and Computer Science, Massachusetts Institute of Technology, March 1982. Also available as MIT VLSI Memo No. 82–114.

[18] Texas Instruments Incorporated, *The TTL Data Book for Design Engineers,* Dallas, Texas, 1976.

[19] P. van Emde Boas, "Preserving order in a forest in less than logarithmic time," *Sixteenth Annual Symposium on Foundations of Computer Science,* IEEE, October 1975, pp. 75–84.

[20] H. Yoshizawa, H. Kawanishi, and K. Kani, "A heuristic procedure for ordering MOS arrays," *Twelfth Design Automation Conference Proceedings,* ACM–IEEE, 1975, pp. 384–393.

Routing and Interconnection

A New Channel Routing Algorithm
W.S. Chan

River Routing: Methodology and Analysis
R.Y. Pinter

Area and Delay Penalties in Restructurable Wafer-Scale Arrays
J.W. Greene and A. El Gamal

A New Channel Routing Algorithm

Wan S. Chan
California Institute of Technology
Pasadena, California 91125

ABSTRACT

This paper presents a new algorithm for solving the
two-layer channel routing problem with doglegging. Based
on a set of intuitive and reasonable heuristics, the
algorithm tries to obtain a channel routing
configuration with a minimum number of tracks. For every
benchmark problem tested, the algorithm gives a routing
configuration with the smallest number of tracks
reported in the literature.

I. INTRODUCTION

In this paper we present a new algorithm for solving the
two- layer channel routing problem with restricted
doglegging, identical to the problem considered in
[1,3]. Although the algorithm was developed as part of a
silicon compiler [5], it can be applied to any channel
routing environment.

Our algorithm is based on the following new improved
ideas:

(a) by viewing each two-pins subnet as a basic
component, we eliminate the need for the "zone" concept
introduced in [1];

(b) at each iteration of our algorithm, we identify the
set of subnets to be considered for assignment, based on
both the vertical and horizontal constraints rather than
on the horizontal constraint alone, as proposed in [1];

(c) to determine the assignment of subnets to tracks, we
attach weights onto the edges of the bipartite graph and
use the max-min matching [4], a refinement of the

maximum cardinality matching [4] used in [1];

(d) by appropriately defining the weight on these edges one can introduce various heuristics into the algorithm. In our algorithm, the weight has been defined to control the growth in the length of the longest path through the vertical constraint graph, which is a lower bound on the number of tracks needed;

(e) instead of starting from a density column and assigning all the subnets on one side of the column before considering any subnet on the other side of the column, our algorithm considers assigning subnets on both sides of the column simultaneously. We also derive an algorithm and a theorem to detect the constraint loops for the bidirectional case, which reduce to those presented in [1] for the unidirectional case.

We tested our algorithm on four channel routing problems taken from the literature and obtained the same best results reported among various papers [1,3].

The paper is organized as follows: in Sec. II we formulate the two-layer channel routing problem with restricted doglegging; in Sec. III we discuss all the important ideas behind our algorithm, using several examples for illustration; in Sec. IV we present our algorithm in high level pseudo code; in Sec. V we summarize the experimental results we obtain; and in Sec. VI we present additional ideas we have on the channel routing problem.

II. PROBLEM FORMULATION

Consider a rectangular routing channel with one row of pins along each of its sides. For each pin, its x-coordinate is fixed but its y-coordinate is to be determined by the router. A net is a set of two or more pins which are to be electrically connected by the router.   As shown in Figure 1a, net 1 consists of two pins, one at the bottom side and the other at the left side of the channel. A net with p pins can be viewed as consisting of (p-1) subnets, each of which connects two neighboring (along x-axis) pins of the net. As shown in Figure 1b, net 2 which consists of three pins, can be viewed as two subnets 2a and 2b.

The router is to connect all the pins of each net using a set of vertical and horizontal segments. In the case of no doglegging, each net consists of a single horizontal segment, as shown in Figure 1a. By introducing doglegs at various columns, a net may consist of several horizontal segments [1,2,3]. As in [1,3], we consider a restricted form of doglegging in this paper: doglegs of a net can be introduced only at the column where the net has a pin. In other words, each subnet will consist of a single horizontal segment and dogleg can be introduced only at the endpoints of a subnet, as shown in Figure 1b.

Although this restricted doglegging will, in general, use more tracks than the unrestricted doglegging of [2], we adopt it for the following reasons. If it is assumed that the given x-coordinates of the pins are such that there is enough distance between each pair of columns to satisfy the process design rules, then introducing doglegs only at the endpoints of subnets does not require checking for enough distance between the columns to insert a dogleg. In addition, this restricted doglegging allows the conceptual simplification of treating each subnet as a basic component.

The router uses two independent layers for interconnection: one used for all horizontal segments and the other for all vertical segments. The electrical connectivity between a horizontal segment and a vertical segment is achieved through a contact located at their intersection.

In order to avoid unintended electrical connection, two horizontal segments which belong to two different nets must not superimpose. Or, equivalently, two subnets which overlap horizontally and which belong to different nets must be assigned to different tracks. For example, in Figure 1b, subnets 2b and 3 must be assigned to different tracks. This type of constraint is refered to as horizontal constraint.

Similarly two vertical segments which belong to two different nets must not superimpose. Hence, if two subnets, which belong to different nets, have a pair of endpoints aligned in the same column, then the subnet whose aligned endpoint is to be connected to a pin at the top side must be assigned a track above the track assigned to the subnet whose aligned endpoint is to be connected to a pin at the bottom side. For example, in Figure 1b, subnet 2a must stay above subnet 1, and subnet 3 must stay above subnet 2b. This type of constraint is refered to as vertical constraint.

Note that the vertical constraint is transitive, i.e. if subnet a must stay above subent b, and subnet b above subnet c, then subnet a must stay above subnet c even though subnets a and c may not have aligned endpoints or may not even overlap horizontally. Unless otherwise stated, the term vertical constraint will include both the direct and indirect (via transitive closure) vertical constraints throughout this paper.

Obviously if there is a loop of vertical constraints, then there exist no solution to the channel routing problem. However such a loop can be broken by inserting a new pin into one of the subnets in the loop, thus splitting the subnet into two subnets which can be independently assigned to different tracks. Hence without loss of generality, we assume that there is no loop of vertical constraints in the given channel routing problem.

A solution to the channel routing problem is to find an assignment of the subnets to the tracks without violating any horizontal or vertical constraints, e.g. Figure 1a and 1b. In other words, viewing a track as a set of subnets, a solution to the channel routing problem is simply a partition of the set of all the subnets, which does not violate the horizontal and vertical constraints.

There are several criteria for evaluating a channel routing configuration [2] but minimizing the number of tracks used, which is proportional to the chip area, is the most important. In this paper we shall present a heuristic algorithm which tries to obtain a channel routing configuration with a minimum number of tracks.

III. UNDERLYING IDEAS

In this section, we will present all the important ideas upon which our channel routing algorithm is based.

A. Density

Given a channel routing problem, the density of the problem is defined as the largest number of different nets crossing a (vertical) column of the channel. For example, the density of the problem in Figure 1 is 2 and that of Figure 2 is 5. As pointed out in Sec. II, in order to satisfy horizontal constraints, all the subnets of a column which belong to different nets must be assigned to different tracks. Therefore the density is a lower bound on the minimum number of tracks needed for a given channel routing problem.

Let a column which has the largest number of different nets crossing be known as a density column. For example, Figure 2 has two density columns next to each other. In our algorithm, we shall first find all the density columns of the given channel routing problem. For reasons that will become obvious later in Sec. III.E, we choose to start with the middle one among all the density columns. We then assign a different track to each subnet in that column which belongs to a different net. For example, in Figure 2, subnets 1,2,3,4 and 5 are each initially assigned to a different track. The reason for starting with a density column is that since we will need at least density number of tracks, we might as well use that many tracks from the beginning.

B. Front Line Subnets (FLS)

After assigning each subnet in the chosen density column to a track, our algorithm enters an iterative loop where at each iteration a set of yet unassigned subnets closest to the tracks is considered for assignment. Such a set of subnets is called the front-line subnets and they are defined below.

Let left (right, respectively) end-point of a track be defined as the left- (right-, respectively) most end-point of all the subnets assigned to the track. Let $lx(b)$ ($rx(b)$, respectively) denote the x-coordinate of the left (right, respectively) end-point of b where b can be either a subnet or a track. For the sake of brevity, we will discuss only the subnets on the right of the tracks. All the ideas below apply similarly to the subnets on the left of the tracks, with obvious modification.

To find the rfls (right-front-line-subnets), we first sort all the right subnets into ascending order of their lx. Following that order, we examine each subnet to see if it is a rfls. A right subnet is said to be a rfls if and only if it has a horizontal or vertical constraint with every one of the rfls selected thus far. Intuitively, at each iteration, the set of rfls is the maximal set of right subnets which are at the front line facing the tracks and should be considered for assignment without waiting until the next iteration.

Consider the example in Figure 2. Suppose we first choose the right density column and initially assign subnets 1-5 to tracks 1-5 respectively. Note that all the subnets have been sorted into ascending order of their lx. At the first iteration, the first right subnet considered is subnet 6. Since at this point no rfls has been selected, subnet 6 satisfies all the requirements of a rfls and is selected as a rfls. Next, subnet 7 is considered and since it does not have any horizontal or vertical constraint with subnet 6, subnet 7 is not a rfls. Similarly, subnets 8-12 do not qualify as rfls at this iteration. Intuitively, subnets 7-12, not having any constraint with subnet 6, can be assigned to the

same track as subnet 6, and hence should wait until
subnet 6 is assigned before they are considered for
assignment. In other words, subnet 6 shields subnets
7-12 from the tracks so that they are not at the front
line facing the tracks and hence need not be considered
for assignment at this iteration. Now suppose subnet 6
is assigned to track 4 as shown in Figure 2. At the
second iteration, the first right subnet considered is
subnet 7 and hence it is a rfls. Next subnet 8 is
considered. Although it does not have horizontal
overlaps with subnet 7, it does have a vertical
constraint with subnet 7. Thus subnet 8 is also a rfls
at this iteration. However subnets 9-11 are not rfls
because they do not have any constraint with subnet 7.
Subnet 12 is a rfls because it has vertical constraint
with both subnets 7 (via transitive closure) and 8.

The above example illustrates an important idea: by
considering the vertical constraint in addition to the
horizontal constraint in the definition of fls, we
enlarge the set of subnets which ought to be considered
at each iteration (e.g. subnet 12 as a rfls). This
increase in the scope of consideration will only help to
reduce the chances of requiring extra tracks because
rfls were unwisely assigned in earlier iterations by
arbitrarily choosing one out of several possible
assignments.

C. Max-min Matching

After we have chosen the set of fls at each iteration,
we must decide how to assign these fls to the tracks so
as to minimize the number of tracks required. Following
[1], we view this problem as a bipartite matching
problem [4, Chap. 5]: each track corresponds to a source
node on the left-hand-side of the bipartite graph; each
fls corresponds to a destination node on the
right-hand-side; and there is an edge between a track
node and a fls node if and only if the fls can be
assigned to the track without violating any horizontal
or vertical constraint. A matching of a bipartite graph
is defined as a subset of the edges where no two edges
in it are incident to the same node. Since every fls not
in a matching given by solving the bipartite matching
problem will require a new track, it is clear that we
must try to find the maximum cardinality matching [4,1],
i.e. the matching with the largest number of the fls
matched to the tracks.

In general there are several maximum cardinality
matchings for a given matching problem. In order to make
an intelligent choice among these maximum cardinality
matchings, we introduce the weight onto the edges of the
bipartite graph. This serves as a convenient mechanism

for incorporating heuristics into our algorithm by an
appropriate definition of the weight. In the following
we shall define the weight we use in our algorithm and
explain the heuristics behind it.

At each iteration of our algorithm, we define a vertical
constraint graph [1] based on the state of the channel
routing problem at that iteration, as follows. A
vertical constraint graph is a directed graph where each
node corresponds to a track or an unassigned subnet, and
each arc (u,v) denotes node u must stay above node v due
to a direct vertical constraint. The indirect vertical
constraint between two nodes, which arises from
transitive closure, corresponds to a directed path
between these two nodes in the graph. Figures 3a and 3b
show a channel routing problem and its vertical
constraint graph. Note that a vertical constraint graph
must be acyclic, i.e. free of constraint loops, for a
channel routing solution to exist. (Throughout this
paper, the term edge is used to refer to the undirected
edge in the bipartite graph, and the term arc is used to
refer to the directed edge in the vertical constraint
graph.)

Let the length of a directed path be defined as the
number of nodes in that path, or equivalently, 1 plus
the number of arcs in that path. Let the critical path
length of a node u in a vertical constraint graph,
denoted by $cpl(u)$, be defined as the length of the
longest path through u in the vertical constraint graph.
Let the critical path length of a vertical constraint
graph, denoted by cpl, be defined as the length of the
longest path through the graph. In other words,

$$cpl = max \; \{cpl(u) \mid all \; nodes \; u \; in \; the \; graph\} \quad (1)$$

In Figure 3b, $cpl=3$.

Since each node in a directed path has a vertical
constraint with all other nodes in the path, it must be
assigned to a track different from all other nodes in
the path. Hence the minimum number of tracks required in
a channel routing problem cannot be smaller than its
cpl. In other words, cpl is a lower bound on the minimum
number of tracks needed for a given channel routing
problem. In Figure 3a, density (which is another lower
bound on the number of tracks required) is 2, cpl is 3,
and it is easy to see that at least 3 tracks are needed
to satisfy the vertical constraints among the subnets.

However, unlike density, which is a fixed quantity for a
given channel routing problem, cpl tends to increase as

the subnets are assigned to the tracks while solving the
problem. It is because after a subnet is assigned to a
track, the resultant new track will have all the
vertical constraints of both the old track and the
assigned subnet since they must remain together on the
same horizontal position hereafter. In terms of
vertical constraint graphs, the resultant graph can be
derived from the original graph by moving all the
constraint arcs incident with the subnet onto its
assigned track, and then deleting the isolated subnet.

Consider the example in Figure 4. Its density=2 as shown
in Figure 4a, and before any subnet is assigned, its
cpl=2 as shown in Figure 4b. Suppose subnets 1 and 2 are
first assigned to tracks 1 and 2 respectively. At the
first iteration, subnets 3 and 4 are the fls whereas
subnet 5 is shielded by subnet 4. Suppose subnets 3 and
4 are assigned to tracks 1 and 2 respectively as shown
in Figure 4c. Then the cpl of the resultant graph is
increased to 3, a number greater than its density, as
shown in Figure 4d (where we use u+v to denote the track
which consists of subnets u and v). On the other hand,
if subnets 3 and 4 are assigned to tracks 2 and 1
respectively, then the cpl remains at 2 and, as shown in
Figure 4e, only 2 tracks are needed to solve the given
channel routing problem. Thus there is a need to choose
among the maximum cardinality matchings the one which
minimizes the increase in cpl.

Based on this observation, we incorporate the following
heuristics into our algorithm. Let the weight $W(u,v)$ of
an edge (u,v) between track u and subnet v be defined as
-1 times the cpl through the track resulting from
assigning subnet v into track u.

$$W(u,v) = -1 * cpl(u+v) \quad (2)$$

From (1) and (2) we have,

        cpl of the resulting vertical constraint graph
GEQ max {cpl(u+v) ¦ all (u,v) in the matching}
   = max {-1*W(u,v) ¦ all (u,v) in the matching}
   = -1 * min{W(u,v) ¦ all (u,v) in the matching}

Hence in order to keep the cpl of the resulting vertical
constraint graph minimum, our problem becomes that of
finding a maximum cardinality matching for which the
minimum of the weights of the edges in the matching is
maximum, i.e. the max-min matching problem [4]. We use
the Threshold Method [4, pp. 198] to find the max-min

matching.

The term cpl(u+v) can easily be computed as follows. Given a vertical constraint graph, let us assign the top-down level numbers (tdl) to the nodes as follows:

Top-down level assignment algorithm:

```
BEGIN
    i:=0;
    WHILE not all nodes are assigned a tdl, DO
    BEGIN
        i:=i+1;
        FOR each node u with no incoming arc, DO
        BEGIN
            set tdl(u):=i;
            delete u and all the arcs incident to it
        END
    END
END
```

In other words, tdl(u) is the length of the longest path from the top to node u. Note that another way of finding tdl is to perform a depth- first search of the vertical constraint graph, and assign tdl of each node to be 1 plus the largest tdl of its immediate descendents. Our experiments show that the depth-first search is faster than above algorithm only for very large graphs.

Similarly, we define the bottom-up level number (bul) of a node. Hence

$$cpl(u) = tdl(u) + bul(u) - 1 \quad (3)$$

Since the longest path from the top to node u+v, resulting from assigning node v into node u, is simply the longer of the two longest paths from the top to node u and to node v,

$$tdl(u+v) = \max \{ tdl(u) , tdl(v) \} \quad (4)$$

Similarly,

$$bul(u+v) = \max \{ bul(u) , bul(v) \} \quad (5)$$

Applying (3) to u+v and substituting (4) and (5), we

have

$$cpl(u+v) = \max\{tdl(u),tdl(v)\}+\max\{bul(u),bul(v)\}-1 \quad (6)$$

We can further refine the definition of the weight by noting that if there is an edge (u,v), and if $rx(u)=lx(v)$, then the right-most subnet assigned to track u and subnet v belongs to the same net (because the existence of edge (u,v) implies that u and v do not have vertical constraint and with $rx(u)=lx(v)$, we can conclude that they share the pin located at $rx(u)$). By adding 0.5 to the weight of such (u,v) to encourage (u,v) to be chosen in the max-min matching, we will tend to reduce unnecessary doglegging.

D. Delayed Assignment

At each iteration of our algorithm, after finding the max-min matching between the tracks and the fls, we assign the fls to the tracks according to the max-min matching. However not all fls need to be assigned at this point for the problem solving process to progress, i.e. to introduce new fls from the remaining subnets. Since the assignment of the remaining subnets depends critically on the past assignments, intuitively it is a good strategy to avoid any premature assignment, (i.e. commiting a subnet to a track before it is necessary for the introduction of new fls,) and thus retain as much flexibility as possible. The following example illustrates the advantage of such a delayed assignment strategy.

Consider the channel routing problem in Figure 5a which has density=3. Suppose subnets 1,2 and 3, which intersect a density column, are initially assigned to tracks 1,2 and 3 respectively. At the first iteration, we have the vertical constraint graph as shown in Figure 5b, and the weighted bipartite graph as shown in Figure 5c. Suppose the max-min matching obtained is $\{(1,4),(2,5)\}$. If we assign both subnets 4 and 5 accordingly, we get the vertical constraint graph shown in Figure 5d and the weighted bipartite graph shown in Figure 5e. Clearly only one of subnets 6 and 7 can be assigned to track 1+4, and a new track is needed for the other subnet. Observe that subnets 6 and 7 are shielded by subnet 4 only and after subnet 4 is assigned to a track, subnets 5, 6 and 7 form a legitimate set of fls. If at the end of the first iteration we assign only subnet 4 to track 1, then we get the vertical constraint graph in Figure 5f and the weighted bipartite graph in Figure 5g. There is a max-min matching $\{(3,5),(1+4,6),(2,7)\}$ which matches all the fls and hence does not require any new track.

Therefore the delayed assignment strategy is incorporated into our algorithm as follows. After finding a max-min matching, our algorithm assigns to the tracks only those fls which are to the left of the first subnet not yet chosen as a fls (recall that those unchosen subnets are sorted in ascending order of their lx). In other words, only those fls with rx<lx of the left-most non-fls subnet are assigned, all other fls remain as fls for the next iteration, and new fls are chosen in the same manner as presented in Sec. III.B. The delayed assignment strategy is also incorporated in Algorithm #2 of [1] where at Step 4 only nets terminating at the current zone are assigned.

E. Bidirectional Search

Thus far we have presented our ideas and examples for the case with no subnet on the left of the chosen density column. Since a density column can be at any x-coordinate of the channel, we must consider the general case where there are subnets on both sides of the chosen density column. One obvious approach, taken in [1], is to consider one side of the chosen density column first, iteratively assign all the subnets on that side, then turn around and iteratively assign the subnet on the other side of the chosen density column. Our experience with this approach indicates that because the second side is completely ignored while assigning the subnets on the first side, it tends to require more tracks when assigning the subnets on the second side. Therefore we adopt the following bi-directional search strategy: at each iteration we consider both sides of the tracks simultaneously by building a weighted bipartite graph and finding a max-min matching for each side. Since a density column is the column where the largest number of subnets intersect, it tends to be the critical area where new tracks are added. To spread these difficult areas evenly on both sides of the chosen density column, we take the chosen density column to be the middle one among all the density columns of a given channel routing problem. With this bidirectional strategy, we do get channel routing configurations with smaller number of tracks.

F. Constraint Loops Detection

Although an edge (u,v) in the bipartite graph guarantees that subnet v can be assigned into track u without violating horizontal or vertical constraints, a simultaneous assignment of two or more subnets to the tracks according to a matching may create a vertical constraint loop. To see this point, consider the example in Figure 6a which has density=2. Suppose subnets 1 and 2 are initially assigned to tracks 1 and 2. For the first iteration, the vertical constraint graph and the bipartite graph are shown in Figure 6b and 6c. Since {(2,4),(1,3)} is the only maximum cardinality matching, it is also the only max-min matching. If we make the assignment accordingly, we find that there is a vertical constraint loop between tracks 1+3 and 2+4 as shown in Figure 6d.

In [1], an algorithm was given to remove the edges in a matching which create constraint loop. However that algorithm cannot be adopted into our algorithm because it works only for a unidirectional search and not a bi-directional search. Instead we use the following algorithm to detect constraint loops. The input to the algorithm consists of a matching of the bipartite graph and the associated vertical constraint graph. The nodes and the edges are deleted from the bipartite graph. For the remainder of Sec. III.F, the term nodes refers to the track and fls nodes in the bipartite graph (a subset of those in the vertical constraint graph). The term nabove of a node is defined to be the number of nodes which is above the node in the vertical constraint graph.

To illustrate the algorithm, consider the matching and the vertical constraint graph shown in Figure 7a and 7b respectively. At the first iteration, no matching edge can be deleted. Although both track 1 and subnet 5 have nabove=0, their matching edge (1,5) cannot be deleted because track 1 is matched to subnet 3 on the other side which has nabove=1. Suppose matching edge (2,4) is chosen over (1,3) as an EX edge and is then deleted. At the second iteration, node 4 is first deleted which causes node 3 to have nabove=0. Next matching edges (1,5) and (1,3) are deleted since all three nodes now have nabove=0. Since there are no matching edges left, all the remaining nodes are deleted in the order of their top-down level: 1,3,5 and 2.

The following theorem guarantees that a matching without any EX edge will not result in a constraint loop after the assignment is made according to the matching. Thus for each set of fls, we iterate on the process of finding the max-min matching of the new bipartite graph, and remove all its EX edges from the bipartite graph, until a max-min matching with no EX edge is found.

THEOREM

A matching has no EX edge if and only if the

Constraint loops detection algorithm:

```
BEGIN
    compute nabove of every node;
    WHILE NOT every node is deleted, DO
    BEGIN
        DO
        BEGIN
            (a)delete every matching edge between nodes
               with nabove=0; (note: if a track has two
               matching edges, one on each side, then all
               three nodes must have nabove=0 in order to
               delete)
            (b)delete every node with nabove=0 and with
               no matching edge;
            (c)decrement nabove of every node which has
               an arc from each node just deleted
        END
        UNTIL no deletion is possible;
        IF every node is deleted THEN done ELSE
        BEGIN
            (d)choose as an EX edge [1], a matching edge
               which has a node with nabove=0 (its other
               node must have nabove>0);
            (e)delete the chosen EX edge
        END
    END
END
```

subnet-to-track   assignment   made   according   to   the
matching will not result in a constraint loop.

PROOF:

We will first prove that if there is no EX edge there is
no constraint loop. Since a constraint loop must consist
of at least two nodes (resulting from the assignment of
two fls nodes to two tracks nodes) which have mutual
vertical constraint on each other; these resultant nodes
must both have nabove>0. Since every node of the
matching edges deleted in Step (a) has nabove=0, the
node resulting from the assignment based on these
deleted edges will have nabove=0 and hence cannot be
part of a constraint loop. Since every node deleted in
Step (b) will only result in a node with nabove=0 after
the assignment, they can be forgotten as far as the
possibility of creating constraint loops is concerned.
Applying the same reasoning in the subsequent iterations
of Steps (a) and (b) until all nodes are deleted, we
conclude that no EX edge implies no constraint loop.

Next we will prove that if there is an EX edge then there is a constraint loop. Let $(u(1),v(1))$ be an EX edge. Without loss of generality, assume that $u(1)$ has nabove=0, and $v(1)$ has nabove>0. Since the vertical constraint graph before assignment is acyclic, there exists a node $u(2)$ with nabove=0 which is an ancestor of $v(1)$. Since $u(2)$ is not yet deleted, there exists a matching edge $(u(2),v(2))$ where $v(2)$ has nabove>0. Applying the same reasoning repeatedly, we have an infinite sequence of matching edges $(u(1),v(1)),(u(2),v(2)),\ldots$ Since the graph is finite and the sequence is not, there exists an k such that $u(1)=u(k)$ and $v(1)=v(k)$. Since $u(i+1)$ is an ancestor of $v(i)$, after assigning according to the matching, $u(i+1)+v(i+1)$ is an ancestor of $u(i)+v(i)$. Hence there is a constraint loop around $u(1)+v(1)$ after the assignment. Q.E.D.

Observe that our constraint loop detection algorithm and the theorem reduce to those in [1] for the unidirectional case.

IV. ALGORITHM

Combining the ideas presented in the previous section, we arrive at the following heuristic algorithm for solving a two-layer channel routing problem with a minimum number of tracks.

V. EXPERIMENTAL RESULTS

The algorithm presented in the previous section has been implemented in the programming language MAINSAIL on a DEC 20 computer and a HP 9826 desk-top computer. The algorithm was tested on four channel routing problems taken from the literature. They range in size and complexity and the last one is the well-known "Difficult Example" [1,3]. For each problem our algorithm uses the same number of tracks as the minimum reported among various papers. Table 1 summarizes for each benchmark problem, its source, the number of subnets it has, its density, the minimum number of required tracks reported in the literature along with its reference, and the number of tracks used by our algorithm. Note that our algorithm uses only the density number of tracks for the first two problems, and one plus density for the other two. The channel routing configuration obtained by our algorithm are presented in Figures 2,8,9 and 10.

In order to determine the effectiveness of our

Channel routing algorithm:

```
BEGIN
    initialize each subnet in the middle density column
      as a track;
    WHILE a subnet is not yet assigned, DO
    BEGIN
        find left and right sets of fls;
        build left and right weighted bipartite graphs;
        DO
        BEGIN
            delete EX edges of last matchings from
              bipartite graphs;
            find left and right max-min matchings of new
              bipartite graphs;
        END
        UNTIL the new matchings have no EX;
        determine which fls should be assigned in order
          to get new fls;
        FOR each of these fls, DO
        BEGIN
            IF it is in the matching,
            THEN assign to track accordingly
            ELSE create a new track for it
        END
    END;
    position the tracks to satisfy vertical constraints
      among them
END.
```

algorithm, it should be tested against many more
benchmark channel routing problems. However they are
not readily available in the machine readable form and
to code them by hand is a tedious and error-prone
process.

VI. ADDITIONAL COMMENTS

A. Wire Length

In addition to the number of tracks used, the total
length of the wire is also an important criterion in
evaluating a channel routing configuration. Given a
channel routing problem, the total length of its
horizontal segments is fixed. Given a channel routing
configuration, the length of a vertical segment can be
reduced in two ways: (a) by reassigning a subnet which
causes a dogleg to a new track to reduce or even
eliminate the length of the dogleg; (b) by reassigning a
subnet, which is not a subnet with one pin at the top

side and another at the bottom side of the channel, to a
new track closer to the (top or bottom) side where the
pins are. Of course, the new assignment must still
satisfy the vertical and horizontal constraints. For
example in Figure 9, we can reassign subnet 1 from track
8 to track 6 to eliminate its dogleg of length=2, and
reassign subnet 2 from track 7 to track 11 to reduce
wire length by 2*(11-7).

B. Contacts

Another important criterion to evaluate a channel
routing configuraton is the number of contacts, or vias,
used to electrically connect the two independent
interconnect layers. Each pin along the top or bottom
sides of a channel needs one contact to connect to its
horizontal segment(s) if there is no dogleg, and it
needs two contacts if there is a dogleg in that column.
Besides, for a vertical segment which is not intersected
by any horizontal segment, it can be routed in the layer
used for horizontal segments and thus will not need a
contact between itself and its horizontal segment(s).
Hence we have

      number of contacts required
    = (number of pins on top and bottom sides of channel)
     +(number of doglegs)
     -(number of vertical segments which are not
       intersected by any horizontal segments)

Given a channel routing problem the number of pins in
the top and bottom sides are fixed. However given a
routing configuration, one can reassign the subnets to
new tracks, within the restriction imposed by the
vertical and horizontal constraints, to decrease the
number of doglegs and to increase the number of vertical
segments not intersected by any horizontal segments.

Therefore by post-processing a channel routing
configuration, one can decrease the wire length and the
number of contacts required.

C. Three-layer Channel Routing

With the recent advance in the process technology, in
particular the second metal layer, it is possible to
have three independent layers for interconnection. For
such process technology the channel routing problem
becomes relatively simple. The three-layer channel
routing problem can be formulated as in Sec. II except
that the first layer is used for the horizontal
segments, the second layer for the vertical segments

connected to pins along the top side of the channel, and the third layer for the vertical segments connected to pins along the bottom. Since the second and the third layers can superimpose without causing any electrical connection, one no longer needs to be concerned about the vertical constraints. Hence the three- layers channel routing problem becomes that of finding the minimum cardinality set of paths through the horizontal constraint graph, where each node corresponds to a subnet and each edge (u,v) indicates that node v is to the left of node u, i.e. v can be assigned to the same track as u. A solution to this problem can be found by using the greedy strategy, i.e. assign each subnet to the track closest to it. It is also easy to show that for any three-layer channel routing problem there always exists an optimal solution which uses the same number of tracks as its density.

VII. CONCLUSION

We have presented a new algorithm for solving the two-layer channel routing problem along with the insight and the experimental results. It will be used as part of a general routing system now being implemented based on the ideas developed by Mr. G. Clow and myself. Although our general routing system was initially intended as part of a silicon compiler [5], our ideas apply just as well in an interactive graphic computer-aided design environment.

ACKNOWLEDGEMENT

This work was carried out during the author's tenure as a visiting associate representing Hewlett-Packard at the Silicon Structures Project in California Institute of Technology. The author wishes to express his sincere gratitude to HP and SSP for providing an excellent research environment.

REFERENCES:

[1] T. Yoshimura and E. S. Kuh, "Efficient algorithms for channel routing," IEEE Transactions on Computer-aided Design of Integrated Circuits and Systems, vol. CAD-1, pp. 25-35, January 1982.

[2] R. L. Rivest and C. M. Fiduccia, "A greedy channel router," Proceedings of 19th Design Automation

Conference, paper 27.2, June 1982.

[3] D. N. Deutsch, "A dogleg channel router," Proceedings of 13th Design Automation Conference, pp. 425-433, 1976.

[4] E. L. Lawler, "Combinatorial optimization: networks and matroids," Holt, Rinehart and Winston, New York, 1976.

[5] T. S. Hedges, K. H. Slater, G. W. Clow and T. Whitney, "The SICLOPS silicon compiler," to appear in Proceedings of International Circuit and Computer Conference, September 1982.

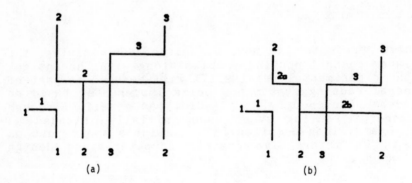

Figure 1. Minimum tracks routing configurations,
(a) without doglegging;
(b) with restricted doglegging.

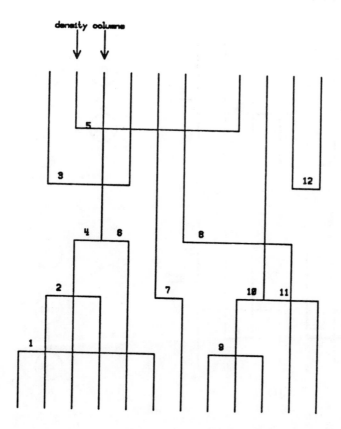

Figure 2. A simple channel routing problem.

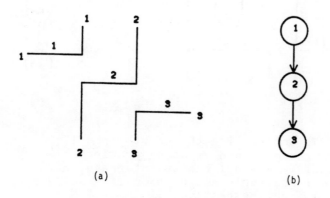

Figure 3. An example of cpl>density,
(a) routing configuration;
(b) vertical constraint graph.

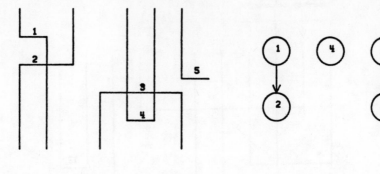

(a)

(b)

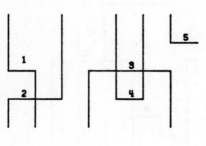

(c)

(d)

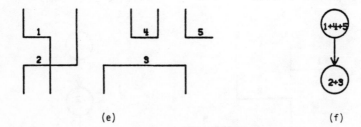

(e)

(f)

Figure 4. Controlling the growth in cpl,
(a) the given channel routing problem;
(b) initial vertical constraint graph;
(c) final routing configuration if 3 and 4 are assigned
    to 1 and 2 respectively;
(d) final vertical constraint graph if 3 and 4 are
    assigned to 1 and 2 respectively;
(e) final routing configuration if 3 and 4 are assigned
    to 2 and 1 respectively;
(f) final vertical constraint graph if 3 and 4 are
    assigned to 2 and 1 respectively.

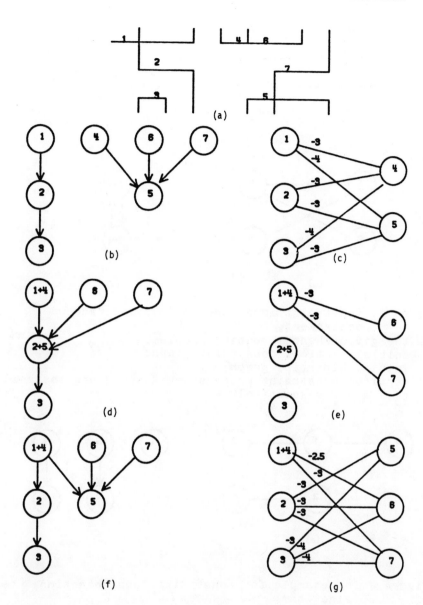

Figure 5. Advantage of delayed assignment strategy,
(a) the given channel routing problem;
(b) initial vertical constraint graph;
(c) initial weighted bipartite graph;
(d) vertical constraint graph if 4 and 5 are assigned
    to 1 and 2 respectively;
(e) bipartite graph if 4 and 5 are assigned to 1 and 2
    respectively;
(f) vertical constraint graph if only 4 is assigned to 1
(g) bipartite graph if only 4 is assigned to 1.

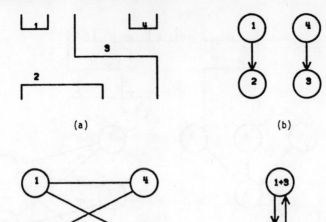

(a)   (b)

(c)   (d)

Figure 6. Constraint loops created by simultaneous
          assignment,
(a) the given channel routing problem;
(b) initial vertical constraint graph;
(c) initial bipartite graph;
(d) vertical constraint graph after 4 and 3 are assigned
    to 1 and 2 respectively.

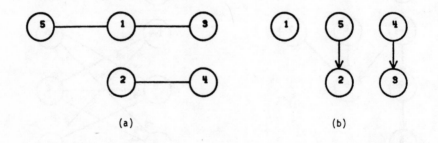

(a)   (b)

Figure 7. An example for constraint loop detection
          algorithm,
(a) bidirectional bipartite graph;
(b) vertical constraint graph.

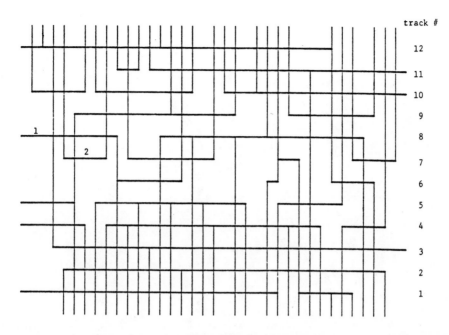

Figure 8. Routing configuration generated for benchmark problem #2.

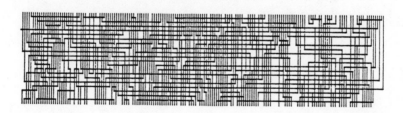

Figure 9. Routing configuration generated for benchmark problem #3.

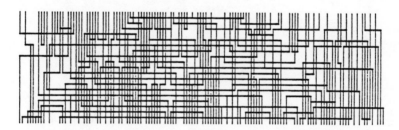

Figure 10. Routing configuration generated for benchmark problem #4.

River Routing: Methodology and Analysis*

Ron Y. Pinter
Bell Laboratories
Murray Hill, New Jersey 07974

Abstract. The problem of river routing across a channel is only a special case of more general routing configurations. Both its methodological and combinatorial characteristics can be extended in useful ways which will be explored in this paper. The two characteristics that we generalize here are planarity and grouping. Planarity means that the connections are realizable in one layer; ie the interconnection pattern of the nets is planar. Grouping means that the connections are made in order, that is to say that the routing of net $i+1$ is adjacent, conceptually and preferably physically, to the routing of net i.

This paper investigates both placement and routing problems. First we provide a graph theoretical model that accomodates the interconnect specifications in a succinct manner, allowing us to find a placement that enables a planar routing pattern in linear time. Second we study problems of detailed routing, namely whether wires fit in the area allotted by a specific placement. Detailed planar routing of two-point nets for an entire chip (with rectangular modules) is shown to be NP-complete, whereas a polynomial time algorithm is given for detailed routing for a simple polygon (ie one without holes). Routability testing is shown to be easier than the actual generation of the wires. Finally we show how to view general channel routing as a superimposed collection of river routing problems.

1. Introduction

Layout systems for digital integrated circuits traditionally decompose the layout problem into two subproblems: placement and routing†. In this paradigm, we first place predesigned pieces of the circuit — called modules — on the chip, and then route wires to interconnect common signal nets. Each module has terminals located along its boundary that must be interconnected properly using a given number of wiring layers which depends on the fabrication technology. The prime objective is to minimize the total area required to realize the circuit subject to various design rules that ensure the feasibility of fabrication.

The complexity of the layout problem is overwhelming. Even when no routing is

---

* This research was conducted at the MIT Laboratory for Computer Science, and was supported in part by an IBM Graduate Fellowship and by the DEC Semiconductor Engineering Group at Hudson, Mass.
† For complicated VLSI circuits, one may need to form a hierarchy of such problems, as in [Pr79].

specified, the placement problem is NP-complete [LaP80]. The intractability is aggravated when routing requirements are being taken into account. Also, most practical routing problems are NP-complete in themselves [LaP80, SaBh80, Sz81].

To deal with this complexity various design methodologies have been suggested that yield tractable layout problems. These methodologies are usually coupled with simplified layer assignment methods, in which the geometric issues are separated from the detailed decisions of layer specifications. Notable examples of this strategy are composition systems such as Bristle-Blocks [Jo79, Jo81] and data-path generators [Sh82]. In both cases modules can be put together in restricted fashion: the wiring is often constrained to one layer, giving rise to river routing situations. The problem of river routing across a channel has been analyzed extensively [To80, DKSSU81, LeiPi81, SieDo81, Pi82a] and efficient algorithms have been devised to solve most interesting placement and routing problems that arise in this context. Little work, however, has been done to extend these successful endeavors, and the prime goal of this paper is to pursue this course.

The term "river routing" embodies two important characteristics: (i) all connections must be made in order among two sequences, and (ii) the interconnection topology is realizable in the plane. We wish to generalize both properties in the chip assembly context and examine their effect on placement and routing and thus establish the power and limitations of a single active layer. In addition, we show how these concepts can be extended in yet another way to solve general instances of channel routing using a newly proposed wiring model called the via-free model.

The issue of planarity is not new. Much of the attention devoted to the subject in the context of the design of electrical circuits dates back to the printed circuit board (PCB) era [AGR70], when layer changes were expensive and degraded the quality of products. But even in the current MOS technologies single layer realizations have non-negligible advantages, because one layer (metal) is highly preferable to the others.

To discuss the grouping issue, we need to define cables and sequences. A sequence of terminals encapsulates the notions of contiguity and geometric order between terminals*. A maximal set of nets that connects one sequence of terminals on a side of a module in order to another sequence on another module is called a cable. Cables occur frequently in many designs, most notably in the design of microprocessors where fields of data have to be transmitted from one place to another. The recognition of routing patterns such as cables constitutes an abstraction mechanism by which some of the low level details are suppressed in favor of a clear perception of the architectural issues. Such an abstraction is also useful in reducing the complexity of solutions to several problems. In addition, it is sometimes easier to control the circuit level performance of a parallel connection if the realization of all signals involved is similar, ie the wires are known to have similar characteristics in terms of layering, length, and jogging.

---

* For sake of presentation we agree that a sequence cannot be split between modules and is lined up along a straight piece of a module's boundary. Both these restrictions are not essential, since more general notions of sequences can be constructed using this basic type without impairing the validity of the technical results presented in this paper.

In [LeiPi81] it was shown that the problem of routing a cable across a channel in a given area can be solved efficiently and optimally. One salient assumption in that setting was that the orderings of terminals on both sides conformed in such a way that a planar embedding of the wires was possible. In the general case, we must decide whether such an embedding exists to start with. This test has two parts. First, "Is there a passage between the locations of the involved sides?", and second, "Are the sides oriented properly with respect to each other?" Figure 1 (a) demonstrates how the orientability issue comes up even in the channel case. When the sides are further apart on the chip this problem may look harder (see Figure 1 (b) and (c)), but in fact it is not. The problem can be formalized as follows*:

[CR]   **Cable Routability**

INSTANCE: A placement of a set of rectangular modules within a bounding box, and a cable connecting two sequences of n terminals each on two (not necessarily disjoint) sides of modules.

QUESTION: Is there a planar (one-layer) realization for this cable?

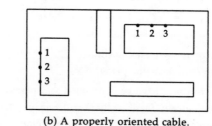

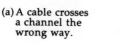

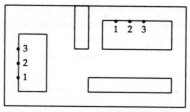

(b) A properly oriented cable.

(a) A cable crosses a channel the wrong way.

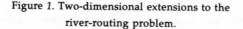

(c) An improperly oriented cable.

Figure 1. Two-dimensional extensions to the
river-routing problem.

In a more general setting, however, we may have the freedom of orienting the modules in such a way that will enable routing in the plane. Reflection of modules is essential in ensuring the feasibility of planar embeddings from a topological point of view, whereas translation and rotation may be required for geometric purposes — to make room for wires. All three operations are realizable at the mask level for integrated circuits without affecting any of the physical aspects of the design. Reflection, however, may not be as easy to realize in a PCB design, which is a major concern

---

* We follow the style of Garey and Johnson [GaJo79] and state problems in their decision version.

since this very operation is the one that is most important for the possibility of planar embeddings.

How can these operations of reflection, translation, and rotation (or a subset thereof, excluding reflection) be used in generating a placement (or modifying an existing one) in which all cables are routable in the plane? First we state the problems concerning the topological issue alone. For the feasibility of planar interconnect we distinguish between the case in which the placement is given and the case where we are free to determine it. First we present the "static" case:

[PR]   **Planar Routability**

INSTANCE: A placement of a set of rectangular modules within a bounding box, numbered terminals on modules' boundaries.

QUESTION: Is there a planar embedding of the interconnect wires for this given placement?

The "dynamic" variant is, in fact, a placement problem:

[PO]   **Placement Orientation**

INSTANCE: A set of (non-oriented, reflectable) rectangular modules, numbered terminals on each modules' boundary.

QUESTION: Is there an embedding of the modules on the plane such that the routing is also feasible in the plane?

Once modules have been oriented and placed in such a way that a planar embedding is topologically feasible, there still remains the question of fitting the actual wires in the given area. Naturally, a placement found to be consistent with a one layer realization induces a set of possible routing paths for the nets. However, such a set may not be uniquely determined by the topology suggested by the placement, but even if the paths are unique, the actual placement may impose some further constraints on the actual routing. Thus, the problem of detailed routing arises at two different levels. First, the more general problem is that of finding a routing for a given placement, where each wire can be routed wherever it fits:

[DR]   **Detailed Routing**

INSTANCE: A placement of a set of rectangular modules within a bounding box, numbered terminals on modules' boundaries.

QUESTION: Is there a one-layer detailed routing for this configuration?

The second detailed routing problem starts off with more information about the routing plan. For every net, we are given the homotopy of its intended wire relative to the placed modules. That is to say, we know how the path taken by the wire is related to the positions of the modules — how it goes "between" pairs, whether it goes to the "right" of or "above" a module etc. Figure 2 exemplifies the notion of a homotopy. A given homotopy of a wire is sometimes called a rough routing of the net (as opposed to detailed routing which entails specifying the exact path). Also, we say that we are given a rough routing for a problem if all its nets are roughly routed. This notion corresponds to that of global routing which is commonly used in the literature in the context of the general (nonplanar) routing problem (see, for example, [SouRo81, Riv82]).

The second detailed routing problem, then, can be formulated as follows:

[DRH]   Detailed Routing given a Homotopy

INSTANCE: A placement of a set of rectangular modules within a bounding box, numbered terminals on modules' boundaries, homotopy (rough routing) for each net.

QUESTION: Is there a one-layer detailed routing for this configuration that conforms with the given homotopy?

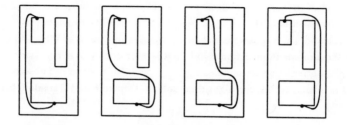

Figure 2. Different homotopies for the same wire.

When specifying a rough routing, the cable abstraction comes in handy. Cables provide a succinct manner by which the homotopy of a set of "parallel" wires can be specified, and moreover, many of the details concerning the relative position between wires in the same cable are rendered unnecessary.

A natural restriction of the planar routing problem is the situation in which there is no difference between DR and DRH. This is the case in which routing is being performed in a simply connected polygon, where there are no "holes" to route around and the rough routing for each net is trivially unique (topologically). Figure 7 shows such a case in which planarity is guaranteed. The formulation of the decision problem is given as follows:

[DRSRP]   Detailed Routing in a Simple Rectilinear Polygon

INSTANCE: A simple rectilinear polygon with terminals on its boundary.

QUESTION: Is there a one-layer detailed routing for this configuration?

The rest of this paper is devoted to investigating and classifying the complexity of the problems stated so far. We show that DR is NP-complete, and present polynomial time algorithms for CR, PR, PO, and DRSRP. The classification of DRH remains open. For those problems that are solvable in polynomial time, there may be a qualitative difference between the time it takes to decide the problem and the time it takes to produce an actual solution (in the affirmative case). Such distinctions will be pointed out as we go along. We conclude with some remarks concerning the applicability of the results in this paper and the lessons that both a circuit designer and the provider of design aids can learn from our discussion.

2. Placement

We start off by studying the problem of cable routability (CR) in Subsection 1. Solving it will provide insight required to study the placement problem, PO. We shall use the solution to CR in formulating the orientability constraints that need to be satisfied by any feasible placement. These constraints, when applied to a given set of modules (with a given orientation), may force some of the modules to be reflected. Such requirements can be straightforwardly accomodated in a graph theoretical framework to be developed in Subsection 2. The case in which modules are not allowed to be reflected is essentially the same as the planar routability problem (PR), since rotation and translation are inconsequential from a topological point of view. PR can be solved within the aforementioned framework, but some extra work is necessary to ensure that the original orientation of modules is preserved. This is done in Subsection 3 by modifying the planarity testing algorithm due to Hopcroft and Tarjan ([HoTa74], [Ev79, Chapter 8]).

2.1. Cable Routability

Two problems need to be solved in order to resolve the cable routability (CR) problem: reachability and orientability. Reachability is the problem of deciding whether the side on which one end of the cable lies on can be connected to the other side by some path travelling through the routing area. This can be solved easily by finding the connected parts of the (not necessarily simple) polygon that comprises the routing area, and testing whether the two sides belong to the same component. The overall complexity of the reachability test is $O(m\alpha(m))^*$, where m is the number of modules. This can be achieved by partitioning the routing area into $O(m)$ stripes and then merging them using a union-find algorithm.

The orientability problem asks whether the sequences at both ends of the cable are oriented in a way consistent with a planar realization. This has been exemplified in Figure 1 above. There is a qualitative difference between cables comprised of two nets on one hand, and cables with three or more wires on the other hand. Two terminals can usually be permuted to obtain the two possible sequences on any side of a module and still be realizable in the plane, but three terminals or more cannot. The reason for this is that whereas two nets can be realized both as one single cable and as two separate cables consisting of one wire each, and still fit in the plane, with three nets any split that maintains the sequence property at both ends of the cable will cause a violation of planarity as a consequence of Kuratowski's Theorem. Figure 3 shows how a two-net cable can be realized in the plane regardless of the relative orientations of the sequences at its ends — once as defined (a) and once by splitting it into two cables (b). A three-net cable, however, must be oriented properly (parts (c) and (d) of Figure 3).

The orientability test for a given cable is easy and takes constant time to perform. Essentially, all we have to do is to find which way each sequence goes relative to the inside of the module it is on, and then compare the orientations of the two sequences

---

$^*\alpha$ is the inverse of Ackermann's function, and can be regarded as a constant for all practical purposes.

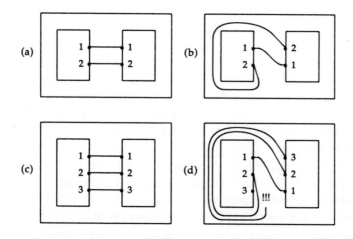

Figure 3. Two wires can be routed as one cable (a) or two cables (b), but three wires are forced into one cable (c and d).

and see if they conform or not. First we look at the first two terminals of each sequence — the rest must follow in the same direction. Then we compare the ways in which the directions of the sequences are related to the inside of the module. When traversed from terminal 1 to terminal n, the inside of the module for one sequence should be on its right, whereas for the other sequence the module should be on its left. In short, the modules should be on opposite sides relative to the sequences, just like the requirements in the channel case.

An explicit implementation of this test can be attained by constructing a 4 by 4 table indexed by the four directions of sides on which terminals can be relative to the module. Each entry tells which way sequences on the corresponding sides must go in order to conform with each other.

If the number of wires in the cable is only two, we can first apply the above test to see if the cable is routable as one unit. If not, we know that we have to split the cable into two wires. Then, however, the reachability test has to be performed twice, since after the first wire has been connected, we must find out whether the second one is still connectable in the new topology created by the connection made.

2.2. Flippable Modules

The problems concerning the routability of two wires are just a special case of Planar Routability: how do we accomodate several cables on the same chip? The layout of one cable may interfere with that of subsequent wires, and thus we need a comprehensive placement method by which individual cables can be properly oriented so they can be each routed in one layer, and at the same time not cross over (at the cable level).

The objective of making a cable properly orientable can be achieved by reflecting modules, ie flipping them over so that some sequences change directions relative to their modules. Obviously, flipping a module once changes the directions of all sequences around it, thus some binding between sequences' orientations is imposed by each module. The second problem, that of placing the modules so that cables do not get into each others' way, seems harder, since now we are trying to resolve some global conflicts. This is where the cable abstraction comes in handy. Using cables as "super-wires" we can model the overall interconnection pattern of the chip by a graph that abstracts all the necessary properties. The graph will be planar if and only if the modules can be placed and oriented in a way that makes a one-layer realization of the interconnect feasible.

Before introducing the way in which cables are used, we present the following definition of the interconnect graph, that is similar to the layout graph concept found in [vanLOtt73]. However, our usage and modification of this concept in this and the next subsections are more elaborate and present a different point of view on the subject.

We model each module by a "wagon-wheel" by placing a vertex in the middle of the module, making a ring out of the terminals around it, and then connect the terminal vertices to the module vertex by spokes. Each net is represented by an edge connecting its two terminals. The correspondence between a layout problem and its interconnect graph is shown in Figure 4.

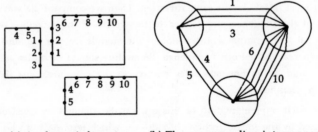

(a) A schematic layout.   (b) The corresponding interconnect graph.

Figure 4. The modules in the schematic layout of (a) are modelled by the wagon-wheel constructions in the interconnect graph (b).

Definition 1: For a given layout problem $L=(M,N)$, consisting of a set of modules M and a set of two-point nets N with terminals on the modules' boundaries, we define the interconnect graph $G(L)=(V,E)$ as follows: $V=M \cup T$ where $T=\{t \mid t \in \nu$ for some $\nu \in N\}$, ie there is one vertex for each module (the module-vertices) and one for each terminal (the terminal-vertices); $E=\{(m,t) \mid m \in M, t \in T,$ and lies on m's boundary$\} \cup \{(t_1,t_2) \mid t_1$ and t_2 are adjacent on the boundary of one module$\} \cup \{(t_1,t_2) \mid \{t_1,t_2\} \in N\}$.

Next we use cables to reduce the size of the interconnect graph by modifying the given layout problem as follows: every cable that consists of 3 or more nets is replaced by exactly 3 nets — the first, last, and one of the interim original nets (lying between the first and the last net). Thus all sequences on modules sides are at most 3 terminals long. Then we use the modified layout to produce the modified interconnect graph. The modified graph for the layout of Figure 4(a) is shown in Figure 5. Notice that the number of vertices has been reduced from $m+2n$ (where $n=|N|$, $m=|M|$) to $m+2c$ where

c is the number of cables ($c \leqslant n$). The number of edges in the graph becomes smaller as well.

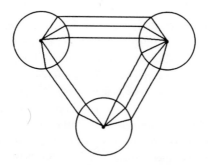

Figure 5. The modified interconnect graph for the
schema in Figure 4.

From here on, the solution of the layout problem can be reduced to the graph planarity problem, since a layout is realizable in one layer (allowing module reflection) if and only if the corresponding (modified) interconnect graph is planar. The correctness of the reduction for the interconnect graph follows from the way modules are modelled, which forces the order of terminals around a module to be preserved by any planar embedding of the graph. The spokes ensure that no connections are being made "inside" modules, ie nets can neither be routed through modules nor can one module be embedded within another. The correctness for the modified graph follows from our discussion on the orientability of cables in the Subsection 1.

Algorithms for planarity testing (and actual generation of an embedding) are known to be fast. Both algorithms presented in [Ev79, Chapter 8] (one is due to Hopcroft and Tarjan, and the other is due to Lempel, Even and Cederbaum) run in time linear in the number of the vertices of the graph. Thus the time complexity of our procedure is $O(m+c)$ once sequences have been formed.

2.3. Unflippable Modules

The planarity testing algorithms used in Subsection 2 cannot be applied verbatim to solve planar routability (PR) if the given modules are not allowed to be reflected. Now the orientation of each module relative to a global frame is fixed a priori (up to rotation and translation). But the wagon-wheel constructs by which modules are modelled can be embedded in the plane in two different orientations. This situation, however, can be remedied by modifying the planarity testing algorithm used. In this subsection we show how to modify Hopcroft and Tarjan's path addition algorithm [HoTa74] so as to handle such extra constraints as the orientation of wagon-wheel subgraphs.

Preprocessing. Before applying the path addition algorithm, some preprocessing is required. Obviously, a necessary condition for embedding the interconnect in the plane for the given orientation of modules is that all cables comprised of three nets or more are routable. This can be tested easily by applying the algorithm given in Subsection 1 for CR (we do not have to check reachability because modules can be translated freely).

Next we look at cables that consist of exactly two nets. If such a pair can be routed as one cable we do nothing at this stage. However, if the pair has to be split into two cables, the implication of connecting both nets has to be taken into account. Figure 6 shows the four ways in which an incompatible pair can be connected as two separate cables. The effect on the planarity of other connections between the two modules is the same in all four cases: the remaining terminals of one module cannot be connected to the remaining terminals of the other module by any path in the plane. Nets 3 and 4 in Figure 6 demonstrate this phenomenon. Terminals whose connection is forbidden in this way are called **separated**. Our modification to the path addition algorithm will be to guarantee that no path (possibly through other modules) connects separated terminals (unless it goes through one of the edges that belong to the pair).

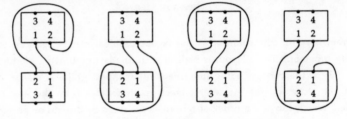

Figure 6. Four ways to connect nets 1 and 2, which
all cut off nets 3 and 4.

For every pair that cannot be routed as a cable, we have to designate the terminals that are separated by it. We generate a pair of labels, l and Γ, for every problematic pair of nets (there is a linear number of such pairs), and assign complementary labels to separated terminals.

Modifications to the path addition algorithm. Now we can start running the path addition algorithm. Every time an edge $e=(u,v)$ is traversed during path generation the labels from the tail, u, are added to the labels of the head, v. There is one important exception: an edge that belongs to the pair of nets that generated the labels pair (l,Γ) does not propagate either of these labels.

The planarity testing is modified as follows: every time a net edge (as opposed to a wagon-wheel edge) is traversed, we check if a conflict has occurred: if we try to add Γ to a set containing the label l the planarity test fails. (This failure condition comes in addition to the failures which may arise from the original parts of the algorithm.)

Correctness. During the path addition algorithm every edge is traversed once. By the definition of the label propagation rule, the only way separated terminals can be connected via a path* in the embedded graph is if they use a net edge that caused their separation, but not through any other net edge.

---

* Notice that a path in the graph may be comprised of parts of several paths that have been added during the algorithm.

Complexity. The modifications required in the path addition algorithm cause only constant overhead per step. We have to assume, however, that set operations such as union and membership can be done in constant time; this can be achieved if a membership vector is used to represent the sets. The size of the sets that have to be kept is the number of problematic pairs, which is linear. The preprocessing that is required is linear as well.

All in all we have

Theorem 1. Planar Routability (PR) can be solved in linear time.

(A detailed proof will appear in a forthcoming paper).

Recently, the result of Theorem 1 has been idependetly obtained using methods different from the one suggested here. In [NaSu82] we find a modification of the node addition algorithm (see [Ev79, Chapter 8]), whereas [ViWi82] has a more or less direct proof (inspired by the path addition algorithm used here).

3. Detailed Routing

So far we have considered only the topological aspects of the layout problem. In this Section we focus on the geometric issues. Design rules (such as in [MeaCo80]) require that each wire has a certain minimal width, and also that different wires are separated by a minimum distance. These two conditions can be combined into the single requirement that a minimal distance is kept between the center lines of disjoint wires. Thus we can abstract a wire by the zero-width path corresponding to its center line; similarly, terminals are represented as points.

In general, the paths corresponding to a wire can be an arbitrary curve in the plane. In layout systems, however, one wishes to constrain the allowable shapes of wires. A variety of one-layer wiring models are supported by exisiting layout packages, and by and large all of them yield to rigorous analysis of their river-routing properties (see [LeiPi81] for a list of wiring models and corresponding results). For didactic purposes, we chose here to deal with the rectilinear wiring model alone. In this model wires can travel vertically and horizontally in a grid-free fashion.

The techniques used in Subsection 3.1 to solve problem DRSRP generalize to the other models, but the complexity of the algorithms may change in certain cases to reflect the intrinsic difficulty of a particular model. All running times for the models listed in [LeiPi81] are polynomial. In Subsection 3.2 problem DR is shown to be NP-complete even for the rectilinear model, which is the simplest one of all.

3.1. Routing in a Simple Rectilinear Polygon

Given an instance of Detailed Routing in a Simple Rectilinear Polygon (DRSRP), we start off by checking whether the connections specified by the nets are at all realizable in one layer from a topological point of view, ie whether the interconnect pattern is planar when constrained to lie within the polygon (ignoring minimum spacing rules). We could have used a test similar to the general test suggested in Subsection 1.2, but here we can use a simpler method. What we have to check is whether the terminals as they appear along the boundary of the polygon match without intersecting. This is analogous to checking whether a set of parentheses is properly balanced, as formulated in the following algorithm:

(1) Initialize the stack S to be empty.

(2) Cut the boundary of the polygon at any (single) point and straighten it out.

(3) Scan the list of terminals from left to right. For each terminal compare the net number (or any other form of a net **id**) to the number at the top of S. If they are equal, pop S, else push the number on S.

(4) If at the end of the scan S is empty, the nets are properly nested. Otherwise, they are not realizable in one layer.

Correctness. We interpret the first occurrence of a terminal in a net as an opening (left) parenthesis, and the second occurrence* as a closing (right) parenthesis. Regardless of where the boundary has been cut, the interconnection pattern is internally planar if and only if the parentheses in the expression (obtained by the above interpretation) are properly balanced. Nets must be nested within each other or be mutually exclusive in their span in order to allow a planar (non-crossing) realization, which is the same as requiring that a set of parentheses be balanced. This correspondence is demonstrated by Figure 7, parts (a) and (b).

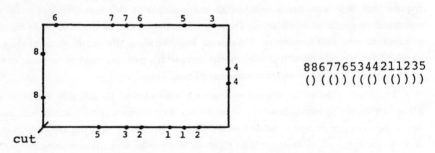

Figure 7. The correspondence between planar routing in a polygon (a) and parentheses expressions (b).

We now turn to the question whether the area of the rectangle is enough to realize the routing of the nets. Sometimes it is faster to check whether a set of nets is routable within a given area than to produce the actual paths taken by the wires. For the problem of river routing across a channel, for example, testing routability takes time $O(n)$ for n nets, whereas producing the layout may take time $\Omega(n^2)$ [LeiPi81]. In what follows we show that the situation for river routing in a polygon is similar. It is easier to test for routability than to produce a layout.

We first provide an algorithm that river routes nets in a rectangle whenever possible. Use the template given in the algorithm for planarity checking (without straightening the boundary): whenever a net is being popped off the stack, simply route it! The routing is done in a "greedy" fashion relative to the contour — the router stays as close

---

* Remember that we are dealing with two point nets only. Multipoint nets can be treated by duplicating the internal terminals, thus inserting a ")(" in the parentheses expression.

as possible to the boundary by initially routing nets along the original boundary (for "innermost" nets), and then staying as close as possible to the contour formed by previous nets towards the boundary. This routing strategy is demonstrated by Figure 8.

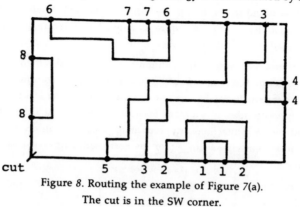

Figure 8. Routing the example of Figure 7(a).
The cut is in the SW corner.

This procedure takes time that is proportional to the number of straight (or elementary) wire segments that are produced; this number can be as large as $\Omega(n^2)$ in the rectilinear model, as shown, for example, in Figure 9 (where n is the number of nets to be interconnected). The worst case time complexity of the algorithm is $O(n^2)$, matching the existential lower bound, thus it is optimal from this point of view. However, it may produce results that are suboptimal in terms of wire length and number of jogs.

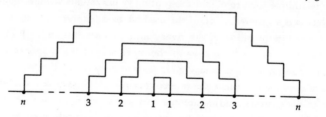

Figure 9. The greedy routing rule may produce $O(n^2)$ segments even when all terminals are on one side.

A similar algorithm has been discovered by C. P. Hsu [Hs82] and has been brought to my attention by David N. Deutsch. In his paper Hsu provides heuristics for reducing the number of jogs and the for making wires shorter without impairing the feasibility of the solutions. In fact, each such algorithm is just one member in a family of algorithms obeying the following greedy routing rule:

"Each net is routed when all nets between its terminals along the boundary (according to the linear order) have already been routed."

The sufficiency of such a rule is quite obvious: all nets are routed in a way that minimizes their mutual interference. Each net is routed as tightly as possible relative to the area requirements of the nets that have been routed so far, and as far away as possible from any subsequent net. The order of routing is dictated by the planarity — up to mutual exclusive nestings.

Now we turn to the question of routability. Unfortunately, the full-fledged algorithm for testing routability in a polygon is too complicated to be described here. This description can be found in [Pi82b, pp. 64--74]. Thus we shall briefly outline the main points of the routability testing procedure as applied to a rectangle when rectilinear wiring is being used.

We classify nets according to the relation between the sides their terminals lie on. If both terminals are on the same side of the rectangle, the net constitutes a **single-sided connection**. If the terminals are on adjacent sides, we say the net is a **corner connection**. If they are on opposite sides, it is a **cross connection**. Notice that although single-sided connections and corner connections can occur at all possible locations in one instance of the problem, cross connections can either go north-south or west-east, but not both ways (see Figure 10).

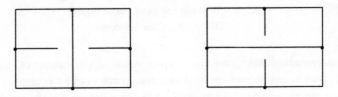

Figure 10. Cross connections can be of one kind only.

The technique used to obtain the routing requirements without generating all the routing details is a generalization of the method used for river routing across a channel. The geometric interpretation of the river routing constraints in [LeiPi81] is a 45° lines drawn from terminal positions towards the other side of the channel. These lines told us how other terminals had to stay clear from the source of the "ray" to accommodate the wires around its net. Since the interconnection pattern was known (all nets went in order across the channel) and monotonic wiring was sufficient, we could make all lines the same length (as a function of the separation) and compare each endpoint statically to the rest of the channel. Here the structure is more complicated, and we have to dynamically change the length of the "ray" we draw in order to establish the routing requirements around the current net relative to the other nets.

We first generate the boundary around single-sided connections. Here there is no problem of routability because the connections can always be realized (assuming the rectangle above the side has unbounded height), and all what we seek is the smallest area required for routing. The output of our procedure will be a description of the routing contour that is necessitated by the given configuration, ie the path closest to the boundary that can be taken by the outermost nets.

The contour is generated by looking at the effect that each terminal has on the routing of the outermost net. This is done by drawing 45° rays from terminals whose directions depend on the nesting of the nets. These rays are clipped according to the depth of the net in the nesting and thus form markers that outline the desired boundary. The nesting structure of the nets is heavily used in making the running time of the whole procedure as low as possible, namely $O(n)$ where n is the number of nets.

The next step is to deal with corners. Notice that a corner connection may have single-sided connections nested within it, as shown in Figure 11, but not vice versa. Thus single-sided connections need not be dealt with any further from this stage on.

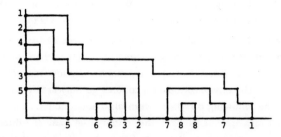

Figure 11. Corner connections (1, 2, and 3) with-single sided (and other corner) connections netsed within them.

Unlike single sides, corners may not always be routable. First, single-sided connections coming off orthogonal sides could interfere with each other, as in Figure 12. Second, connections to terminals may be blocked by narrow passages between the single-sided contours.

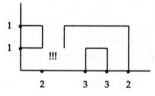

Figure 12. Interference between single-sided connections coming off adjacent sides.

To solve these interference problems we maintain a separate contour for each side, and then check whether the contours intersect. Thus all single-sided connections are processed first, setting up contours for subsequent analysis. We also process those corner connections that do not have single-sided connections embedded in them, which is an easy extension to the single-sided algorithm.

The essence of the procedure for testing the routability of corners is finding out whether the wires that must be connected to the sides have enough room to pass through the "straits" formed by the single-sided contours. The test we devise applies not only to making corner connections per se, but also to the ability of terminals that are on the sides of a corner to "get out of the corner" and be connected to their mates regardless whether the mates reside on the other side of the corner or on another side of the polygon altogether.

The technique for testing routability in a corner is to draw 45° "rays" from all convex (protruding) corners of single-sided contours and find their intersection with the rest of the contour. Recall that the contour consists of original sides as well as pieces generated by single-sided connections. Each ray is trimmed at its first intersection with the rest of the boundary. If the ray hits the boundary from the "inside" then we know that two contours collide and the test fails. If, on the other hand, the ray has positive length, we compute how many wires can cross it. The crossing test for a ray succeeds if this number is not smaller than the number of nets that must cross it.

The next step is a generalization of the corner routability test so that it can be used to test routability in a rectangle. Three rays (rather than one) are being drawn from each corner, and also rays are being generated by terminals. The three rays that are generated at each convex corner of a contour go outwards relative to the inside of the contour in the following directions: vertically, horizontally, and in a 45° angle. The first and second rays are extensions to the contour itself, and the third bisects the right angle created between them, as shown in Figure 13. What is not shown in this figure (so as not to impair its clarity) is that now we also draw two 45° rays from each terminal — one to each direction.

The test for routability is the same as in the corner case: all trimmed rays must be longer than the number of nets that have to cross it. A ray that starts from within the contour of another side has negative length, thus the test as stated covers the contour intersection problem as well. The necessity of the three rays is due to the interaction between single-sided contours that are generated on opposite sides. Forty-five degree rays are not guaranteed anymore to be trimmed by a perpendicular side, thus we need an explicit density test between opposite contour. This is exemplified in Figure 13, where the forty-five degree rays miss the contour on the opposite side, but the passage between the contours is too narrow to permit four nets to be routed across it.

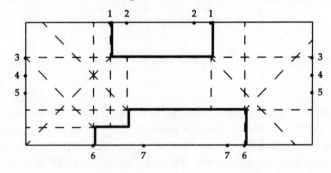

Figure 13. The necessity of rectilinear rays is demonstrated by the narrow passage between the single-sided contours.

When routing in a rectilinear polygon, rather than a rectangle, the corners of the polygon have to be treated as contour-corners, generating three rays each. If the polygon is known to be simple, all routability tests can be done in linear time. So all in all

Theorem 2. Detailed planar routability in a simple rectilinear polygon (DRSRP) can be decided in linear time.

3.2. Routing around Modules

The problem of detailed routing around modules when no homotopies of wires are given (DR) is NP-complete. In this subsection we show how to modify a recent result of Kramer and van Leeuwen [KrvanL82] to obtain the claimed statement. The intractability of the problem is due to the possibilities of routing wires around modules in more than one way. Unlike the situation in Section 2, where planarity was the only concern, here the spacing requirements play a key role. Routing a wire using a certain path may render other wires unroutable due to the consumption of routing area by the first wire. Thus some global decisions need to be made in a way that is so typical of most known NP-complete problems.

The two problems discussed in [KrvanL82] concern one-layer routing of two-point nets on a square grid. In the first problem ("Routing") routing is done in the absence of modules altogether. In the second problem ("Obstacle Routing") modules are allowed as part of the problem specification, but they are viewed only as forbidden areas rather than pieces of logic with connections on their boundaries. Thus all terminals are grid-points that can be routed from in all four directions (unless one of the immediate neighbours is occupied by a terminal of a different net). This deviation from our notion of terminals as residing on modules' boundaries can be remedied easily by providing a simple construct for local replacement [GaJo79, Section 3.2.2] to be used in our transformation.

The transformation can be made from either of the problems discussed in [KrvanL82]. We prefer here to use the problem in which modules (obstacles) are present. Notice that in all the constructions of the proof in [KrvanL82], a path leaves a terminal in one of two opposite directions. Thus all arguments can be retained if we construct a layout in which the paths are constrained a priori to leave a point in one of two opposite directions (or even in one of three directions). This can be done by using the replacement of Figure 14: terminal P in the original construction is replaced by terminal P' on the side of a module that is parallel to the direction in which the paths leaving P go in the original construction. The point P has to coincide with P'', the immediate neighbor of P', rather than with P' itself.

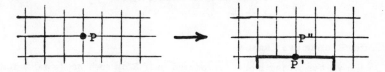

Figure 14. A pin in the Kramer and van Leeuwen construction can be replaced by a terminal on a module's side.

Since most terminals in the construction of [KrvanL82] are located next to modules (obstacles) anyway, and their routing direction is parallel to the side of the module, this replacement can be performed by just moving P to the boundary. In the few cases where this is impossible, we splice a new module into the layout, thereby expanding the grid by the necessary dimensions without affecting the area available for routing. Thus the feasibility of a solution remains invariant under this transformation, proving

Theorem 3. Detailed planar routing is NP-complete.

The major question that remains open concerning detailed routing is that of finding the detailed routing when the homotopy for all wires is given a priori (DRH). Then the decision of how to route around the modules topologically is solved, but we still have to find exact paths for the wires. When no modules (holes) were present, nets could be routed using the greedy routing rule defined in the previous section. Now, however, there is no unique contour with which nets can be associated. To start with, it is unclear "how long" a wire has to stay close to the side of one terminal and when to start moving towards the side of the other terminal in the net (not to mention sides that the net passes by on the way).

This problem manifests itself even in the simple situation of a doughnut router, which is typical for routing to pads (Figure 15). Johannsen [Jo81, Appendix 3] provides a heuristic for solving the routing to pads problem that is based on the greedy routing rule for a channel, but his strategy does not produce optimal results. Routing in a doughnut and the associated placement problem (where to put a module within a "frame" so as to ensure routability) have been both solved recently: placement and routing each take quadratic time. These results will be described in a forthcoming paper [BaPi83].

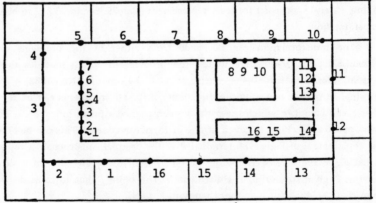

Figure 15. River routing to pads.

4. Via-free Routing in a Channel

In the via-free wiring model (introduced in [Pi82b]) each net (wire) is routed in one layer and may not change layers along the path. We assume that all terminals are available in all layers, so each net can be assigned a layer by the routing algorithm. This approach alleviates the problems caused by using contacts, at the expense of not being a "general" wiring model. Figure 16 shows that not every interconnection pattern can be realized using the via-free wiring model, but it can still be used to embed many useful networks, such as the shuffle-exchange graph. This "one layer per net" model is in many ways a natural generalization of one-layer routing, as we shall see here.

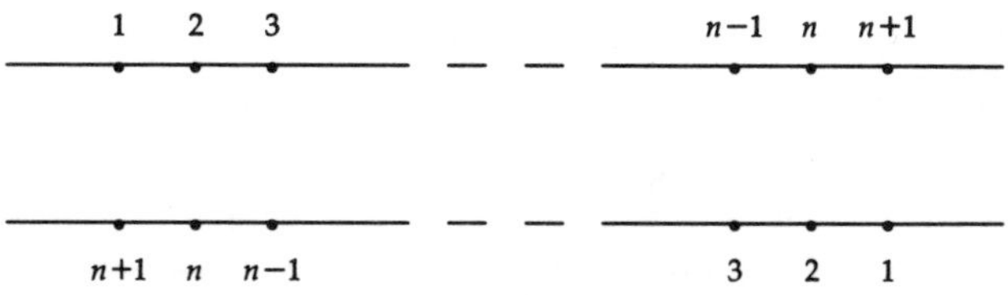

Figure 16. A channel that cannot be routed in the via-free
model using n layers.

We first show how to efficiently determine the minimum number of layers required to route two-point nets across a channel using the via-free wiring model. Then we show how to find the minimum width of a channel when the smallest number of layers is being used in the via-free wiring model. This is done by combining the framework developed for the layer minimization problem with results on river-routing across a channel from [LeiPi81].

In the via-free model two nets have to be assigned different layers if and only if they cross each other, that is if the straight lines drawn to connect their terminals intersect. In symbols, the crossing condition for nets N_i and N_j is $(a_i-a_j)(b_i-b_j)<0$. A set of nets going across a channel can be assigned l layers so that no two crossing nets will be assigned the same layer if and only if the following coloring problem can be solved: Given are l colors and the graph $G=(V,E)$ where $V=\{N_i, i=1,...,n\}$ is the set of nets, and $E=\{(N_i,N_j)|N_i$ crosses $N_j\}$ represents crossings. The coloring problem in general is NP-complete, but the kind of graphs set up by the channel routing problem in the via-free model is a special class called **permutation graphs**. Such graphs can be colored in polynomial time for any number of layers l. In case, however, single-sided connections are allowed, the channel routing problem becomes NP-complete since it is equivalent to the problem of coloring a circle graph [GaJo79].

Even, Pnueli, and Lempel [EPL72] have provided an $O(n^2)$ algorithm for coloring a permutaion graph with n nodes. In fact, they recognized the applicability of such graphs to the very same problem discussed here (cast in a PCB context). A faster algorithm, running in time $O(n \log l)$, where n is the number of nets and l is the number of layers, is derived in [Pi82b] as a modification to the algorithm in [EPL72].

Now we turn to the subject of minimizing the channel's width in the via-free model. The interaction between wires as far as spacing is concerned occurs in each layer independently of the other layers. Thus we can partition the wires into set according to the layer they are routed in. We can exploit the river routing constraints as developed in [LeiPi81] in order to determine the width that is required for routability: whenever a layer is being assigned to a net, we test whether it will fit in the specified (or current, depending whether we are solving the decision or the minimization problem) width.

Unfortunately, in order to obtain the minimum width attainable with the minimum number of layers required, we have to resort to the slower layer assignment algorithm from [EPL72], rather than the new algorithm devised in [Pi82b]. Figure 17 shows two solutions to a channel routing problem using the via-free model, one using the slow layer assignment algorithm and the other using the fast one.

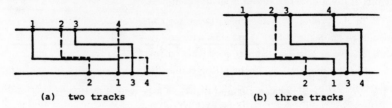

(a) two tracks   (b) three tracks

Figure 17. Two solutions to the same channel routing problem using two layers in the via-free model. Each has minimum width relative to its layer assignment, but (a) is narrower than (b).

Another point concerens the relationship between the minimum width and the minimum number of layers. The width minimization works only when the routing is done with the minimum number of layers needed. If more layers are available, the extra degree of freedom makes the optimization problem too compicated to handle with the same technique. The only situation for which this problem is solved is river routing (see [LeiPi81]).

5. Conclusions

One-layer realizations are limited in their capability of handling arbitrary wiring patterns. The interconnect graph must be planar, and if modules are not flippable, the situation is even more restrictive. On the other hand, the algorithms for testing whether the strategy is applicable for a given situation, as well as routing algorithms for producing truly optimal results themselves, are generally efficient with minor exceptions. Thus it is reasonable to spend time in trying to use a one-layer method before going on to more complicated strategies.

It is, however, naive to believe that the interconnect of an arbitrary circuit could be realized in the plane. But this is exactly where our advice comes in. Because the actual routing is so much easier when doing it in one layer, try to design in such a way that will make use of as many cables as possible. This may not be so hard when designing certain kinds of circuits, such as microprocessors, in which a fair amount of the data is being grouped in sequences. In other cases, the order of signals coming off certain logical modules, such as PLAs, can be reordered at will (without changing the functionality of the circuit) so as to generate conformable sequences.

In general, it is hard to find a maximal subset of planar interconnections: this is equivalent to the maximum planar subgraph problem, and is known to be NP-complete [GaJo79, p. 197]. But the planarity of a single cable is easy to check both for orientability and connectivity. First, the proper orientation of a cable can be checked in constant time, so we have a fast way to detect planarity of such subgraphs. Second, it is easy to check whether the insertion of a new cable into an existing layout will violate its planarity or not. Thus a simple hill-climbing heuristic can be used with many start positions in trying to find a reasonably large set of cables all realizable in one layer. The reduction of complexity achieved by the cable abstraction can, in fact, be exploited by any other heuristic for the maximal planar subgraph or even by an exhaustive procedure if the number of cables is small enough.

To the provider of support software for computer aided design (CAD), our advice would be to make cables "first-class citizens" in the design environment. The user should have at her/his disposal easy ways to specify and manipulate cables. Such aids should include operations for handling connectivity as well as tests for orientability and routability. In this way, the designer can test the feasibility of her/his design while creating it, exploiting planarity as one goes along, without the fear that at the end of the process she/he will find out that the planar realization failed without being able to trace it to a specific culprit.

Similarly, one would like to have support for parts of the design as well. We may partition the nets into sets that will be each realized in one layer. The planarity tests will apply to one set at a time, and the user should be able to specify the identity of such sets. There should be an easy way of moving around members of these sets without changing their connectivity data, but in a way that will make the necessary geometric adjustments.

Another consideration that is relevant to the design of microprocessors is the possibility that some of the signals "traveling" on a cable need to be fanned-out to more than one terminal. These are special kinds of multipoint nets that occur rather frequently in structured design such as microprocessors. We propose a generalization of cables that encapsulates this routing situation, namely **cable busses**. A cable bus is a construct in which two identifiable sequences must be connected in one layer, and some of the nets have more connections other than within the sequences. These extra connections could be realized by attachments to the main cable that are river routed in themselves on another layer.

Cable busses pose quite a few problems, such as how to bias the routing of the main cable so as to make the routing of the attachments easier, and how to conduct detailed river routing to targets that are "floating" along a wire, rather than being fixed terminals. These problems remain open, but the methodological framework exists for some support at the design aids level.

We summarize by giving a list of "rules" for good design in one layer:

(1) Use cables, since they reduce complexity.

(2) Provide design aids that know about cables and can manipulate them well.

(3) Assign the same layer to cables that run in "parallel".

(4) Provide subsetting facilities that maintain association with layers.

(5) Extract cables from multipoint nets if they can be used in a natural way.

(6) If the genus of the interconnect pattern is low, via-free routing is a good strategy, although it may require larger width than necessary.

Acknowledgements

I would like to thank Alain Hanover, Charles Leiserson, Ron Maxwell, and Ron Rivest for many helpful discussions. Al Aho provided numerous comments on an earlier version of this paper.

References

[AGR70]   Akers, S. B., J. M. Geyer, and D. L. Roberts: "IC Masks Layout with a Single Conductor Layer"; Proceedings of the Seventh Design Automation Workshop, June 1970, pp. 7--16.

[BaPi83]   Baker, B. S., and R. Y. Pinter: Optimal Placement of a Circuit Within a Ring of Pads; in preparation.

[DKSSU81] Dolev, D., K. Karplus, A. Siegel, A. Strong, and J. D. Ullman: "Optimal Wiring Between Rectangles"; Proceedings of the Thirteenth Annual ACM Symposium on Theory of Computing, May 1981, pp. 312--317.

[Ev79]   Even, S.: Graph Algorithms; Computer Science Press, Potomac, MD, 1979.

[EPL72]   Even, S., A. Pnueli, and A. Lempel: "Permutation Graphs and Transitive Graphs"; Journal of the Association for Computing Machinery, Vol. 19, No. 3 (July 1972), pp. 400--410.

[GaJo79]   Garey, M. R., and D. S. Johnson: Computers and Intractability: A Guide to the Theory of NP-completeness; W. H. Freeman & Co., San Francisco, CA, 1979.

[HoTa74]   Hopcroft, J. E., and R. E. Tarjan: "Efficient Planarity Testing"; Journal of the Association for Computing Machinery, Vol. 21 (1974), pp. 549--568.

[Hs82]   Hsu, C. P.: General River Routing; unpublished manuscript, Dept. of EECS, Univ. of California, Berkeley, CA, 1982.

[Jo79]   Johannsen, D. L.: "Bristle Blocks: a Silicon Compiler"; Proceedings of the Caltech Conference on VLSI, January 1979, pp. 303--310. Also appears in the Proceedings of the Sixteenth Design Automation Conference, June 1979, pp. 310--313.

[Jo81]   Johannsen, D. L.: Silicon Compilation; Technical Report No. 4530 (Ph.D. dissertation), Dept. of Computer Science, Caltech, Pasadena, CA, 1981.

[KrvanL82] Kramer, M. R., and J. van Leeuwen: Wire-Routing is NP-complete; Technical Report RUU--CS--82--4, University of Utrecht, the Netherlands, February 1982.

[LaP80]   LaPaugh, A. S.: Algorithms for Integrated Circuit Layout: An Analytic Approach; MIT/LCS/TR--248 (Ph.D. dissertation), MIT, Cambridge, MA, November 1980.

[LeiPi81]   Leiserson, C. E., and R. Y. Pinter: "Optimal Placement for River Routing"; Proceedings of the CMU Conference on VLSI Systems and Computations, October 1981, pp. 126--142. A revised version will appear in SIAM Journal on Computing, Vol. 12, No. 3 (August 1983).

[MeaCo80] Mead, C., and L. Conway: Introduction to VLSI Systems; Addison-Wesley

Publishing Company, Reading, MA, 1980.

[NaSu82]   Nakajima, K., and M. Sun: "On an Efficient Implementation of a Planarity Testing Algorithm for a Graph with Local Constraints"; Proceedings of the Twentieth Annual Allerton Conference on Communication, Control, and Computing, October 1982, pp. 656--664.

[Pi82a]   Pinter, R. Y.: "On Routing Two-Point Nets Across a Channel"; Proceedings of the Nineteenth Design Automation Conference, June 1982, pp. 894--902.

[Pi82b]   Pinter, R. Y.: The Impact of Layer Assignment Methods on Layout Algorithms for Integrated Circuits; Ph.D. dissertation, Dept. of Electrical Engineering and Computer Science, MIT, Cambridge, MA, August 1982.

[Pr79]   Preas, B. T.: Placement and Routing Algorithms for Hierarchical Integrated Circuit Layout; Computer Systems Laboratory Technical Report No. 180/SEL--79--032 (Ph.D. dissertation), Stanford University, Stanford, CA, August 1979.

[Riv82]   Rivest, R. L.: "The 'PI' (Placement and Interconnect) System"; Proceedings of the Nineteenth Design Automation Conference, June 1982, pp. 475--481.

[SaBh80]   Sahni, S., and A. Bhatt: "Complexity of the Design Automation Problem"; Proceedings of the Seventeenth Design Automation Conference, June 1980, pp. 402--411.

[Sh82]   Shrobe, H. E.: "The Data Path Generator"; Proceedings of the MIT Conference on Advanced Research in VLSI, January 1982, pp. 175--181.

[SieDo81]   Siegel, A., and D. Dolev: "The Separation for General Single-Layer Wiring Barriers"; Proceedings of the CMU Conference on VLSI Systems and Computations, October 1981, pp. 143--152.

[SouRo81]   Soukup, J., and J. Royle: "On Hierarchical Routing"; Journal of Digital Systems, Vol. 5, No. 3 (Fall 1981), pp. 265--289.

[Sz81]   Szymanski, T. G.: Dogleg Channel Routing is NP-complete; unpublished manuscript, Bell Laboratories, Murray Hill, NJ, September 1981.

[To80]   Tompa, M.: "An optimal solution to a wire-routing problem"; Proceedings of the Twelfth Annual Symposium on Theory of Computing, April--May 1980, pp. 161--176.

[vanLOtt73] van Lier, M. C., and H. J. M. Otten: "On the Mathematical Formulation of the Wiring Problem"; International Journal on Circuit Theory and Applications, Vol. 1, No. 2 (1973), pp. 137--147.

[ViWi82]   Vijayan, G., and A. Wigderson: Planarity of Edge Ordered Graphs; Technical Report No. 307, Dept. of Electrical Engineering and Computer Science, Princeton University, Princeton, NJ, December 1982.

Area and Delay Penalties in Restructurable Wafer-Scale Arrays

Jonathan W. Greene* and Abbas El Gamal**
Information Systems Lab
Stanford University

Abstract: The penalties for restructuring wafer-scale arrays for yield enhancement are assessed. Each element of the fabricated array is assumed to be defective with independent probability p. A fixed fraction R of the elements are to be connected into a prespecified defect-free configuration by means of switched interconnections. The area penalty is determined by the required number of tracks per wiring channel t. Propagation delay is determined by the required maximum connection length d. It is shown that: Connection of RN fixed I/O ports to distinct nondefective elements from an N-element linear array requires $d, t = \Theta(\log N)$; Connection of RN pairs of elements from two N-element linear arrays requires only constant d and t; Connection of a chain of RN^2 elements from an $N \times N$ array also requires only constant d and t; Connection of a $\sqrt{R} N \times \sqrt{R} N$ lattice from an $N \times N$ array requires $d = \Omega(\sqrt{\log N})$. Constant t suffices to connect a lattice if $d = \Theta(\log N)$. Algorithms are presented that connect any fraction $R < 1-p$ of the elements with probability approaching one as N increases. It appears that these results hold even under actual defect distributions.

1. Introduction

Recent efforts at wafer-scale integration rely on the use of redundancy to maintain reasonable yields. Lincoln Labs [LI] and McDonnell Douglas [HF] are currently experimenting with restructurable whole-wafer processor arrays. After testing the array elements, on-chip switches are programmed to connect nondefective elements into the desired configuration. Switches that irreversibly open [AB] [SM], irreversibly close [MI], or reversibly latch [HF] are available. These techniques have already found widespread application in more conventional forms of

---

*   Work partially supported by an NSF Graduate Fellowship and under Air Force contract F 49620-79-C-0058.

**   Work partially supported under DARPA contract MDA 903-79-C-0680 and Air Force contract F 49620-79-C-0058.

VLSI, such as 64K RAMs containing spare rows and columns; yield improvements of a factor of five to eight have been reported [SM].

This improvement in yield is achieved at the expense of overhead area occupied by the switches and extra interconnections, and an increase in signal propagation delay. Also, determining the switch settings may entail a non-trivial computation. In this paper, we investigate these penalties for several regular array configurations. Bounds on the severity of the penalties as a function of array size are derived and algorithms for restructuring the wafer are given. Our approach is best illustrated through the following simple example.

A linear array of K identical processors, connected in a chain, is to be implemented on a single wafer. Assume that each processor, or more generally, circuit *element*, has an independent probability p of being *defective* and $1-p$ of being *active*. Then the yield, or probability that a wafer is functional, is $(1-p)^K$, approaching zero exponentially as K increases.

In order to maintain good yield, the number of elements on the wafer is increased to $N=K/R$ for some $R<1-p$ and switches are provided to insert the elements in the chain. After manufacture, the elements are tested. If the number of active elements is less than K the wafer is discarded. Otherwise, K elements can be connected as shown in Figure 1.1a. Since $R<1-p$, the probability that the wafer will have sufficient active elements approaches one exponentially as K increases. In this example, only one track is required in the wiring channel. A serial memory wafer fabricated as an early demonstration of this concept is shown in Figure 1.1b.

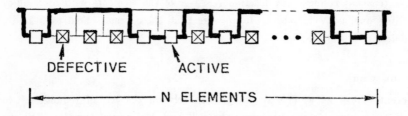

DEFECTIVE   ACTIVE

|←————————— N ELEMENTS —————————→|

Fig. 1.1a: Connection of a chain from an N-element linear array.

Unfortunately, signals from one element to another can now encounter additional propagation delay since the connections between elements are longer than before. (The exact relationship between wire length and delay is not important here). Suppose that the maximum tolerable connection distance is fixed at d, where the elements are spaced at unit length. There are fewer than N places at

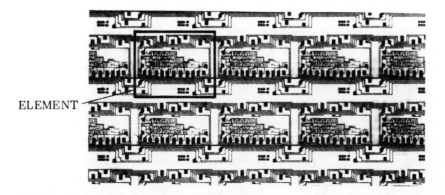

ELEMENT

Fig. 1.1b: A portion of the metallization pattern of a serial memory wafer using the data path structure of Fig. 1.1a. Interconnections are controlled by non-volatile MNOS memory cells. (From Fig. 6 of [HCE]).

which a chain can start, and probability p^d that there will be too many defects before the next active element. Since there are K active elements in the chain, the yield is less than $N(1-p^d)^K < N\exp(-Kp^d)$. This approaches zero unless the size of the chip, N, grows exponentially with K—a very unsatisfactory situation.

We require here and throughout that the fraction of elements connected, R, must be held constant as N grows. Then the only way to maintain good yield is to permit d to grow with $N=K/R$. For fixed $R<1-p$, it is easily shown that $d=O(\log N)$[†] permits connection of the chain with probability approaching one polynomially in N.

Our analysis is based on two assumptions, both of which have been used in practice [LI]. The first is that faults in the interconnect, including power distribution, are taken to be much less likely than element defects. Since the interconnect requires fewer fabrication steps than the active circuitry and is not affected by variations in electrical parameters, this is expected to be the case. Furthermore, the interconnect we propose is in many cases flexible enough to route around faulty wiring as well as defective elements. The second assumption is that defects are independent and identically distributed. Though perhaps not completely accurate, this enables us to derive analytical results. Arguments for the general validity of these results under actual defect distributions are presented later in the paper.

---

†   We use $O(\cdot)$ to denote an upper bound, $\Theta(\cdot)$ to denote an exact bound, and $\Omega(\cdot)$ to denote a lower bound, all to within a constant factor. Natural logarithms are used throughout.

Our wiring model is as follows. A prespecified number t of wiring tracks is provided in each channel between elements. At each point where tracks intersect, programmable switches are provided. The switches enable the intersecting tracks to be connected in any way, for example in a crossover or elbow. It is required that the length of each resulting electrically connected segment be no greater than d. The *yield* is the probability that the defects are distributed in such a way that a configuration of the desired size can be connected under these constraints. For upper bound results, we assume elements and tracks are spaced at unit width.

We bound the parameters d and t for the four configurations listed below.[‡] Upper bounds are demonstrated by presenting algorithms that connect any fraction $R < 1-p$ of the total number of elements with a yield approaching one polynomially in the size of the array. The running time of the algorithms is linear in the number of elements.

Sec. 2: $K = RN$ fixed input/output ports are to be connected to distinct, active elements from a linear array of N elements. (This might be necessary to connect bit slices to a bus, for example). This is accomplished using a channel containing t wiring tracks between the ports and the array. A connection may not run along the channel for a distance greater than d. We term this arrangement, shown in Figure 1.2, a *selector*. Lower and upper bounds of $t, d = \Theta(\log N)$ are obtained. These are due to the high probability that the longest run of consecutive defects is $\Theta(\log N)$ elements.

Sec. 3: The connection of K pairs of active elements from two parallel N-element arrays, shown in Figure 1.3, is surprisingly easier than the task of the selector. A run of defects does not necessarily cause a problem here because there are no fixed ports—alternate pairs of elements can be connected from other parts of the array. Only constant d and t are required.

Sec. 4: The connection of a chain from a two-dimensional array, as shown in Figure 1.4, can be achieved with $t = 2$ wiring tracks between elements and constant d. This result is closely related to the percolation problem of statistical physics. A different scheme achieving similar results has been found independently by Leighton and Leiserson [LL].

Sec. 5: The connection of a $K \times K$ square lattice from an $N \times N$ array of elements spaced at unit length (Figure 1.5) requires wire length $d = \Omega(\sqrt{\log N})$ to prevent yield from approaching zero, no matter how many tracks are provided. This is due to the high probability that there is a block of $\Theta(\log N)$ defective elements which

---

[‡]   In several cases, this paper gives only proof sketches for simplicity. Fully detailed proofs appear in [GE].

K PORTS

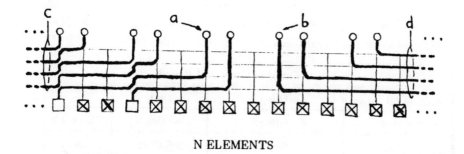

N ELEMENTS

Fig. 1.2: A section of a selector. $K=(2/3)N$ and $t=4$ wiring tracks are provided.

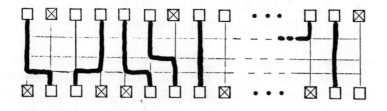

Fig. 1.3: Pairwise connection of two parallel linear arrays of N elements.

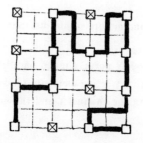

Fig. 1.4: Connection of a chain of $K=11$ elements from a 4×4 array.

Fig. 1.5: Connection of a 3×3 square lattice from a 4×4 array.

must be crossed by a connection. We conjecture that only a constant number of wiring tracks between elements is necessary, though the best distance we have achieved with constant t is $d=O(\log N)$. Leighton and Leiserson [LL] have proposed a connection scheme with $d=O(\sqrt{\log N}\,\log\log N)$ and $t=O(\log\log N)$.

However, their algorithm may require an impractically large look-up table.

The following table summarizes the bounds. It appears that the additional interconnect area and delay necessitated by the use of redundancy will grow slowly enough not to be a significant impediment to wafer-scale integration.

TABLE: Order of Growth of Bounds		
Chain in 1D array	$d = \Theta(\log N)$	$t = 1$	-
Selector	$d = \Theta(\log N)$	$t = \Theta(\log N)$	Thms. 1,2
Pairing	$d \leq O(1)$	$t \leq O(1)$	Thm. 3
Chain in 2D array	$d \leq O(1)$	$t \leq 2$	Thm. 4
Lattice	$d \geq \Omega(\sqrt{\log N})$	-	Thm. 5
	$d \leq O(\log N)$	$t \leq O(1)$	Thm. 6
	$d \leq O(\sqrt{\log N} \log\log N)$	$t \leq O(\log\log N)$	[LL]

The problems of connecting a chain and a lattice from a two-dimensional array of faulty elements were first studied by Manning [MN] and Aubusson and Catt [AC], and more recently by others [FV] [HS] [KO] [LM]. Though in most cases the algorithms are studied empirically, analytical performance estimates are lacking. A different approach using one dimensional arrays is considered in [RO].

2. Selectors

We begin by proving a lower bound on the maximum connection distance and number of tracks required in a selector.

Theorem 1: For any $0 < \delta < 1$, the probability that $K = RN$ ports, aligned parallel to a linear array of N elements, can be connected to distinct active elements tends to zero as $\exp(-N^{1-\delta}\overline{p}/p)$ [†] unless the number of tracks in the channel t and the maximum connection distance d both satisfy

$$d, t > \frac{\delta R \log N}{-2\log p} - \frac{3}{2} = \Omega(\log N).$$

Proof sketch: For any $0 < \delta < 1$, let m be the largest odd integer less than or equal to $-\delta R \log(N)/\log(p)$ and suppose that either $t < (m-1)/2$ or $d < (m-1)/2$. Divide the array and the ports by perpendicular cuts into $\lfloor (NR-1)/m \rfloor - 1$ blocks containing at least m ports and a block at each end containing at least $(m+1)/2$ ports.

---

†   For ease of notation, $\overline{x}$ denotes $1-x$.

There are at least NR/m blocks. If there is a block containing only one or no active element, then all but one of its ports must be connected to elements in other blocks. This would require that more than t tracks leave a side of the block and that at least one of these tracks travels a distance greater than d before reaching an element in another block. Thus under the above constraints on t and d, the yield is bounded by the probability that all blocks contain at least two active elements, which can be shown to be less than $\exp(-N^{1-\delta}\bar{p}/p)$. Substituting the definition of m into the assumptions on d and t yields the result.□

Theorem 1 is true because of the high probability of a run of $\Theta(\log N)$ defects. Fortunately, this is also the longest run that is likely to occur, permitting us to prove the following theorem.

Theorem 2: For any $R<\bar{p}$, let $z>0$ be any constant such that $\Phi(z) = p\exp(zR)+\bar{p}\exp(-z\bar{R}) <1$. Then for any $0<\delta<1$ and arbitrarily large N it is possible to connect $K=RN$ ports to distinct active elements of a linear array with yield $1-O(N^{1-1/\delta})$, using maximum connection distance $d=\lceil\log(N)/2^z\delta R\rceil = O(\log N)$ and a number of tracks $t=\lfloor Rd\rfloor = O(\log N)$.

Proof sketch: The ports are spaced nearly evenly above the array as follows. For $i\in\{1,2,...,N\}$, let $P_i=1(\lfloor Ri\rfloor > \lfloor R(i-1)\rfloor)$ [†]. A port is placed above element i if $P_i=1$. Let d and t be defined as in the statement of the theorem.

The optimal connection procedure is phrased in terms of (+)-wires propagating from a port rightward to an active element and (-)-wires propagating from an active element rightward to a port. The procedure moves element by element along the array. If an active element is encountered, the longest (+)-wire is terminated at it or else a new (-)-wire is started. If a port is encountered, the longest (-)-wire is terminated at it or else a new (+)-wire is started. If more than t (-)-wires are stacked up, the longest one is removed and its element remains unused. Note that all wires passing any given element must be of the same type.

For $1\leq i\leq N$, let $W_i\in\{0,1,...\}$ equal t plus the number of (+)-wires less the number of (-)-wires passing to the right of element i. Define $W_0=t$. Under the above procedure, $W_i = \max\{P_i-A_i+ W_{i-1}, 0\}$ where $A_i = 1(\text{element } i \text{ is active})$.

The procedure can fail only in one of the following four circumstances.

1. There are more than t (+)-wires at any point in the array.

[†] $1(\cdot)$ is the indicator function: one if the argument is true, and zero otherwise.

2. There are any (+)-wires left at the end of the array.

3. A (+)-wire propagates from an as yet unconnected port past $d+1$ elements. Then the wire passes at least $\lfloor Rd \rfloor = t$ ports connected to subsequent (+)-wires, implying that there are at least $t+1$ (+)-wires are stacked up. (For example, in Figure 1.2 the (+)-wire leaving port (b) has traveled distance 6, so there are at least $\lfloor (2/3) \cdot 5 \rfloor + 1 = 4$ (+)-wires at (d)).

4. A (-)-wire propagates past $d+1$ elements. Then the wire passes at least $\lfloor Rd \rfloor = t$ ports connected to previous (-)-wires, implying that there are at least $t+1$ (-)-wires at the element where the wire began. (For example, in Figure 1.2 the (-)-wire terminating at port (a) has traveled distance 6, so there are at least $\lfloor (2/3) \cdot 5 \rfloor + 1 = 4$ (-)-wires at (c)). But the procedure never stacks up $t+1$ (-)-wires, so this cannot occur.

Thus the array is connected if the number of (+)-wires stacked up never exceeds t and no (+)-wires are left after the last element; that is, $W_i < 2t+1$, $1 \le i \le N-1$, and $W_N < t+1$. We can upper bound W_i by a queuing variable and, using some methods of queuing theory [FE] and the Chernoff bound [GA], show that the probability that the procedure fails is $O(N^{1-1/\delta})$ for the given parameter values.□

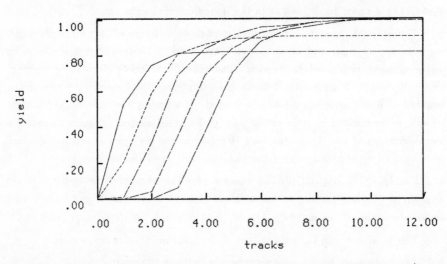

Fig. 2.1: Yield over 500 trials for an N element selector with $p=0.5$, $R=3/7$, t as indicated on the horizontal axis, and $d=2t$. $N=21$ (solid line), 63 (- - -), 189 (— - — -), 567 (- - -), and 1701 (— —).

We have now demonstrated that the maximum connection distance and number of tracks must grow asymptotically as $\Theta(\log N)$. For example, if $p=0.5$ and $R=3/7 \approx 0.4286$, Theorem 1 indicates that $t > (0.309) \log N$ is required while

Theorem 2 shows that $t \approx (0.865)\log N$ will suffice. Figure 2.1 gives simulation results for these values of p and R and four fixed values of N.

3. Pairing of Two Parallel Arrays

We now demonstrate that parallel connection of two linear arrays requires only constant d and t, independent of N. However the value of the constant does depend on the fraction of active elements that are connected; as R approaches $\bar{p}$, the constant becomes larger.

Theorem 3: For any $R < \bar{p}$, $0 < \delta < 1$ and arbitrarily large N it is possible to connect $K = RN$ pairs of active elements from two N-element arrays with yield $1 - O(1/N)$ using a constant number of tracks and constant maximum distance

$$d = \left\lceil p\bar{p}/2(\bar{p}-R)^2 \, \delta^2 - 1 \right\rceil \qquad t = \left\lceil p\bar{p}/4(\bar{p}-R)^2 \, \delta^2 - 1/2 \right\rceil.$$

Proof: We propose a scheme under which the expected fraction of elements that can be connected is $R + \delta(\bar{p}-R)$, just a bit more than is required.

Let $b = \left\lceil p\bar{p}/2(\bar{p}-R)^2 \, \delta^2 \right\rceil$ and let d and t be defined as above. Construct two linear arrays of N/b blocks, with b elements per block. Provide $\lfloor b/2 \rfloor = t$ tracks between the arrays. Let X be the number of active elements in a typical block of one array and let Y be the number of active elements in the facing block of the other array. No matter where defects occur in the block, we can connect $\min\{X,Y\}$ pairs of active elements with the tracks provided since at most $\lfloor b/2 \rfloor$ of the elements to be connected in each block face defective elements. Only distance $b-1 = d$ is required.

Since X and Y are independent and both binomially distributed with parameters $[b, \bar{p}]$,

$$E^2(|X-Y|) \leq E(X-Y)^2$$

$$= E(X-EX)^2 - 2E(X-EX)E(Y-EY) + E(Y-EY)^2$$

$$= 2 \, Var(X)$$

$$= 2bp\bar{p}.$$

Then the expected number of pairs which can be connected in each block is

$$E(\min\{X,Y\}) = EX - E(|X-Y|)/2$$

$$\geq b\bar{p} - \sqrt{bp\bar{p}/2}$$

$$\geq b\left(R + \delta(\bar{p}-R)\right) \qquad\qquad (3.1)$$

which is the desired value.

Since the number of connected pairs in each of the blocks are independent, Chebyshev's inequality [GA] can be applied to bound the probability that at least RN pairs are connected, showing that a yield $1-O(1/N)$ is achieved.□

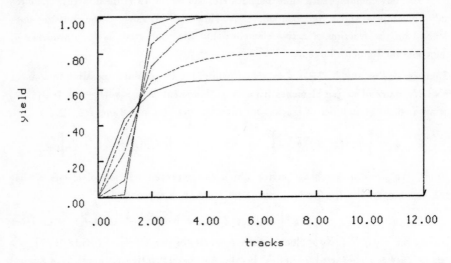

Fig 3.1: Yield over 500 trials for pairing of two N element arrays with $p=0.5$, $R=3/7$, t as indicated on the horizontal axis, and $d=2t$. $N=21$ (solid line), 63 (- - -), 189 (— · — -), 567 (- - -), and 1701 (— —).

As an example, for the values $p=0.5$ and $R=3/7$, Theorem 3 indicates that $t=12$ and $d=24$ will suffice to achieve yield approaching one as $N\rightarrow\infty$. This can be compared to the simulation results shown in Figure 3.1. These were obtained using an optimal scheme which proceeds from left to right along the array connecting each active element whenever the constraints permit.

4. Chains Connected From a Two-dimensional Array

The problem of connecting a chain of active elements from a two-dimensional array is closely related to percolation theory. Percolation processes have been studied extensively since they were first defined by Broadbent and Hammersley [BH]. (See [WI] for a recent survey).

The site percolation problem concerns an infinite lattice of sites which are independently empty with some probability q or occupied with probability $\bar{q}$. A site is said to *percolate* if it is connected to (or is a member of) an infinite connected cluster of occupied sites.[†] The probability that a site percolates is the same for any site, and so can be expressed as a percolation probability function $P(\overline{q})$, which is monotonic increasing and attains the value 1 at $\overline{q}=1$. Broadbent and Hammersley demonstrated that $P(\overline{q})=0$ for $\overline{q}$ less than some critical value characteristic of the lattice. Monte Carlo estimates have established empirical curves for various lattices [FHW]. The curve for a square lattice is reproduced in Figure 4.1.

Fig. 4.1: The percolation function $P(\overline{q})$ for the square lattice as determined by Monte Carlo estimates. (From Fig. 6 of [FHW]).

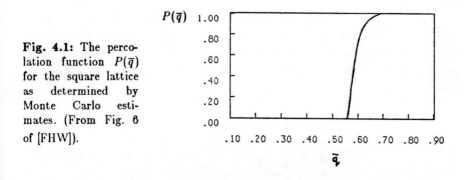

One can consider the percolation lattice formed by the elements of our array and all possible connections between them of distance up to some constant d. Suppose the probability that an element is active, $\overline{p}$, is less than the critical probability for this lattice. Then it will not be possible to connect active elements into an arbitrarily large cluster, or in particular, a chain.

Conversely, for a given p, connections of a certain distance d will be required for percolation to occur. The following theorem upper bounds the required constant d and provides a scheme for connecting a chain from the array.

Theorem 4: With probability $1-O(N^{-2})$, a chain of length $K=RN^2$ can be connected from an $N\times N$ array with maximum distance

$$d = 6\left[\sqrt{\log((\overline{p}-R)/c) / 3\log(p)}\right]-3,$$

for some constant $c>1$. No more than two tracks are required in any channel.

Proof sketch: Group the elements into N^2/b square blocks of b elements each. Choose b large enough that each block has only a small probability q of containing fewer than 4 active elements. We now consider percolation on an $N/\sqrt{b}\times N/\sqrt{b}$

---

†   This definition of percolation is taken from [FHW]. It differs slightly from the more common one in [WI] but is more useful here.

square lattice of blocks, rather than elements. Blocks with at least four active elements correspond to occupied sites and other blocks to empty sites. The probability that any given site is adjacent to a single large cluster of occupied sites can be estimated from the value of $P(\overline{q})$ given in Figure 4.1. This probability approaches one rapidly as b and hence $\overline{q}$ are increased.

A tree of maximum degree 4 can be constructed which spans the cluster of occupied sites and all sites adjacent to the cluster, with all non-leaf nodes situated on occupied sites. This can also be considered as a spanning tree on the blocks. Since all "non-leaf" blocks have at least 4 active elements, a chain of active elements can be formed by looping around the tree without ever having to connect two elements from non-adjacent blocks. (See Figure 4.2). The construction of the spanning tree requires $O(K)$ steps and the connection of subchains in the blocks also requires $O(K)$ steps since there are only a constant number of elements in each block.

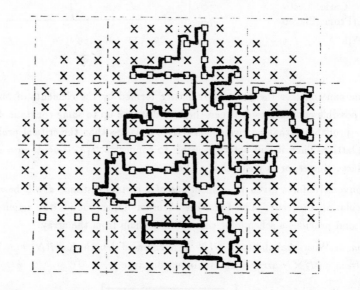

Fig. 4.2: The connection of a chain on a wafer. Each full block contains $b=16$ elements. $p=.71$, $q=.27$, $P(\overline{q})\approx1$. A total of 48 active elements are connected, achieving $R=.21$. The distribution of defective elements is taken from an actual 1K RAM wafer [HA].

Only two tracks are needed between blocks since only two connections are made between adjacent blocks. The connection of all active elements within a block to the tracks between blocks requires only one track between elements. This is straightforward to prove by induction from an $i \times j$ block to an $(i+1) \times j$ or $i \times (j+1)$ block. The average number of tracks per channel is $t = 1 + 1/\sqrt{b}$. The maximum Manhattan connection distance required is

$$d = 6\sqrt{b} - 3. \tag{4.1}$$

The bound on d given in the theorem is established analytically through several intermediate steps. First we establish that the fraction of sites connected to the cluster converges to $P(\overline{q})$. Then the correlation inequality [WI] is applied to show that the fraction of active elements in the cluster converges to $\overline{p}$, so that the fraction of array elements connected, R, converges to $P(\overline{q})\overline{p}$. Next we derive a lower bound on $P(\overline{q})$ in terms of q and then b. Finally we apply (4.1) to relate b to d.□

The following example illustrates the practicality of this scheme. Suppose each element is defective with probability $p = 0.5$. Choose $b = 9$, and hence $d = 7$. The probability that a block of 9 elements has at least 4 active elements is $\overline{q} \approx .7461$. From Figure 4.1, we see that for an infinite square lattice and this value of $\overline{q}$, any block is practically certain to be a member of or adjacent to a cluster. Even for a finite 30×30 element array, and accordingly a 10×10 lattice of blocks, an average of 94.9 blocks were members of or adjacent to the largest cluster of blocks with 4 or more active elements in 500 random trials.

5. Lattices Connected from Rectangular Arrays

The following theorem gives a lower bound on the required maximum connection distance. Note that no assumption is made about the number or width of wiring tracks and that connections are not required to be placed in rectilinear fashion. Only the distance between the elements to be connected is considered.

Theorem 5: Consider an $N \times N$ square array with elements spaced unit distance apart. Let $K^2 = RN^2$. Then for any $0 < \delta < 1$, the probability that a $K \times K$ lattice can be connected tends to zero subexponentially in N unless the maximum connection distance satisfies

$$d \geq \sqrt{\frac{\delta \log(N)}{-2\log(p)}} = \Omega(\sqrt{\log N}). \tag{5.1}$$

Proof sketch: It turns out that blocks of $O(\log N)$ defective elements, which occur with high probability, make this distance necessary. The proof proceeds in three steps. First we define sets of array elements composed of such blocks called *grids*. Then we show that a grid of defective elements exists with probability approaching one. Finally, we assume the existence of a defective grid and show that if (5.1) is violated it is not possible to connect a lattice.

For any given N, R, and p, we carefully choose certain values $L = \Theta(1)$, $m = \Theta(\sqrt{\log N})$. Identify L bands of elements across the array as shown in Figure 5.1. Each band can contain several $m \times m$ blocks of defective elements. The regions between the left side of the array and the first defective block, between each pair of adjacent blocks, and between the last block and the right side of the array are called *gaps*. If there is no gap wider than $2N/Lm^2$ elements in any band, the set of elements in the defective blocks is called a *grid*.

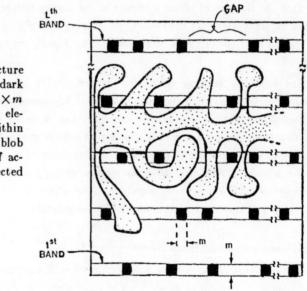

Fig. 5.1: The structure of a grid. The dark squares represent $m \times m$ blocks of defective elements arranged within the L bands. The blob represents the set of active elements connected around the grid.

Because there are so many acceptable positions for the defective blocks, it is highly likely that a grid of defective elements exists. It may be shown that this happens with probability $1 - O\left(N \exp\left(-\dfrac{2RN^{1-\delta}/(2+R)}{(\sqrt{-\delta \log N/\log p} + 1)^3} \right) \right)$.

The active elements must be connected despite the arrangement of defects in the grid. Suppose that the maximum connection distance $d < (m+1)/\sqrt{2}$. Then it is not possible for the connected active elements to enclose a defective block. (See Figure 5.2). This means that the set of connected elements must be distributed like the blob shown in Figure 5.1, consisting of a central subset positioned entirely between two bands and disjoint subsets extending like fingers, each connected through their own gap.

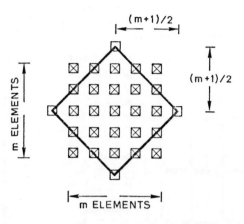

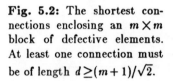

Fig. 5.2: The shortest connections enclosing an $m \times m$ block of defective elements. At least one connection must be of length $d \geq (m+1)/\sqrt{2}$.

We proceed to upper bound the number of elements in each subset. The size of the central subset is obviously less than N^2/L since it is contained between two of the L bands. The size of the fingers can be bounded as follows. The given distance d is not sufficient to penetrate through a gap, and so every connection through a gap must stop at an element within the gap. This means that the set of elements within the gap is a cutset: if all the elements in a gap are removed, the elements in the associated finger are disconnected from the rest. Since we know how wide and deep the gap can be, we can upper bound the size of the cutset. Furthermore, the connectivity of a square lattice is such that if a cutset of size c is removed, at most $(c^2 + c)/2$ elements are disconnected. This allows us to upper bound the size of each finger and thus the total size of all the fingers and the central subset. It turns out that this sum must be less than K^2. Thus it is not possible to connect a $K \times K$ lattice with the above maximum connection distance d. For the chosen value of m, this implies that (5.1) must be satisfied to connect the array.□

The theorem readily generalizes to triangular and hexagonal lattices.

Theorem 5 offers no lower bound on the number of wiring tracks t required between elements. We see no reason why more than $t = O(1)$ should be required. However, the best result we have been able to obtain using a constant number of tracks is

Theorem 6: For any rational number $R < \bar{p}$, let b be any fixed integer such that

$$\alpha \triangleq \frac{(\bar{p}/R)^R (p/\bar{R})^{\bar{R}}}{1 - 2(pR/\bar{p}\bar{R})^b} < 1.$$

Then for any $0 < \delta < 1$, arbitrarily large N, and $K = RN$, it is possible to connect a

$K \times K$ lattice from a $K \times N$ array with yield $1-RN^{2-2/\delta}$ using a maximum connection distance $d = 2 + \dfrac{8\overline{R}\log(2N)}{-\delta\log(\alpha)} = O(\log N)$ and $t = b+1 = O(1)$ tracks between rows of elements.

Proof sketch: Note that b exists since $(\overline{p}/R)^R(p/\overline{R})^{\overline{R}}$ and $pR/\overline{p}\overline{R}$ are both less than one for $R < \overline{p}$. We now describe the array.

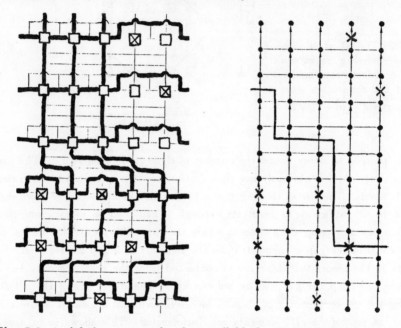

Fig. 5.3. At left is an example of a small block. $K=6$, $m=5$, $b=2$, $t=3$ and $d=10$. Three chains are connected through the block. At right is the corresponding flow graph. Edges with arrows are directed; the others are bidirectional. The zero capacity edges corresponding to defective elements are marked with an "$\times$". The solid line crossing the graph bisects the edges in a typical cutset.

For a carefully chosen value $m=\Theta(\log N)$ we construct a $K \times N$ array of N/m blocks, as shown in Figure 5.3. Each block has K rows of m elements. Between each pair of adjacent rows run $t=b+1$ tracks. Of these, b are used to connect Rm chains twisting through each block from bottom to top, stepping from an active element in one row to an active element in the next row. A horizontal chain is then connected along each row of the array using one additional horizontal track and one vertical track in the manner of Figure 1.1. The greatest horizontal separation between chains in adjacent blocks is $4\overline{R}m+2 \leq d$ for the chosen m.

Together, the vertical and horizontal chains form a $K \times K$ square lattice.

It must be demonstrated that, with high probability, the defects fall in such a way that the chains can be connected. To do this, we describe the formation of the vertical chains more precisely in graph theoretical terms.

The directed flow graph in Figure 5.3 represents a block of the array. The points where tracks cross correspond to vertices. Each segment of a horizontal track corresponds to a bidirectional edge of capacity one. Each segment of a vertical track corresponds to a directed edge of capacity one. Each element corresponds to a directed edge of capacity one if the element is active, and zero otherwise. A row of source vertices is added at the bottom and a row of sink vertices at the top.

If there is a flow of capacity Rm (or equivalently, Rm edge-disjoint paths of capacity one) from the source vertices to the sink vertices, then Rm non-crossing chains can be connected through the block from bottom to top as desired. Define a cutset C to be a set of edges whose removal separates the sources from the sinks. The capacity of a cutset $F(C)$ is the sum of the capacities of its edges. The Max-flow Min-cut Theorem, extended to the case of multiple sources and sinks [DE], states that: A flow of Rm exists if (and only if) all cutsets C have capacity $F(C) \geq Rm$.

We can restrict consideration to the set S of cutsets including exactly m vertical edges, all of which correspond to elements, since for every cutset $C_1 \notin S$, there exists a cutset $C_2 \in S$ such that $F(C_1) \geq F(C_2)$. Thus $F(C) \geq Rm$ for all $C \in S$ implies that $F(C) \geq Rm$ for all C, and in turn by the Max-flow Min-cut Theorem, that the chains can be connected. The probability that any such cutset, in any of the N/m blocks, has capacity less than Rm can be upper bounded by $RN^{2-2/\delta}$. Thus the chains can be connected in all the blocks, with yield approaching one.□

Although Theorem 6 shows only that the chains exist, it is a simple matter to find them. There need not be exactly Rm chains in each block; if there is at least one chain in each block, and a total of K in the array, the wafer is connected satisfactorily. Each block is done independently. Starting at the bottom left source vertex, one searches for the leftmost path to a sink vertex. Chains are completed in turn until no more can be formed in the block. Since the chains never cross, the algorithm need never examine those vertices positioned to the left of the already completed chains. Since each vertex is visited only a constant number of times, the task is completed in $O(KN)$ steps.

Leighton and Leiserson [LL] have proposed a graph embedding scheme which can connect any fraction $R < \bar{p}$ of the elements with probability $1 - O(1/N^2)$ using $d = O(\sqrt{\log N} \log\log N)$ and $t = O(\log\log N)$, where wires have width one. They partition the array into blocks of size $c \log N$ elements, where c is chosen sufficiently large that with high probability every block has at least $Rc\log N$ active elements. A sublattice can then be connected in each block using $O(\log\log N)$ tracks per channel. However, doing this may require a look-up table of size $N^{c/\log 2}$. For the previous example of $p=0.5$ and $R=3/7$, $c\approx 68$ so the table is impractically large.

The lower bound of Theorem 5 can be extended to incorporate the physical assumption that the area of an element must increase linearly with the length of the longest wire it drives [TH]. Since the capacitance of a wire increases linearly with its length, the drive current needed to charge or discharge the wire in a given time, and hence the area of the driving transistor, must increase linearly as well. We therefore suppose that the elements may be rectangular but must occupy area $A_e=\Omega(d)$. The proof of Theorem 5 can then be modified to show that $A_e=\Omega(\log N)$ and $d=\Omega(\log N)$. These bounds can be achieved using elements of height $O(\log N)$ and constant width in the scheme of Theorem 6.

6. The Distribution of Defects.

The results of Sections 2-5 all rely on the assumption that element defects are independent and identically distributed. In reality, this may not be true, and problems could arise in three ways.

1. If there are clusters of defective elements within a region of good yield, long connections may be required. This difficulty would particularly affect the selector and lattice configurations—recall that Theorems 1 and 5 both rely on the existence of wholly defective regions within the array to establish the impossibility of using only short connections.

2. If the active elements form separate clusters within a large region of zero yield, they cannot be connected into a single array without several especially long connections.

3. The element yield $1-p$ may vary over the wafer and among wafers. If the variation over a wafer is predictable, different regions can be equipped with interconnect designed to accommodate different values of p. To the extent that p is not predictable from wafer to wafer, the interconnect must be designed conservatively.

Clusters of defective elements would most likely result from single point defects or clusters of defects which overlap two or more elements. However, since the size of the elements contemplated for whole-wafer integration is rather large, such defects are improbable. Thus problem 1 should not be a source of concern.

Photomask alignment errors are probably the most significant cause of large regions of zero yield. An analysis by Kim and Ham [KH] finds that the unaffected circuits should be concentrated in a convex region of the wafer bounded by several contours associated with the various misalignments. This conclusion was born out by a 1978 study of several lots of 1K RAM wafers at RCA, which found no cases of multiple isolated high yield regions [HA]. Thus problem 2 should not be significant either.

The same study found "reasonable uniformity of local yield" within the high yield area of a given wafer. However the value of the local yield varied widely from wafer to wafer. This would indicate that problem 3 may be significant. More appropriate and recent data will tell.

References

[AB]   Abbott, R., et. al. Equipping a Line of Memories with Spare Cells. Electronics (July 28, 1981) pp. 127-130.

[AC]   Aubusson, R., and Catt, I. Wafer-Scale Integration—A Fault Tolerant Procedure. IEEE J. Sol. State Circuits SC-13, 3 (June 1978) pp. 339-344.

[BH]   Broadbent, S., and Hammersley, J. Percolation Processes. I. Proc. Cambridge Phil. Soc. 53 (1957) pp. 629-641.

[DE]   Deo, N. Graph Theory with Applications to Engineering and Computer Science. Prentice Hall, Englewood Cliffs, NJ (1974) pp. 390-391.

[FE]   Feller, W. An Introduction to Probability Theory and Its Applications. Vol. II. 2nd ed. Wiley, New York (1971) pp. 194-198.

[FHW]   Frisch, H., Hammersley, J., and Welsh, D. Monte Carlo Estimates of Percolation Probabilities for Various Lattices. Physical Rev. 126,3 (May 1, 1962) pp. 949-951.

[FV]   Fussell, D., Varman, P. Fault Tolerant Wafer-Scale Architectures for VLSI. 9th Ann. Symp. on Comput. Arch. (1982) pp. 190-198.

[GA]   Gallager, R. Information Theory and Reliable Communication. Wiley, New York (1968) pp. 126-128.

[GE]   Greene, J., El Gamal, A. Area and Delay Penalties for Restructurable VLSI Arrays. To appear, J. of ACM.

[HA]   Ham, W. Yield-Area Analysis: Part I--A Diagnostic Tool for Fundamental Integrated-Circuit Process Problems. RCA Rev. 39 (June 1978) pp. 231-249.

[HCE]   Hsia, Y., Chang, G., and Erwin, F. Adaptive Wafer Scale Integration. Japanese J. of Appl. Phys., Vol. 19 (1980) Supplement 19-1, pp. 193-202.

[HF]   Hsia, Y., and Fedorak, R. Impact of MNOS/AWSI Technology on Reprogrammable Arrays. Symposium Record, Semi-Custom Integrated Circuit Technology Symposium (May 26-27, 1981) Washington, DC.

[HS]   Hedlund, K., Snyder, L. Wafer Scale Integration of Configurable, Highly Parallel (CHiP) Processors. Extended Abstract. Proceedings of International Conference on Parallel Processing, IEEE (1982) pp. 262-264.

[KH]   Kim, C., Ham, W. Yield-Area Analysis: Part II--Effects of Photomask Alignment Errors on Zero Yield Loci. RCA Rev. 39 (Dec. 1978) pp. 565-576.

[KO]   Koren, I. A Reconfigurable and Fault Tolerant VLSI Multiprocessor Array. Proc. 8th Ann. Symp. on Comput. Architecture, Minneapolis, MN (May 12-14, 1981) pp. 425-442

[LI]   Lincoln Laboratories. Semiannual Technical Summary: Restructurable VLSI Program. ESD-TR-81-153 (Oct. 1981).

[LL]   Leighton, F.T., and Leiserson, C. Wafer Scale Integration of Systolic Arrays. (Extended Abstract). 23rd Ann. Symp. on Foundations of Comput. Sci. (Nov. 1982) pp. 297-311.

[LM]   Lowry, M., Miller, A. Analysis of Low-level Computer Vision Algorithms for Implementation on a VLSI Processor Array. Proc. ARPA Image Understanding Workshop, (September 1982).

[MI]   Minato, O. et.al. HI-CMOS II 4K Static RAM. Digest of Technical Papers, IEEE Solid State Circuits Conf. (1981) pp.14-15.

[MN]   Manning, F. An Approach to Highly Integrated Computer-maintained Cellular Arrays. IEEE Trans. Computers C-26, 6 (June 1977) pp. 536-552.

[RO]   Rosenberg, A. The Diogenes Approach to Testable Fault-Tolerant Networks of Processors. Duke Univ. Comp. Sci. Dept TR CS-1982-6.1 Rev. May 1982.

[SM]   Smith, R., et.al. Laser Programmable Redundancy and Yield Improvement in a 64K DRAM. IEEE J. Solid-State Circuits SC-16, 5 (Oct. 1981), pp. 506-513.

[TH]   Thompson, C. A Complexity Theory for VLSI. (Dissertation). Carnegie-Mellon University (August 1980) pp.41-44.

[WI]   Wierman, J. Percolation Theory. Ann. Prob. 10,3 (1982) pp. 509-524.

Formal System Models

Verification of VLSI Designs
R.E. Shostak

A Hierarchical Simulator Based on Formal Semantics
M.C. Chen and C.A. Mead

Trace Theory and the Definition of Hierarchical Components
M. Rem, J.L.A. van de Snepscheut and J.T. Udding

Deriving Circuits from Programs
J.L.A. van de Snepscheut

Verification of VLSI Designs

Robert E. Shostak
Computer Science Laboratory
SRI International
333 Ravenswood Avenue
Menlo Park, CA 94025

Abstract

We present a method for formal proof of correctness of VLSI designs. Our approach is an adaptation of software verification techniques that works on circuit networks rather than program flowgraphs. The method may be viewed as a dual of Floyd's method for program verification in which the roles of state and control are interchanged. We illustrate the approach with the semi-automatic verification of a simple NMOS design, and discuss its application to large scale VSLI designs. A proof of soundness of the method is presented in a forthcoming paper [ShS 83].

1 Introduction

Though much emphasis has been placed on methodologies for rapid development of VLSI designs in the last few year years, relatively little effort has been applied to the problem of verifying the correctness of such designs. The need for formal verification methods is especially acute for the complex circuits made possible by very large scale integration. The high gate counts and gate-to-pin ratios that characterize these circuits make the classical techniques of simulation and prototype testing far more time-consuming and less apt to be reliable.

While the bulk of research in verification has been pursued in connection with software, some attention has been given to its application to hardware designs. Wagner [Wag 77], for example, has used symbolic manipulation to detect errors in digital circuits at the register transfer level. The approach taken in that work involves specification of designs in a non-procedural register transfer language. The verification process results in a proof that a higher-level abstract description of a circuit is correctly implemented in terms of a lower-level description. This notion of abstract machine simulation has also been used in the verification of microcode [Bir 74], and

---

This research was supported in part by NSF grant MCS-7904081, and by AFOSR contract F49620-79-C-0099.

underlies the hierarchical design methodologies (such as HDM [LRS 79]) that have been used for software system design and verification. Some more recent work [Fos 81] that addresses itself directly to VLSI uses ideas from formal language theory to verify designs obtained by combining standard cells [Fos 81]. This approach uses context-free grammars to specify compositions of standard cells, and hinges on showing that the correctness of the composition follows from that of the primitive cells. The use of denotational methods, which focus on the semantics of designs rather than their structure, has also been recently suggested [Gor 81].

The approach described here is closely related to the assertional method described by Floyd [Flo 67] for the verification of sequential programs. Assertional methods depend on the annotation of the program to be verified with predicates that describe what must be true whenever control reaches the point in the program with which the predicate is associated. The proof of correctness rests on the application of an induction principle to demonstrate the truth of the predicates.

Our method may be viewed as a dual of Floyd's in which the roles of *state* and *control* are interchanged. Reflecting the duality, assertions that annotate the circuit graph to be verified describe local state at each point in time rather than global state at particular points in time. The Hoare axioms that characterize the behavior of primitive programming language constructs in Floyd's method are replaced here by *transfer predicates* that describe the relation among inputs and outputs of circuit elements as functions of time.

The method has been tested on a number of examples using the STP system [ShS 82] for deductive support[1].

The presentation of the method is organized as follows. Section 2 discusses the formal circuit model on which the proposed method is based. The next two sections formally define correctness with respect to that model and give an example. The proof method itself is described in Section 5. Section 6 illustrates the method with the proof of correctness of a VLSI implementation of the Muller C-element, using the STP system to provide mechanical support. The last section discusses issues to be addressed in the application of the method to the verification of large VSLI systems. A proof of soundness of the method is given in [Sh 83].

---

[1]STP is a mechanical verification system originally intended for software proof. It has been used in the design proof of the SIFT fault-tolerant operating system [ScM 82].

2 The Circuit Model

A system to be verified is formally modeled in our method as a graph whose nodes correspond to circuit elements and whose edges correspond to signal (or information) paths that connect these elements. While this nomenclature is suggestive of electronic devices, the concepts are meant to apply to arbitrary concurrent systems– individual processors, for example, in a distributed network. In the context of hierarchical design, a given circuit element could abstract a complex system that might itself be represented by a circuit graph at a lower level of the hierarchy.

Associated with each circuit element is a set of *input* and a set of *outputs*. Each input and output has a formal *signal type* that represents the domain of values it may assume. A signal type may either represent some physical quantity, such as a voltage or current, or may be more abstract, as in the case of "array of integers" or "ordered pair of voltage, current". A circuit element is interpreted as forcing some relation among its inputs and outputs, considered as functions of time. Its behavior is formally modeled by a *transfer predicate* that expresses this relation.

A resistor of resistance r, for example, might be modeled as a circuit element with an input x of signal type $< voltage, current >$ and an output y of the same type. The transfer predicate P_t might be given by

$$y_t^{current} = x_t^{current}$$
$$\wedge \quad y_t^{voltage} = x_t^{voltage} + r x_t^{current}$$

A *circuit graph* is formally defined as a directed graph with a circuit element associated with each node. A node must have exactly one incoming edge corresponding to each input of the associated circuit element, and one outgoing edge corresponding to each output. Moreover, the signal types of the input and output associated with the two ends of each edge must be identical.

As a convenient means of modeling inputs and outputs of the circuit as a whole, we allow dangling edges– we speak of a dangling edge with an initial node but no destination node as an *output edge*, and one with a destination node but no initial node as an *input edge*.

Roughly speaking, each nondangling edge represents a connection between the circuit elements on either end. In the case of an electronic circuit, the edge may be thought of as a wire; for a distributed computer network, it may represent a data link. In any case, the connection is considered to be perfect in the sense of forcing the values of the input and output at each end to be equal. This assumption does not entail any loss of generality, of course, since imperfect transmission paths may be modeled as circuit elements. (Note that circuit element outputs that drive several inputs can be modeled in the same manner- one merely introduces a *junction* circuit element with one input and several outputs.)

Each edge of a circuit graph may thus be viewed as representing a *signal* whose value is that of the input and output bridged by the edge. For convenience, we will assume in the following that each edge of a circuit graph is labeled with a unique signal name.

3 The Meaning of Correctness

The goal of the method detailed in the following sections is to show that a design operates in accordance with a given characterization of its intended behavior. The characterization has three components: an *input specification*, which assigns an assertion to each input edge; an *output specification*, which assigns one to each output edge; and an *initial condition* specification, which assigns assertions to arbitrary edges of the graph. Initial conditions are useful to capture initial state information, such as the initial settings of registers. Each input, output, and initial condition assertion is a predicate on the signal that labels the edge to which the assertion is assigned.

Intuitively, *correctness* means that if the circuit inputs satisfy the input specification, if the initial condition assertions are assumed to hold, and if each circuit element operates in faithfulness to its transfer characteristic, then the circuit outputs are guaranteed to satisfy the output specification.

In order to give correctness a precise meaning, we need first to formalize the notion of time. For this purpose, it is convenient to introduce the concept of a *time domain*, by which we will mean any well-founded, partially-ordered set. As will be clear later, the well-foundedness is necessary for the induction principle that underlies the proof method. In our examples, the time domain T given by the set of real numbers greater or equal to some arbitrary real t_0 will be used for T. One might think of t_0 as the time at which the circuit in question is powered up, or as the first time of interest. In other contexts, a discrete time domain such as that underlying temporal logic specifications may be appropriate. For some applications, it may be useful to specify a nontotally-ordered domain.

To formalize the notion of correctness with respect to a given specification, we define the concept of *interpretation*. An interpretation for a given circuit graph G assigns a function of time to the signals associated with the edges in such a way that the transfer predicate of each circuit element is satisfied. Intuitively, an interpretation is just a possible behavior of the circuit, expressed in terms of the values of the signals at each time t. An interpretation of a graph containing a resistor node, for example, must assign voltage-current pairs to the incoming and outgoing signals that have identical current components and that obey Ohm's Law.

The correctness of a circuit design can be stated formally as follows. We will say that a graph G is *correct* with respect to a given input, output, and initial condition specification if every interpretation of G that satisfies the input and initial condition specifications also satisfies the output specification.

Figure 1a., for example, shows a schematic for a simple negative feedback amplifier implemented by means of an operational amplifier with nonzero open loop gain $-g$. Figure 1b. shows a corresponding circuit graph. Note that in addition to resistor nodes, the graph has nodes associated with three other kinds of circuit element: an operational amplifier, a "wire-or", and a "signal splitter". These latter two are just junctions that force the voltage components of their inputs and outputs to be equal and the sum of their current components to be zero.

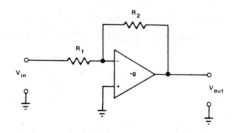

(a) SCHEMATIC OF FEEDBACK AMPLIFIER

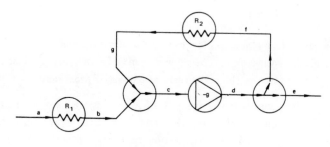

(b) CORRESPONDING CIRCUIT GRAPH

FIGURE 1

Note that each edge is labeled with a signal name. All signals are of type $<$ *voltage, current* $>$.

The op amp circuit element has one input x, one output y, and transfer predicate P_t given by

$$y_t^{voltage} = -gx_t^{voltage}$$
$$\wedge \quad x_t^{current} = 0$$

The second condition of the predicate gives the op amp an infinite input impedance.

The interpretations of the graph are the assignments of $<$ *voltage, current* $>$ functions (of time) to the signals labeling the edges that cause the transfer predicate of the circuit element associated with each node to be satisfied. It is easy to see, for example, that the assignment giving all signals the constant value $< 0,0 >$ is one (uninteresting) such interpretation.

Where V_t denotes the input voltage, the circuit can be shown correct with repect to the specification that assigns the assertion $a_t^{voltage} = V_t$ to the input edge and assigns the following assertion to the output edge (no initial condition is needed).

$$e_t^{voltage} = \frac{-R_2 V_t}{R_1\left(1 + \frac{1}{g}\right) + R_2}$$

The output assertion defines the closed-loop gain of the amplifier.

4 Proving Correctness

We now give a formal method for proving the correctness of a circuit design with respect to a given specification. The approach may be viewed as a dual of Floyd's method [Flo 67] for proving the correctness of sequential programs.

Floyd's method entails the association of assertions with certain points in the flowgraph of a program in such a way that each loop of the flowgraph is "cut" (i.e., intercepted) by an assertion. Each assertion captures a property of the program state (defined by the instantaneous values of the variables) that must hold each time the flow of control reaches the point in the flowgraph with which the assertion is associated. The essence of the proof strategy is to show that the program semantics along any execution path from an assertion point A to an assertion point B are such as to guarantee that if the assertion at A holds just prior to execution of the path, then the one at B must hold just afterwards. Once this has been established for each straight-line execution path connecting two assertion points, it automatically follows by an induction argument that program inputs satisfying the assertion associated with the entry point must produce outputs (given that the program terminates) satisfying the exit assertions.

The differences between Floyd's method and the technique we will describe owe chiefly to two distinctions between program flow graphs and circuit graphs. First, circuit graph semantics do not depend on any notion of flow of control. Second, circuit state is associated with the collective condition of localized signals, rather than that of global variables. These differences may be viewed in terms of a duality between state and control: circuit *state* is associated with particular localities, whereas circuit *control* is global; for program graphs, exactly the reverse is true.

Like Floyd's method, our approach depends on the association of assertions with a subset of the edges in the graph. These include the input and output assertions, together with a set of additional assertions (called *loop assertions*) sufficient to cut each cycle exactly[2] once. Reflecting the duality, these assertions characterize particular signals at arbitrary points in time rather than arbitrary signals (program variables) at particular points in time.

The crux of the proof strategy is to show through an induction argument that each assertion must hold at each time t, assuming that the input and initial condition assertions hold. As will be clear momentarily, the induction is carried out simultaneously over *time* and over the *structure* of the graph.

Once the circuit graph is annotated with assertions, the proof of correctness proceeds in three stages. First, a kind of path analysis is carried out, resulting in a set of acyclic subgraphs of the original graph. Second, a set of mathematical formulas called *verification conditions* is derived, one from each subgraph. Finally, the verification conditions are each proved, either manually or using a mechanical theorem prover.

The path analysis phase produces an acyclic subgraph for each noninput edge e of the graph labeled by an assertion. The subgraph for e is "rooted" at the initial node for e, and is formed by tracing backwards from this root, branching at nodes along the way with more than one incoming edge. Incoming edges labeled by assertions, however, are not traversed, so that the backwards trace terminates at nodes all of whose inputs edges are labeled by assertions. These edges are left dangling in the subgraph, and are referred to as the *inputs* of the subgraph. Note that, since each loop of the original graph must be cut by an assertion, the tracing process must terminate and the resulting subgraph must be acyclic.

---

[2]Section 6 describes a modification of the method in which more than one assertion may cut each loop.

Figure 2a., for example, shows an annotated circuit graph with a single input assertion ϕ_t, output assertion ψ_t, and loop assertion ω_t. Figure 2b. shows the subgraph for the edge labeled by ω_t.

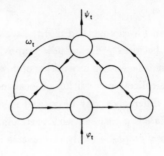

(a) CIRCUIT GRAPH

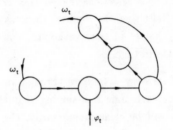

(b) ACYCLIC SUBGRAPH FOR ω

FIGURE 2

The verification condition for the subgraph generated from an edge e with noninput assertion $\mathcal{A}_t$ is constructed in the following way.

First, for each node n in the subgraph, let p_t^n denote the associated transfer predicate, expressed in terms of the signals that label the inputs and outputs of that node. Let P_t denote the conjunction of the p_t^n's.

Next, let D_t denote the conjunction of all assertions labeling dangling inputs of the subgraph other than e itself. For example, in Figure 2b., D_t includes ϕ but not ω, even though ω labels a dangling input.

Finally, let I_t denote the conjunction of all initial condition assertions.

Then the verification condition is given by the formula

$$\forall t \in T \, (P_t \wedge I_t \wedge D_t) \quad \supset \quad \forall t \in T \, [(\forall u < t \, \mathcal{A}_u) \supset \mathcal{A}_t]$$

The formula states that if the transfer predicates, initial conditions, and assertions labeling subgraph inputs other than e hold at all times t, and if $\mathcal{A}$ holds at all times *prior* to t, then A must also hold at time t.

As we will show in section 7, the correctness of the circuit follows from the validity of the verification conditions. Note that the assertions included in D_t are assumed to hold at *all* times rather than just times prior to t. The justification for this assumption is somewhat subtle, and has to do with the fact that the induction argument is carried out over the structure of the graph as well as over time.

The last step in the process, and the most difficult one, is the proof of the verification conditions. As in software verification, the size and complexity of verification conditions arising from even simple systems dictates the use of automatic or semi-automatic proof techniques. The following example illustrates the verification process from beginning to end, using the STP verification system to prove the verification conditions.

5 A Sample Proof

Figure 3a. shows an abstract representation of Muller's *C-element*, also known as a *rendezvous* circuit [Mil 65]. The C-element is a logic circuit invented by D. Muller during the 1950's as part of a speed-independent digital design discipline. Speed independent circuits have more recently been used in connection with self-timed VLSI designs [Sei 79].

The operation of the C-element is easily stated. The device has two binary input signals (a and b) and one binary output signal (c). The output becomes 1 whenever both inputs become 1; it becomes 0 whenever both inputs become 0. Otherwise, the output remains in whatever condition it was in before.

In order to formalize this specification, of course, one needs to be very careful about what is meant by "becomes" and "before". Given that time is represented by a continuous domain (such as the reals), the exact meaning of "just before" is by no means clear. It is interesting to note, in this connection, that manufacturers of digital devices often hedge this question in their specifications through the use of *pseudo-discretization*. As a notation to indicate that a flip-flop output does not change for a given combination of inputs, for example, one often sees

$$Q_{n+1} = Q_n.$$

This convention is often used in contexts in which there is no clock refererence. This is not to say, of course, that proper circuit operation depends on quantization of time in any way. The notation is used merely to trade off precision for simplicity of expression.

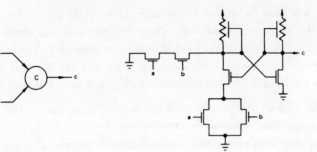

(a) MULLER C-ELEMENT AND NMOS REALIZATION

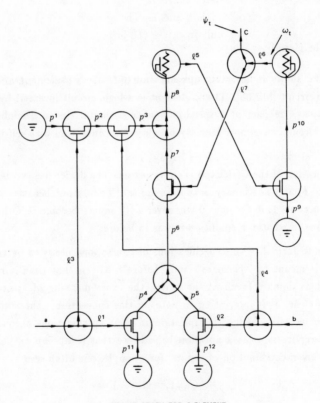

(b) CIRCUIT GRAPH FOR C-ELEMENT

FIGURE 3

In any case, the exact meaning of "the time just prior to a given event" can often be resolved to a specific time that depends on propagation delays of certain signals internal to the specified device. In the case of the C-element, the behavior of a typical implementation can be formally specified in the following way:

$$c_{t+\delta} = \text{if } (a_t = b_t) \text{ then } a_t \text{ else } c_t$$

Here, δ is a positive constant that reflects propagation delay through the C-element.

The right half of Figure 3a. shows the schematic diagram of an nMOS implementation of the C-element given by Seitz in [MeC 80]. Figure 3b. shows a corresponding circuit graph. The signals that label the graph are of two types. The inputs and outputs (a, b, and c) and signals l^1 through l^6 are of type *logic level*. The values of this type are just the logic levels 0 and 1. The remaining signals are all of type *pull-down*. The values of type pull-down are the logic level 0 and a special tri-state (high-impedance) value *hi-z*.

The circuit elements of the graph are of five kinds, representing *ground, pull-down transistors, pull-up transistors, pull-down wire junctions*, and *signal splitters*. Table 1 gives the signal types and the transfer predicate for each type of circuit element.

The ground circuit element, for example, has no inputs, and one output of type pull-down; it merely forces its output to 0 at all times.

The pull-down transistor has a pull-down input s (for *source*), a logic-level input g (for *gate*), and a pull-down output d (for *drain*.) The operation is similar to that of a switch: if the gate input is high, the source signal is passed through; otherwise the output is forced to the *hi-z* state.

The pull-down junction, with two pull-down inputs and one pull-down output, captures the effect of "wire-or'ing" two pull-down signals together. Its output is low if and only if one of its inputs is low.

The signal splitter (used to split off the output signal c) provides fan-out but otherwise has no effect.

Finally, the pull-up transistor, with one pull-down input and one logic-level output, converts a pull-down signal to a logic level; 0 is mapped to 0, and *hi-z* is mapped to 1. In addition, a propagation delay of ϵ is introduced.

The association of delay with the pull-up element alone is an idealization of actual behavior in a number of respects. Delays in ratio logic are assymmetric (with

Table 1

TRANSFER PREDICATES FOR CIRCUIT ELEMENT TYPES

ELEMENT TYPE	SIGNAL TYPES	TRANSFER PREDICATE
GROUND x	PULL-DOWN x (OUTPUT)	$x_t = 0$
PULL-DOWN TRANSISTOR 	PULL-DOWN s (INPUT) LOGIC-LEVEL g (INPUT) PULL-DOWN d (OUTPUT)	$d_t = $ IF $g_t = 1$ THEN s_t ELSE hi – z
PULL-DOWN JUNCTION 	PULL-DOWN x (INPUT) PULL-DOWN y (INPUT) PULL-DOWN z (OUTPUT)	$z_t = $ IF $(x_t = 0 \lor y_t = 0)$ THEN 0 ELSE hi – z
PULL-UP TRANSISTOR 	PULL-DOWN s (INPUT) LOGIC-LEVEL g (OUTPUT)	$g_{t + \epsilon} = $ IF $s_t = 0$ THEN 0 ELSE 1
SIGNAL SPLITTER 	LOGIC-LEVEL x (INPUT) LOGIC-LEVEL y (OUTPUT) LOGIC-LEVEL z (OUTPUT)	$y_t = x_t \quad z_t = x_t$

respect to rise and fall times) and are associated with all capacitances in the circuit including wire and output loads. In a typical layout of the implementation given in this example, the charging of the gates of the pull-down transistors through the effective resistance of the pull-ups would likely dominate, and so we have have found it convenient to lump all delay in the pull-ups. One could, of course, model this behavior in much finer detail, and we expect to do so in the proposed work. For present purposes, however, our simple delay assumptions suffice to characterize the overall behavior of the circuit.

The specification with respect to which we wish to prove correctness is as follows. First, as no particular assumptions are made about the input signals a and b, the corresponding input assertions are just *true*. Because no assumptions need be made about the initial condition of any signal, no initial condition assertions are specified. Finally, the output assertion ψ_t associated with signal c is given by

$$c_{t+2\epsilon} = \text{if } (a_t = b_t) \text{ then } a_t \text{ else } c_t$$

Note that the parameter δ has been particularized to 2ϵ, where, once again, ϵ is one pull-up delay.

As the graph has only one directed cycle, a single loop assertion suffices. In principle, the assertion can be placed on any edge of the cycle. The signal l^β has been chosen in this instance. Since l^β must equal the output c, an appropriate loop assertion ω is easily formulated:

$$l^\beta_{t+2\epsilon} = \text{if } (a_t = b_t) \text{ then } a_t \text{ else } l^\beta_t$$

The path analysis process gives rise to two subgraphs: one for the edge labeled l^β, and one for the output edge.

It is easy to see that the first of these includes every node in the graph, and has dangling input edges corresponding to a, b, and l^β itself. The conjunction P_t for this subgraph thus contains the transfer predicate instance corresponding to each node, and the conjunction D_t is just the formula *true*.

The subgraph generated for the output edge, on the other hand, contains a single node (the one at the output) with one dangling input edge corresponding to l^β. The conjunction D_t is thus composed of the single loop assertion ω.

An appendix shows the sequence of commands, or *events*, used to specify the circuit and carry out the proof of the two verification conditions in the STP system. Most of the events declare symbols or definitions that are used in the proof. (STP is a strongly typed system, so that every symbol used in a formula, and every formula used in a proof, must first be declared.) Other events cause the system to attempt to prove a formula, or query the system as to the status of a given formula. We give only a summary explanation of the event sequence here. The reader is referred to [ShS 82] for a more detailed description of STP.

The first thirty-four events give declarations relating to the types logic level and pull down and defining the signals associated with the edges of the graph. Note the axiom named *DISTINCT* declared in event 7 that states that the three values *ONE*, *ZERO*, and *HI-Z* are different. The axiom *LLAXIOM* declared in event 9 asserts that *ONE* and *ZERO* are the *only* values of type logic level.

The input assertion ϕ_t, output assertion ψ_t, and loop assertion ω_t that annotate the circuit graph are declared in events 35 through 37.

Events 38 through 50 define the transfer predicates of the different circuit elements used in the design. For example, the transfer predicate for the pull-down transistor (abbreviated *PDT*) is defined in event 47. *PDJ* and *PUT* are used to designate the predicates for the pull-down junction and pull-up transistor, respectively.

The formulation and proof of the first verification condition are given in events 52 through 61. The formulas A_t, D_t, I_t, and P_t are defined in events 53 through 56, while 57 and 58 declare the hypothesis and conclusion of the verification condition itself. A lemma used to prove the verification condition is formulated in event 59 and then proved from the hypothesis of the verification condition in event 60. The system proves the lemma in 170 cpu seconds. The conclusion of the verification condition is then proved in event 61 using the hypothesis, the lemma, and two instances of *LLAXIOM* declared in event 9. That proof requires 70 seconds. Events 62 through 68 show the formulation and proof of the second verification condition, which requires only 7 seconds.

The last two events query the status of the two verification conditions. In each case, of course, the system responds that the condition has indeed been proved.

6 Research Issues

The sample proof given in the last section illustrates the method for a relatively simple design specified at a single level of abstraction. The greatest pay-off of formal verification, however, lies in its application to large, complex systems. Successful application of the method to such systems will depend on its integration into the kind of hierarchical design paradigms that are currently being used to specify and verify large software designs, and will increasingly be used for VLSI design. A hierarchically structured design is specified at several levels, each at a higher level of abstraction than the ones below it. For VLSI designs, these range from high-level description of the functional behavior of the system down to the stick-diagram level (or even the layout level), with intermediate descriptions (such as the register transfer architecture) in between. The verification process for hierarchical designs has two facets. First, the design at each hierarchical level must be proved to be consistent with its specifications. Second, the concepts at each level must be shown to be correctly implemented in terms of those at lower levels.

Practical application of the method will also require further study of specification issues. Owing to the major role played by concurrency and synchronization in practical hardware systems, problems of specification can be particularly subtle in this context. In the examples presented here, time is represented explicitly by variables ranging over a time domain, and time behavior is described by conventional first-order formulation. This approach works well for describing the behavior of simple circuit elements, and may prove to be the best way of specifying propagation delay phenomena. For specifying the behavior of systems in which synchronization plays an important role, however, it may prove to be too cumbersome.

An alternative to explicit modeling of time is the application of *temporal logic* techniques. The idea of applying temporal logic to specification and verification was introduced by Pnueli [Pnu 77] in the context of concurrent programs. A number of researchers have subsequently investigated the use of temporal logic for this purpose and for specification and proof of arbitrary concurrent systems [Hai 80,Lam 81,Map 81,ScM 81,Wol 82,Mao 81]. A particular strength of this technique is that it permits one to talk about the *order* in which events occur without having to be specific about the particular times at which they occur. Specifications (such as descriptions of liveness and safety properties) that require a good deal of syntax to formulate in conventional first-order logic with explicit time can often be stated quite tersely using temporal connectives. A traditional weakness of temporal logic for real applications has been the lack of mechanical support for temporal reasoning. Just recently, however, we have begun to use the STP system to mechanize the correctness proof of a communications protocol specified using temporal assertions. In this proof, the semantics of the temporal connectives are encoded axiomatically and invoked where appropriate by the user.

At low levels of description, the difficulty of specifying VLSI designs is also exacerbated by the need to model analog behavior. In order to account for such phenomena as meta-stability and charge-sharing problems, it may sometimes be necessary to develop fairly complex specifications of device behavior. In principle, it is possible to model such behavior in arbitrary detail. In practice, however, a considerable degree of idealization of device behavior is necessary to prevent the combinatorics of the specification and verification process from becoming unmanageable.

We feel confident that many of the errors that designers tend to make can be caught using relatively abstracted device semantics. For example, novice designers of combinatorial logic in NMOS often violate the requirement that each input combination cause exactly one of several wire-or'ed outputs to be driven. Such mistakes can probably be detected using no more complicated models of device behavior than have been given in our examples. Rules regulating the structure of pass-transistor logic, including those regarding pull-up/pull-down ratios should also be easy to treat with simple models.

More subtle design errors, such those associated with charge isolation, hazards, and race conditions, may be more difficult to deal with. Fortunately, the use of design disciplines (such as that proposed by Mead and Conway [MeC 80]) make these aspects of device function more manageable. Nevertheless, such problems account for many of the errors designers of VLSI circuits tend to make, and the question of how to capture the relevant aspects of device behavior in formal specifications is an important one.

7 References

[Bir 74] Birman, A.,"On Proving Correctness of Microprograms", IBM J. Research and Development, vol.14, no. 3, May 1974, pp. 250-266.

[Flo 67] Floyd, R. W., "Assigning Meanings to Programs", Proc. Amer. Math. Soc. Symp. in Applied Math. 19 (1967), pp. 19-31.

[Fos 81] Foster, M. J., "Syntax-Directed Verification of Circuit Function", VLSI Systems and Computations, H. Kung, B. Sproul, G. Steele, ed., Computer Science Press, Carnegie-Mellon University, 1981, pp. 203-212

[Gor 81] Gordon, M., "Two Papers on Modelling and Verifying Hardware", Proc. VLSI 81 International Conference, Edinburgh, Scotland, August 1981

[LRS 79]Levitt, K., L. Robinson, B. Silverberg, "The HDM Handblook", SRI International, 1979

[MeC 80]Mead, C., L. Conway, Introduction to VLSI Systems, Addison-Wesley, Philipines, 1980

[Mil 65] Miller, R., Switching Theory, vol. 2., Wiley, New York, 1965

[ScM 82] Schwartz, R.L., P.M. Melliar-Smith,"Formal Specification and Mechanical Verification of SIFT: A Fault-tolerant Flight Control System", TR CSL-133, SRI International, Menlo Park California, January 1982

[Sei 79] Seitz, Charles L., "Self-timed VLSI Systems", Proc. Caltech Conference on VLSI, January 1979

[Sho 82] Shostak, R. E., "Deciding Combinations of Theories", Proc. Sixth Conference on Automated Deduction, Courant Institute, New York, June 1982.

[ShS 82] Shostak, R. E., R. L. Schwartz, P. M. Melliar-Smith, "STP: A Mechanized Logic for Specification and Verification", Proc. Sixth Conference on Automated Deduction, Courant Institute, New York, June 1982

[ShS 83] Shostak, R. E., "Formal Verification of Circuit Designs", 6th International Symposium on Computer Hardware Description Languages and their Applications, Carnegie-Mellon University, Pittsburg, May 83

[Wag 77] Wagner, T., "Hardware Verification". Ph.D. Th., Dept. of Computer Science, Stanford University, September 1977

Appendix.   STP Proof for Muller C-Element

Theory of Logic Levels and Pull Downs

```
1.    (DT TRI.STATE NIL)
2.    (DST LOGIC.LEVEL TRI.STATE NIL)
3.    (DST PULL.DOWN TRI.STATE NIL)
4.    (DS TRI.STATE ZERO NIL)
5.    (DS LOGIC.LEVEL ONE NIL)
6.    (DS PULL.DOWN HIZ NIL)
7.    (DA DISTINCT (AND (NOT (EQUAL (ONE)
                                    (ZERO)))
                        (NOT (EQUAL (ONE)
                                    (HIZ)))

                        (HIZ)))))
8.    (DSV LOGIC.LEVEL L)
9.    (DA LLAXIOM (OR (EQUAL L (ZERO))
                      (EQUAL L (ONE))))
```

Circuit Specification

```
10.   (DS LOGIC.LEVEL A (NUMBER))
11.   (DS LOGIC.LEVEL B (NUMBER))
12.   (DS LOGIC.LEVEL C (NUMBER))
13.   (DS NUMBER TO)
14.   (DSV NUMBER T)
15.   (DS PULL.DOWN P1 (NUMBER))
16.   (DS PULL.DOWN P2 (NUMBER))
17.   (DS PULL.DOWN P3 (NUMBER))
18.   (DS PULL.DOWN P4 (NUMBER))
19.   (DS PULL.DOWN P5 (NUMBER))
20.   (DS PULL.DOWN P6 (NUMBER))
21.   (DS PULL.DOWN P7 (NUMBER))
22.   (DS PULL.DOWN P8 (NUMBER))
23.   (DS PULL.DOWN P9 (NUMBER))
24.   (DS PULL.DOWN P10 (NUMBER))
25.   (DS PULL.DOWN P11 (NUMBER))
26.   (DS PULL.DOWN P12 (NUMBER))
27.   (DS LOGIC.LEVEL L1 (NUMBER))
28.   (DS LOGIC.LEVEL L2 (NUMBER))
29.   (DS LOGIC.LEVEL L3 (NUMBER))
30.   (DS LOGIC.LEVEL L4 (NUMBER))
31.   (DS LOGIC.LEVEL L5 (NUMBER))
32.   (DS LOGIC.LEVEL L6 (NUMBER))
33.   (DS LOGIC.LEVEL L7 (NUMBER))
34.   (DS NUMBER EPSILON NIL)
35.   (DD BOOL PHI (T)
            (TRUE))
36.   (DD BOOL PSI (T)
            (EQUAL (C (PLUS T (TIMES 2 (EPSILON))))
                   (IF (EQUAL (A T)
                              (B T))
                       (A T)
```

```
                              (C T))))

37.      (DD BOOL OMEGA (T)
              (EQUAL (L6 (PLUS T (TIMES 2 (EPSILON))))
                     (IF (EQUAL (A T)
                                (B T))
                         (A T)
                         (L6 T))))
```

Circuit Element Semantics

```
38.      (DSV PULL.DOWN X)
39.      (DSV PULL.DOWN S)
40.      (DSV LOGIC.LEVEL G)
41.      (DSV PULL.DOWN D)
42.      (DSV PULL.DOWN Y)
43.      (DSV PULL.DOWN Z)
44.      (DSV LOGIC.LEVEL V)
45.      (DSV LOGIC.LEVEL W)
46.      (DD BOOL GROUND (X)
              (EQUAL X (ZERO)))
47.      (DD BOOL PDT (S G D)
              (EQUAL D (IF (EQUAL G (ONE))
                          S
                          (HIZ))))
48.      (DD BOOL PDJ (X Y Z)
              (EQUAL Z (IF (OR (EQUAL X (ZERO))
                               (EQUAL Y (ZERO)))
                          (ZERO)
                          (HIZ))))
49.      (DD BOOL PUT (S G)
              (EQUAL G (IF (EQUAL S (ZERO))
                          (ZERO)
                          (ONE))))
50.      (DD BOOL SPLITTER (L V W)
              (AND (EQUAL L V)
                   (EQUAL L W)))
51.      (DA EPSILON.POS (GREATEREQP (EPSILON)
                                     0))
```

VC1 and its Proof

```
52.      (DSV NUMBER U)
53.      (DD BOOL ALPHA.VC1 (T)
              (OMEGA T))
54.      (DD BOOL D.VC1 (T)
              (TRUE))
55.      (DD BOOL I.VC1 (T)
              (TRUE))
56.      (DD BOOL P.VC1 (T)
              (AND (SPLITTER (L6 T)
                            (L7 T)
                            (C T))
                   (PUT (P10 T)
                        (L6 (PLUS T (EPSILON))))
                   (PDT (P9 T)
                        (L5 T)
```

```
                        (P10 T))
                  (GROUND (P9 T))
                  (PDT (P9 T)
                       (L5 T)
                       (P10 T))
                  (PDT (P6 T)
                       (L7 T)
                       (P7 T))
                  (PUT (P8 T)
                       (L5 (PLUS T (EPSILON))))
                  (PDJ (P3 T)
                       (P7 T)
                       (P8 T))
                  (PDT (P2 T)
                       (L4 T)
                       (P3 T))
                  (PDT (P1 T)
                       (L3 T)
                       (P2 T))
                  (GROUND (P1 T))
                  (SPLITTER (A T)
                            (L3 T)
                            (L1 T))
                  (SPLITTER (B T)
                            (L4 T)
                            (L2 T))
                  (PDJ (P4 T)
                       (P5 T)
                       (P6 T))
                  (PDT (P6 T)
                       (L7 T)
                       (P7 T))
                  (PDT (P11 T)
                       (L1 T)
                       (P4 T))
                  (GROUND (P11 T))
                  (PDT (P12 T)
                       (L2 T)
                       (P5 T))
                  (GROUND (P12 T))))
57.   (DF HYPO.VC1 (IMPLIES (GREATEREQP T (TO))
                            (AND (P.VC1 T)
                                 (I.VC1 T)
                                 (D.VC1 T))))
58.   (DF CONCLU.VC1 (IMPLIES (AND (GREATEREQP T (TO))
                                   (FORALL U (IMPLIES (AND (GREATEREQP
                                                            U
                                                            (TO))
                                                        (LESSP U T))
                                                   (ALPHA.VC1 U))))
                              (ALPHA.VC1 T)))

59.   (DF LEMMA (IMPLIES (GREATEREQP T (TO))
                         (AND (GROUND (P9 T))
                              (PDT (P9 T)
                                   (L5 T)
                                   (P10 T))
```

```
                                   (PUT (P10 T)
                                        (L6 (PLUS T (EPSILON)))))))
60.     (PR LEMMA (HYPO.VC1 ((T T))))

----------Proving----------

170.25 seconds
Proved

61.     (PR (CONCLU.VC1)
            (HYPO.VC1 ((T T)))
            (LEMMA ((T (PLUS T (EPSILON))))))
            (LLAXIOM ((L (A T))))
            (LLAXIOM ((L (B T))))
            (LLAXIOM ((L (L6 T))))
            DISTINCT EPSILON.POS)

----------Proving----------

70.55 seconds
Proved

VC2 and its Proof

62.     (DD BOOL ALPHA.VC2 (T)
            (PSI T))
63.     (DD BOOL D.VC2 (T)
            (OMEGA T))
64.     (DD BOOL I.VC2 (T)
            (TRUE))
65.     (DD BOOL P.VC2 (T)
            (SPLITTER (L6 T)
                      (L7 T)
                      (C T)))
66.     (DF HYPO.VC2 (IMPLIES (GREATEREQP T (TO))
                              (AND (P.VC2 T)
                                   (I.VC2 T)
                                   (D.VC2 T))))
67.     (DF CONCLU.VC2 (IMPLIES (AND (GREATEREQP T (TO))
                                    (FORALL U (IMPLIES (AND (GREATEREQP
                                                                 U
                                                                 (TO))
                                                            (LESSEQP U T))
                                                       (ALPHA.VC2 U))))
                              (ALPHA.VC2 T)))
68.     (PR (CONCLU.VC2 ((U T)))
            (HYPO.VC2 ((T T)))
            (HYPO.VC2 ((T (PLUS T (TIMES 2 (EPSILON)))))))
            EPSILON.POS)

----------Proving----------

6.55 seconds
Proved

Query of Status of VC1 and VC2

69.     (QF CONCLU.VC1)
```

```
***********************************************

CONCLU.VC1

(IMPLIES (AND (GREATEREQP T (TO))
              (FORALL U (IMPLIES (AND (GREATEREQP U (TO))
                                       (LESSP U T))
                                  (ALPHA.VC1 U))))
         (ALPHA.VC1 T))

PROVED

Premises used:
HYPO.VC1
   T/ T
LEMMA
   T/ (PLUS T (EPSILON))
LLAXIOM
   L/ (A T)
LLAXIOM
   L/ (B T)
LLAXIOM
   L/ (L6 T)
DISTINCT
EPSILON.POS

69.    (QF CONCLU.VC2)

***********************************************

CONCLU.VC2

(IMPLIES (AND (GREATEREQP T (TO))
              (FORALL U (IMPLIES (AND (GREATEREQP U (TO))
                                       (LESSEQP U T))
                                  (ALPHA.VC2 U))))
         (ALPHA.VC2 T))

PROVED

Premises used:
HYPO.VC2
   T/ T
HYPO.VC2
   T/ (PLUS T (TIMES 2 (EPSILON)))
EPSILON.POS
```

A Hierarchical Simulator Based on Formal Semantics*

Marina C. Chen and Carver A. Mead†

Simulation consists of exercising the representation of a design on a general purpose computer. It differs from programming only because the ultimate implementation will be in a different medium, say a VLSI chip. In order for simulation to be in any sense effective, the simulated system must perform the same function as the ultimate implementation. A VLSI chip is a highly concurrent object; the simulation of such a chip amounts to programming a highly concurrent system. It follows that any demonstrably correct simulation technique will be one of the two types:

(1) The entire design is represented as an implementation with objects which are abstract models of the medium at the bottom level (e.g. transistor model). The simulation operates on a representation which is a direct image of the fully instantiated implementation in the medium.

(2) The design is represented as a hierarchy of implementations. Each level of implementation is constructed of objects which are abstract models of the implementation at the level below it. The simulation operates on a hierarchical representation where each level is refined by the level below it.

The first approach requires only a model of the implementation medium. The second approach requires, in addition, a general principle for obtaining an abstract model from a given implementation of objects at the lower level. The second approach is similar to using induction as a proof technique. Instead of proving all possible cases embodied in a

*This work is sponsored by the System Development Foundation.

†Computer Science Department, California Institute of Technology, Pasadena, California.

theorem, one proves the base case, and shows the hypothesis at level n is true with the assumption that the hypothesis is true at level $n-1$. In the context of simulation and modeling, it is necessary to first establish the transistor level model, then show how to obtain a model at level n given a model at level $n-1$. The second approach yields a more efficient simulation and a clearer conceptualization of the design much the same way the inductive technique does to a proof.

In [4], we described a model of computation with formal semantics for highly concurrent systems. This method of obtaining semantics for a concurrent system is exactly the same as the principle used to obtain an abstract model for a design. In this paper, we present a hierarchical simulator which uses this composition and abstraction principle together with the formal model for MOS switch-level circuits developed by Bryant [2] as the basis.

Switch-Level Simulation

In [2], a circuit is approximated by a series of *logical conductance networks*. Transistors that are *on* are represented as a high conductance and those that are *off* as a zero conductance. The simulation of a circuit consists of obtaining the steady state solution for each conductance network and then updating the transistors whose gates have changed to obtain a new conductance network. The process continues until the topology of the conductance network itself no longer changes. The transition from one conductance network to the next is based on the unit-delay model of the switching of the transistors with respect to their gate voltage. Therefore, attaining the steady state for a transistor network involves obtaining two nested levels of steady state, the inner *conductance-level* and the outer *transistor-level*.

Multi-level and Mixed-level Simulation

In the design of a VLSI system, the traditional levels of hierarchy are circuit level, gate-level, and register transfer level. This partitioning helps designers focus on one particular level of design at any given time. When they focus on the register transfer level, for example, they can reason about the overall design in terms of the functionality of the inter-connected blocks and a given timing scheme, without worrying about the details inside each block. On the other hand, when they are designing at the circuit-level, the focus is on one functional block at a time rather than the whole system. Ideally, if the overall system design is shown to be correct in terms of the functional blocks, and each functional block is shown to be correct in terms of its circuit-level or gate-level implementation, the designers need not examine the correctness of the detailed implementation across two different functional blocks, i.e., each of these hierarchical levels provides an abstraction of the level below it. The functionality of the overall design will always be preserved when the designers cross the different levels. The complexity of a large system design can only be effectively managed through these levels of abstraction. Preserving the functionality, i.e., maintaining consistency between hierarchical levels, is the most important property of a hierarchical simulator, and the most difficult to achieve.

A multi-level simulator, when used as a tool to verify a hierarchical design, should provide a way to ensure the consistency of the design throughout all levels. On the other hand, the simulator should allow blocks of different levels to be connected through proper interfaces which handle the timing and the matching of various input/output data types. The key issue in such a multi-level simulator is the interface mechanism. In this paper, we present a simulator in which a uniform representation is used at all levels of the design. We show a method of abstraction with which the consistency can be maintained, and describe how the timing and data types can be properly interfaced.

Semantic Hierarchy and Syntactic Hierarchy

In programming languages, *macros* and *procedures* or *functions* have long been recognized as two different ways to facilitate programming. Macros are used only at a syntactic level to ease the specification, and do not provide any semantic abstraction, since they are expanded during compilation. The object code of a program using macros is exactly the same as its counterpart without using macros. However, procedures and functions are used not only to facilitate specification, but to encapsulate a piece of code with a well-defined interface to other parts of a program. Ideally, a function should not allow any side-effects and therefore provides a semantic abstraction. We make the distinction of the *syntactic hierarchy* vs. the *semantic hierarchy* in a simulator in a way analogous to the distinction between macros and procedures in a programming language. The syntactic hierarchy in the simulator serves two purposes: One is ease of specification, just as macros in a programming language; the other is that it contains information about spatial locality. Since it has been observed that activities in circuits tend to be local [2], this information can be exploited by the simulation algorithm to achieve better performance. Unlike simulators which take a flat network specification where the locality information has been thrown away by the preprocessing to be recovered later on by topological analysis, this simulator takes advantage of user's hierarchical specification and requires neither preprocessing nor topological analysis.

Syntactic Cells.   In the context of a switch-level simulator, transistors and nodes are objects from which everything else is constructed. They are bottom level cells of the semantic hierarchy. We call this bottom level the *conductance-level* since an active device (transistor) is approximated by a passive conductance. A *syntactic cell* is a circuit consisting of an interconnection of transistors and nodes where there is no restriction on the input nodes or the output nodes of the cell. The state of a node in a syntactic cell can directly or indirectly influence nodes in the others and thus the cell provides no abstraction for the behavior of a circuit.

Figure 1 shows an nMOS *exclusive nor* cell which is an example of a syntactic cell.

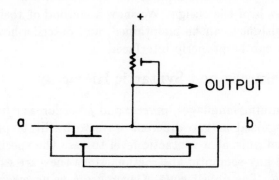

Figure 1. A non-restoring exclusive nor circuit XNOR.

Although a syntactic cell like the one shown in Figure 1 is a composition of transistors and nodes, it is at the same semantic level (or of the same semantic type) as transistors and nodes. A syntactic cell can also be a composition of other syntactic cells of the same semantic type, although these cells can be nested at an arbitrary depth in the syntactic hierarchy. No separation of the hierarchy into leaf cells and composition cells [10] is required, i.e., a transistor or a node can be composed with a syntactic cell directly without making it into a syntactic cell by itself. The simulation of such a cell produces exactly the same result as the circuit represented without hierarchy. The equivalence of Bryant's model of a flat network with our hierarchical represented network can be shown by straightforward induction on the level of the syntactic hierarchy.

The semantic hierarchy is constructed for abstracting the behavior of circuits. A syntactic cell is made into a semantic cell if an abstraction of its behavior is desired. This new *primitive* semantic cell is used as an "atom" in a new syntactic hierarchy which in turn is used within each semantic level in order to clarify the specification and express the locality of cells.

For simulator based on a switch-level model, the semantic levels can be the bottom level transistors and nodes (conductance-level), gate-level, clocked cell level (in a more general sense including a self-timed [11] module with request and acknowledge signals), register transfer level, and other higher levels. We call all levels at and above the clocked cell level the *functional level* since the functionality of cells at those levels can be abstracted.

Gate-level Cells. A *gate-level* cell is a composition of conductance-level cells (transistors, nodes and syntactic cells composed of them) with the restriction that the input nodes be uni-directional, i.e., the input nodes are gates of transistors. Figure 2 shows a gate-level cell. Note that the XNOR circuit in Figure 1 is not a gate-level cell.

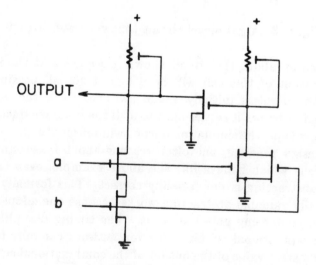

Figure 2. A restoring exclusive or gate XOR.

According to the unit-delay model of a transistor, the gate node of a transistor will not affect the state of the transistor until a given conductance network is settled. Since each of the input nodes of a gate-level cell is the gate node of a transistor, no intermediate state on that node will be seen by the cell until the node reaches the inner conductance-level steady state described above. The gate-level cell is affected by each of its inputs when each of these gate nodes reaches its steady state and causes the corresponding transistor to change state. Notice that to abstract the behavior of a gate-level cell, we can only discard the intermediate states on all of its output nodes (prior to its reaching the inner conductance-level steady state). We cannot just keep the outer transistor-level steady state and discard all the conductance level steady states. Therefore, the conventional way of thinking about a gate-level cell as a functional block (in which all intermediate states before reaching the outer transistor-level steady-state are discarded) is not formally correct. This incorrect abstraction has to be remedied by some other analysis in the design process. For a simulator, such incorrect abstraction will miss, for example, the glitch in the circuit shown in Figure 3.

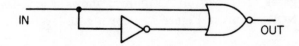

Figure 3. A gate-level circuit that generates logic 0

If we reason about the circuit shown in Figure 3 at the functional level, the output of this cell will be a logic 0 for all possible inputs. Yet when the input is initially 1 and switches to 0, the output will become a logic 1 before it settles back to 0. If the inner conductance-level steady state is kept, the simulation output will reflect the glitch properly. In actual design practice, one often reasons about interconnections of gate-level cells which behave like the above example, even though the functional abstraction is not formally correct. This formally incorrect but practically valuable abstraction works because one adopts a timing discipline in composing gate-level cells. The timing discipline ensures that each combinational circuit in a given network can only be affected by the steady state value of the output of the combinational cell to which it is connected. Familiar examples of this discipline are a two phase non-overlapping clock scheme where the clock period of each phase is long enough for a combinational circuit to settle, and a self-timed request-acknowledge signaling. An intermediate value on an output node will not propagate because of the timing discipline in the same way that the intermediate value of a local variable will not be returned by a function in a programming language. Therefore the timing discipline provides a semantic abstraction similar to a function in a programming language.

Clocked Cells. A clocked-cell is a composition of conductance level cells with the restriction that all input nodes must satisfy a timing discipline which insures that only the steady-state of the output nodes can be seen by other cells to which they are connected. (Figure 4 shows a clocked cell formed by two other clocked cells.)

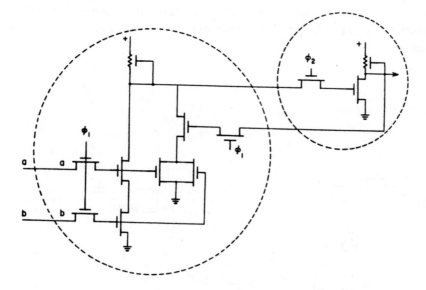

Figure 4. A clocked Muller C-element containing two sub-cells

The steady state of a primitive clocked cell (which does not contain any other clocked cell as a component, for example, as each of the sub-cells shown in Figure 4) is the steady state at transistor-level for that cell. Each of these cells can only see the steady state of other cells and therefore provides an abstraction. The gate-level abstraction is at a lower level than the clocked cell level since not only the steady state of the transistor-level is kept but all the steady states at the conductance level are kept as well. Bryant [2] points out that the gate-level cells provide a useful modeling abstraction since it is not necessary to keep track of the *signal strength* used in the conductance-level. Experience with MOS design has shown that specifying a chip in terms of gate-levels cells is not convenient because the restriction on inputs does not allow effective use of pass transistors. Since clocked cells are usually small, we use them as the semantic level immediately above the conductance-level cells. Sub-circuits within them are represented as syntactic cells. In this simulator we therefore do not support the gate-level abstraction. In a technology other than MOS, the gate-level may well be an appropriate level of abstraction, and is very easy to implement within our simulator.

We will now proceed to illustrate the model and its use in specification and simulation by way of a simple example — a pipelined inner product element similar to that described in [8]. Figure 5 shows an example of a hierarchical partition of a single bit inner product cell IPB described in [12] which will be used later in the pipelined inner product element.

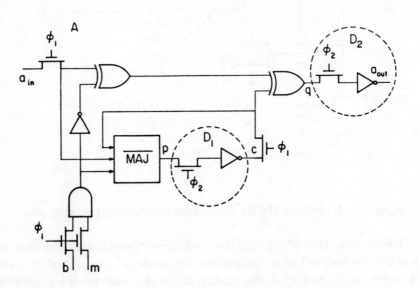

Figure 5. A single bit inner product element IPB

This cell is of type clocked-cell and consist of three clocked-cells, A which is not circled in the Figure, D_1 and D_2 which are circled. Cell A contains five syntactic cells, $\overline{\text{MAJ}}$, two XNOR gates, an AND gate, and an inverter. It also contains four transistors and four gate nodes each with ϕ_1 on it, four input nodes a_{in}, b, m and, c and finally, two output nodes p and q.

Cells D_1 and D_2 are identical. Each consists of one syntactic cell, the inverter, a transistor and a gate node with ϕ_2 on it, an input node p for D_1 and q for D_2, respectively and finally an output node c for D_1 and a_{out} for D_2, respectively. The inputs to the IPB are a_{in}, b, m and the clocks. The output is a_{out}. There is one bit of internal state in this cell, namely the carry c. The current state of c is denoted by c_{cur} and next state by c_{next}.

The simulation of this syntactic cell proceeds by (1) obtaining the steady state of each clocked-cell (A, D_1 and D_2) using its inputs and state, independently of the other cells and, (2) transferring the outputs of one cell to the inputs of the others. Notice that a IPB cell is a syntactic cell, and therefore each of the three sub-cells is only invoked once. The IPB cell can be further abstracted to be a bit-level cell. However, it is then necessary to obtain the steady state of the IPB cell. The above procedure is iterated until the outputs of all three sub-cells do not change any more. This iteration yields the steady state of the IPB cell at the bit-level. This example shows that each semantic level requires an iteration to obtain its steady state. In the example of constructing an n bit pipelined element below, we do not use the bit-level abstraction, and so the above described procedure is invoked only once.

To verify the correctness of this circuit, the three clocked-cells are replaced by their functional specification, shown in Figure 6. The cell specified in a functional form is simulated and compared with the one (with detailed implementation) described above.

```
IF  φ₁ THEN
BEGIN
        q ← (aᵢₙ ⊕ (NOT(b AND m))))⊕c_cur,
        p ← MAJ(aᵢₙ, b AND m, c_cur);
END ELSE
IF  φ₂ THEN
BEGIN
        a_out ← NOT q,          c_next ← NOT p;
END;
```

Figure 6. The functional specification of a 1-bit inner product element

Notice that the data type in the functional specification is *boolean* rather than *signal* (which has two components, the signal strength and the signal state) used in the conductance-level representation, and that the algebra of signals [2] is different from ordinary boolean algebra. In many cases, such as this inner product cell, a correct signal on the output of each clocked cell can only be a logic 0 or a logic 1 which falls into the domain of boolean algebra, although the internal nodes or even the intermediate state of the outputs can be in state X. The algebra of signals contains the signal state X, which represents an intermediate voltage between 1 and 0. The steady state output of a proper clocked cell will always be a binary value. If the output of a clocked-cell is an X, then either an error has occurred in the implementation or functional abstraction at the clocked cell level should not be applied. It is possible, in fact, very common, for the output of a clocked cell to be *unknown*, for example, when the output is a function of some state variable which has not been initialized. At every level we use the symbol $\perp$ (pronounced *bottom*) to represent an undefined value. It must not be confused with X which represents a voltage between logic 0 and logic 1 at the conductance-level. Since an intended output will never depend on an unknown value $\perp$, a possible way of handling a $\perp$ in the input of a function would be to define the corresponding output also to be $\perp$. An algebra extended this way is called *naturally extended* [9] (although there exists other possible extensions, the natural extension is used in this simulator).

Notice that even at the level of a simple 1-bit cell, we can see the kind of data abstraction that always accompanies functional abstraction. Each level has its own algebra for manipulating data and functions. A formal treatment of the model of computation that allows such abstractions (input/output mapping functions for data abstraction and fixed-point semantics for functional abstraction) is given in [4].

Word-level Cells. The functional specification given in Figure 6 is used in the next level of composition, in this example the pipelined inner product element. The performance of the simulation using this specification will be drastically improved in comparison with the circuit-level specification. Although exhaustive checking is possible for verifying the consistency between the two specifications for a single cell like IPB, it becomes very rapidly impractical as the size of the cell grows. Then it becomes necessary for the notation or language in which the design is described to have formal semantics in order to allow verification of the consistency between two different levels of specification. Space-time recursion equations described in [4] are an example of such a notation. Although the current simulator is implemented as an embedded language in an ordinary programming language, the primitives for specifying the design are a direct mapping of the above formal notation.

An n-bit pipelined inner product element can be composed by connecting n IPB cells serially (the specific scheme is shown in [12]). We call this cell IPE. It can be viewed as a word-level cell where a pulse input *lsb* indicates the start of a word (say, least significant bit of an n bit word). We can adapt the interface of this cell so that the clocks can be hidden inside the cell and the bit serial input and output can be abstracted as words. Figure 7 shows the data abstraction which maps input words $\hat{a}_{in}$, $\hat{b}$ and $\hat{m}$ into series of bits and collects output bits to be the word $\hat{a}_{out}$. The bits $\hat{a}_{in}[i]$, $\hat{b}[i]$, and $\hat{m}[i]$ are put in one by one to the IPE at its input ports $ipe_{a_{in}}$, ipe_b, and ipe_m. A suitable clock is also generated for the cell. Once the inputs and the clocks are valid, IPE is called upon to compute its result (written as IPE.*compute* in Figure 7). It computes by invoking each of the IPB cells (which invokes once each of its sub-cells A, D_1 and D_2 as described above), transfers the results of one IPB to the next and repeats until each of the result returned by all IPB cells becomes steady.

The functional abstraction of the IPE is shown in Figure 8. Again, the consistency of the two different levels can be verified by the combination of formal verification and simulating both specifications.

```
IF lsb THEN
FOR i := 0 TO n − 1 DO (loading phase)
BEGIN
        ipe_{a_in} ← â_in[i], ipe_b ← b̂[i], ipe_m ← m̂[i]; (input mapping)
        φ_1 ←high, φ_2 ←low; IPE.compute;
        φ_1 ←low, φ_2 ←high; IPE.compute;
END;
FOR i := 0 TO n − 1 DO (unloading phase)
BEGIN
        φ_1 ←high, φ_2 ←low; IPE.compute;
        φ_1 ←low, φ_2 ←high; IPE.compute;
        â_out[i] ← ipe_{a_out}; (output mapping)
END;
```

Figure 7. Interface of an n bit-serial element to higher-level specification

$$\text{IF } lsb \text{ THEN } \hat{a}_{out} \leftarrow \hat{a}_{in} + \hat{b} \times \hat{m}$$

Figure 8. The functional specification of the inner product element

The n bit inner product element can be used to construct, for example, a systolic array performing matrix multiplication [6]. At the level of a systolic array, the inner product element is used in the functional form as in Figure 8 regardless of its implementation as bit serial or word parallel. We have to be careful however, since the mapping functions that constitute the data abstraction are never unique. The mapping may be from n serial bits to a word or from n parallel bits to a word. The mapping in the former is from time domain to an abstract word and the latter is from space domain to an abstract word. In connecting two inner product elements at this level, one needs to make sure that the output mapping function of one element is the inverse of the input mapping function of the element to which it is connected. Two IPE elements as shown in Figure 7 can be connected since the output mapping of one (from bits $ipe_{a_{out}}[i]$ for $0 \leq i < n$ to an n bit word $\hat{a}_{out}$) is the inverse of the input mapping of the other (from n bit word $\hat{a}_{in}$ to bits $ipe_{a_{in}}[i]$ for $0 \leq i < n$). The lower level has the same interface problem in a different form. It is the timing discipline used in the design that allows abstraction to the clocked cell level. This discipline can be one of several kinds. For example it may be two, three or four phase clock, or a two or four cycle request-acknowledge protocol. Once again the condition imposed on the output of one cell must match that of the input of the cell to which it connects.

The Hierarchical Design Method

The systolic array can also be abstracted from its implementation and used as an abstract machine performing matrix multiplication. The detailed procedure of abstracting a systolic array is presented in [4]. Notice that the abstraction mechanism is precisely the same at all levels. Considering the bottom-up approach, we summarize the two essential steps in the hierarchical design method.

(1) Use a cell defined by its implementation in the context of cells defined by their functional specifications. In order to adapt the interface between the cell and its context, a new cell containing the following three parts is constructed:

   i) A function which maps inputs in the data representation of the high level to the inputs at the lower level.

   ii) The implementation at the lower level.

   iii) A function which maps the outputs at the lower level to outputs in the data representation at the higher level.

This cell is now a proper cell with exactly the same interface as a higher level cell. We can use it as such or

(2) Replace such an implementation with its functional specification. These two versions shall be verified to be consistent either by simulating them against each other, by formal verification or by a combination of the two. Once the results are identical, the first cell can be replaced by its functional equivalent.

The order of these two steps can be reversed for the top-down approach. In the top-down approach, instead of an implementation of lower level functions being abstracted to be a higher level function, a higher level function is implemented by some lower level functions. Such refinement of design can be carried out until the implementation at the bottom is completed.

What we have shown is a method for constructing systems from switch-level circuits to functional blocks through successive semantic abstractions or, stated the other way around, starting with functional specification at the top level and successively refining the specification until it is implemented in the bottom level representation of the medium (in this case, switch-level model of transistors and nodes). In contrast to the conventional view of fixed hierarchical levels, the partitioning of a semantic hierarchy is flexible and problem oriented. Designers can partition each system in the way that is most natural to the design, rather than fit into rigid pre-defined levels which are not necessarily appropriate.

Implementation of the Simulator

The example shown above is very specific, and one can obviously write a multi-level simulator for this particular system. Such a simulator would be of limited use since it handles only designs partitioned in the same way. A general purpose simulator which must support arbitrary levels and mixed-level seems at first unrealistic. We approach this problem by:

1. Separating out the part that is universal to all system levels

2. Using the power of an embedded language [7].

Embedded languages. It has been observed in integrated circuit layout languages that an embedded language — a language supporting graphics primitives in an existing programming language — has the generality and flexibility in the specification of designs that an interactive graphic layout system usually lacks. The effort of making a graphic system as powerful as an embedded language is essentially that of supporting a general purpose programming language. It is much more sensible to let the compiler of an existing language do the work. The same philosophy applies to a specification language for simulation. We build into a programming language the simulation algorithm and an interactive user interface (corresponding to the debugger in a programming environment) for testing the design. One specifies cells in an embedded simulation language by invoking primitives for transistors, nodes, syntactic cells and semantic cells. These primitives are pre-defined in the language. With the power of a general programming language, users can then specify functional abstractions, the data abstractions, and various data types at any level according to their conceptualization of the design.

Representation of Cells and the Dynamics of the Simulation. Cells are represented as *modules* in a programming language supporting separately compilable modules. The language we have used for this specific implementation is Mainsail. A cell has in it the specification of its constituents, i.e., the implementation in terms of lower level cells, and the procedure for computing its result. The former is supplied by user and the latter is incorporated automatically. A semantic cell computes by causing each of its constituents to compute until the steady state is reached; syntactic cells only step through each of the constituents once. Transistors, nodes and cells with functional specification that models their behavior compute directly. Composition cells (cells composed of sub-components) cause their constituents to compute in a recursive manner until one of the primitive functions or abstracted function is encountered. Each of these different ways of computing a cell is made into a template and the template is compiled together with the specification of its construction in the case of a composition cell. A module is typed according to the template incorporated into it. The net effect is that each module contains the functions necessary to compute its own behavior.

We use a module instead of a procedure to represent each cell because a cell can be mapped directly into a module which has a data section to represent the cell's inputs, outputs and states, and a procedure section for the specification of its constituents and computation of its behavior. Modules also have one bit of state which records whether the inputs are the same as those of the last invocation. No computation is necessary if they are the same since the outputs of the last computation are still valid. This "change bit" allows a truly efficient implementation.

Representation of Connections. The computational aspects of a system have attracted much more attention than the communication aspects partly because in a von Neumann architecture, the cost of accessing the random access memory is always the same and partly because historically, variable names cost nothing and furthermore, substitution or elimination of redundant variables has attracted relatively little attention in mathematics. Since the transfer of information does cost energy and resources (wires), it must be taken into account in any model of a computational system. We represent the communication among modules also as a module which contains the connectivity information and various ways to transfer information, such as a uni-directional connection or a bi-directional connection.

The Universal Fixed-Point Algorithm. By viewing a computational system as an ensemble of cells and connections, we devise a fixed-point algorithm to find the steady state of a cell. The fixed-point algorithm performs a task similar to the so-called "relaxation" algorithm. It takes the implementation of a cell in terms of connections and cells in a bipartite format. The algorithm invokes each cell, i.e., causes each cell to compute independently, and after all have been invoked, then invokes each connection to transfer the data and thereby bring about the interactions among connected cells. This procedure is iterated until the steady state is reached. The bipartite arrangement of cells and connections results in the property that the order in which each cell is invoked is immaterial in the algorithm. This algorithm is shown in [3] to yield the steady state of a system. Since the above model is uniformly applied to all semantic levels, and all cells and connections are represented uniformly, only a single universal fixed-point algorithm is necessary.

Elements of a Multi-level Simulator. The following primitives, embedded into Mainsail, serve as templates for user defined circuits:

1. Cells of various types: transistors, syntactic cells, conductance networks, clocked cells, and functional cells.

2. Connections of various types: nodes, bi-directional connections for syntactic cells below clocked cell level, conductance network formation, and uni-directional connections for functional cells. The transition from one conductance network to the next is also represented as a connection.

In the object-oriented view of computation such as Simula [1], Smalltalk [5], etc., these templates are the superclasses of the user defined classes (cells). These templates are the only structure we build into the simulator. Instead of building and maintaining data structures that represent a design, the structure is embedded in user's specification of interconnected modules. Hence, no global simulation algorithm is necessary to traverse the data structure.

Conclusions

We have described a multi-level simulator which allows user-defined levels instead of rigidly pre-defined levels. We draw a clear distinction between the modularization for ease of specification and for semantic abstraction — the syntactic hierarchy and the semantic hierarchy, respectively. An example of multi-level simulation is given which spans from circuit-level up to the abstract function of an inner product element.

With a formal model as a basis, the implementation of the simulator is simple and uniform at all levels. A single universal fixed-point algorithm is used. This approach raises the activity of simulation from a low level corresponding to macro assembly level in a programming language to a hierarchical specification corresponding directly to the conceptualization of user's design. We show how functional abstraction and data abstraction (interfaces between two different levels) can be made. These abstractions are the key to the consistency and efficiency of a multi-level simulator.

We demonstrate the importance of the formal semantics which allows functional abstraction and thereby enables an efficient simulation methodology. In simulation, showing that the specification and the implementation are equivalent is not merely desirable but absolutely essential. We hope that this working example has shown that formal semantics is an essential feature of any design tool as well as any concurrent programming language.

Acknowledgments

We wish to thank Randy Bryant for insightful discussions and suggestions on the subject of simulation, and Alain Martin for his valuable comments in the preparation of this paper.

References

[1] Birtwhistle, G. M., Dahl, O-J, Myhrhaug, B., and Nygaard, K., *Simula Begin*, Petrocelli, New York, 1973.

[2] Bryant, R.E., *A Switch-level Simulation Model for Integrated Logic Circuits*, Massachusetts Institute of Technology, Cambridge Massachusetts, March, 1981.

[3] Chen, M.C., Doctoral Dissertation, Computer Science Department, California Institute of Technology, 1983.

[4] Chen, M.C. and Mead C.A., *Concurrent Algorithms as Space-time Recursion Equations*, Proceedings of USC Workshop on VLSI and Modern Signal Processing, pp. 31-52, November, 1982.

[5] Ingalls, D., *The Smalltalk 76 Programming System: Design and Implementation*, Proceedings of the Fifth ACM Conference on Principles of Programming Systems, pp. 9-16, January 1978.

[6] Kung,H.T. and Leiserson C.E., *Algorithms for VLSI Processor Arrays*, in C. Mead and L. Conway, Introduction to VLSI Systems, Addison-Wesley, 1980, chapter 8.3.

[7] Locanthi, B., *LAP: A Simula Package for IC Layout*, Caltech SSP Report #1862, California Institute of Technology, 1978

[8] Lyon R. F., Two's Complement Pipeline Multipliers, *IEEE Trans. on Communications* COM-24, pp. 418-425, 1976.

[9] Manna, Z., *Mathematical Theory of Computation*, McGraw-Hill, New York, 1974.

[10] Rowson, J. A., *Understanding Hierarchical Design*, Doctoral Dissertation, California Institute of Technology, April, 1980.

[11] Seitz, C., *System Timing*, in C. Mead and L. Conway, Introduction to VLSI Systems, Addison-Wesley, 1980, chapter 7.

[12] Wawrzynek J. and Lin T. M., *A bit Serial Architecture for Multiplication and Interpolation*, 5067:DF:83, Computer Science Department, California Institute of Technology, January, 1983.

Trace Theory and the Definition of Hierarchical Components

Martin Rem, Jan L.A. van de Snepscheut
and Jan Tijmen Udding
(Department of Mathematics and Computing Science, Eindhoven University of Technology, Netherlands)

Abstract

The relevance of trace theory to the design of VLSI circuits is discussed. We present an introduction to trace theory and to regular trace structures in particular. We show, in a number of examples, how trace theory can be used to reason about and to prove properties of hierarchical components.

1. INTRODUCTION

It has been argued elsewhere [8] that one of the most important properties of VLSI is its potential as a carrier of concurrent computations. Concurrent computations require a very careful design technique, for, as we know, uncontrolled concurrency results in uncontrollable complexity. This observation makes complexity control a conditio sine qua non for VLSI design. We know of only one effective technique of complexity control: modular design. Using this technique, the design of a component amounts to the choice of subcomponents and relations. The relations express how the parts (the subcomponents) constitute the whole. Designing the subcomponents in a similar fashion, we obtain hierarchical components.

We shall present our components in the form of program texts. With each such text we associate a meaning: its semantic contents. From the meanings of the subcomponents and the way in which the subcomponents constitute the component the meaning of the whole component can be deduced. Such a hierarchical technique will constrain the complexity of the design task only if the meaning associated with a component does not reflect the component's internal structure. The meaning of a component, consequently, must comprise its net effect only, i.e., it must express all possible communication patterns between the component and its environment. These communication patterns will be formalized as a (possibly infinite) set of finite-length sequences of symbols. Such a sequence is called a trace. The meaning of a component then consists of a trace set (a set of traces) and an alphabet (a set of symbols). Each trace may be interpreted as a possible "external behaviour" of the component. The

225

symbols in the traces are chosen from the alphabet. They may be interpreted as the different "types of communication". The combination of a trace set and an alphabet is called a trace structure.

Example 1.1 A binary semaphore [2], initialized at 0 , would be a component with {v,p} as its alphabet and, using regular expressions [5], the set generated by the expression $(vp)^* + (vp)^*v$ as its trace set. The latter is the set of all finite-length alternations of v and p that do not start with a p . (End of example)

When forming a component from subcomponents, the trace structure of the component will be a function of the trace structures of the subcomponents. Components, trace structures, and composition functions are formally introduced in Section 2.

One may wonder how concurrency is incorporated in our formalism. Given a trace, we can talk of the i-th occurrence of a symbol b (denoted by b_i) in that trace. Given a trace set, we say that b_i precedes c_j if there exists a trace containing b_i and c_j and in every such trace occurrence b_i comes before occurrence c_j . The reflexive-transitive closure of the relation "precedes" is a partial order, and occurrences of symbols that are not ordered by it may be called concurrent.

Others [6] have suggested to define the meaning of a component as a set of vectors of traces, rather than as a set of traces. Concurrency then gives rise to symbols occurring in different elements (traces) of the same vector. The lengths of the vectors is the maximum degree of concurrency with which the component may be executed. There are two reasons why we have not adopted the vector approach. Firstly, the vectors reflect the internal structure of a component. Secondly, the vectors do not lend themselves well to the assignment of meanings to recursive components. (Recursive components are briefly discussed in Section 3.) Our experience is that meanings based on trace sets can be elegantly generalized to recursive components [10].

Brock and Ackerman [1] demonstrate that the meaning of a component should not be defined as a function from input traces to sets of output traces. The reason is that such a formalism cannot express order between occurrences in input traces and occurrences in output traces. Our approach does not have this defect.

We would like to stress that we do not intend to model the physical properties of computing engines. We assign meanings to components. These meanings should be observed by the physical realizations of the components. A realization that is not in accordance with the component's meaning is an erroneous realization. A possible realization method is discussed in [9].

2. TRACE THEORY

A *trace structure* is a pair $\langle T,A \rangle$, in which A is a finite set of symbols and $T \subseteq A^*$. A^* denotes, as usual, the set of all finite-length sequences of elements of A , including the empty sequence which is denoted by ε . A is called the *alphabet* of the

trace structure, and T its *trace set*. The elements of T are
called *traces*.

Let t be a sequence of symbols and B an alphabet. The *projection* of t on B , denoted by $t{\restriction}B$, is defined as follows.

if $t = \varepsilon$ then $t{\restriction}B = \varepsilon$
if $t = ub \land b \in B$ then $t{\restriction}B = (u{\restriction}B)b$
if $t = ub \land b \notin B$ then $t{\restriction}B = u{\restriction}B$

(Concatenation is denoted by juxtaposition.) The projection of a
trace set T on alphabet B , denoted by $T{\restriction}B$, is the trace set
$\{t{\restriction}B \mid t \in T\}$ and the projection of trace structure $\langle T,A \rangle$ on B ,
denoted by $\langle T,A \rangle {\restriction} B$, is the trace structure $\langle T{\restriction}B , A \cap B \rangle$.

Property 2.1
    (i) $(T{\restriction}A){\restriction}B = (T{\restriction}B){\restriction}A = T{\restriction}(A \cap B)$
    (ii) $\langle T,A \rangle {\restriction} B = \langle T,A \rangle {\restriction} (A \cap B)$

Property 2.2 Let A and B be two sets of symbols. Then

$$(\exists s,t: s \in A^* \land t \in B^*: s{\restriction}B = t{\restriction}A)$$
$$\equiv (\exists u: u \in (A \cup B)^*: u{\restriction}A = s \land u{\restriction}B = t)$$

In this section we define two composition operators for trace
structures. The first operation is p-composition. The *p-composite*
of two trace structures $\langle T,A \rangle$ and $\langle U,B \rangle$, denoted by $\langle T,A \rangle \underline{p}$
$\langle U,B \rangle$, is the trace structure

$$\langle \{t \in (A \cup B)^* \mid t{\restriction}A \in T \land t{\restriction}B \in U\} , A \cup B \rangle$$

Whenever obvious from the context, the alphabets are not explicitly
mentioned and we simply talk of the p-composition of trace sets.
For disjoint alphabets p-composition amounts to the shuffle operation [3].

Property 2.3 P-composition is symmetric and associative.

Note One might be tempted to introduce the convention that the alphabet of a trace structure contains exactly those symbols that oc-
cur in the trace set of the trace structure. Such a convention would, however, destroy the associativity of p-composition, as the
following example demonstrates:

$$(\{ba,\varepsilon\} \underline{p} \{ab,\varepsilon\}) \underline{p} \{ab,\varepsilon\} = \{\varepsilon\} \underline{p} \{ab,\varepsilon\} = \{ab,\varepsilon\}$$
$$\{ba,\varepsilon\} \underline{p} (\{ab,\varepsilon\} \underline{p} \{ab,\varepsilon\}) = \{ba,\varepsilon\} \underline{p} \{ab,\varepsilon\} = \{\varepsilon\}$$

(End of note)

Property 2.4 Let $\langle T,A \rangle$ and $\langle U,B \rangle$ be two trace structures.
(i)   If $A \subseteq B$ then $T \underline{p} U = \{t \in U \mid t{\restriction}A \in T\}$;
(ii) If $A = B$ then $T \underline{p} U = T \cap U$.

Property 2.5 Let $\langle U,B \rangle$ be a trace structure and let $\langle T,A \rangle$ and
$\langle T',A \rangle$ be two trace structures with equal alphabets. Then

    (i) $T \subseteq T' \Rightarrow T \underline{p} U \subseteq T' \underline{p} U$

    (ii) $U \underline{p} (T \cup T') = (U \underline{p} T) \cup (U \underline{p} T')$

    (iii) $U \underline{p} (T \cap T') = (U \underline{p} T) \cap (U \underline{p} T')$

We can extend $\subseteq$, $\cup$, and $\cap$ to operate on trace structures with equal alphabets. The properties above can then be reformulated for trace structures.

<u>Property 2.6</u> Let $<T,A>$ and $<U,B>$ be two trace structures and C a set of symbols. Then

$$(T \underline{p} U) \lceil C \subseteq (T \lceil C \underline{p} U) \lceil C$$

<u>Proof</u> Let $x \in (T \underline{p} U) \lceil C$ and $y \in (A \cup B)^*$ be such that $y \lceil C = x \wedge y \in T \underline{p} U$. Hence, $y \lceil A \in T \wedge y \lceil B \in U$. Since $y \in (A \cup B)^*$,

$$y \lceil (B \cup C) \in ((A \cap C) \cup B)^*$$

Using Property 2.1 we find

$$(y \lceil (B \cup C)) \lceil (A \cap C) \in T \lceil C$$

Combined with

$$(y \lceil (B \cup C)) \lceil B \in U$$

this shows that

$$y \lceil (B \cup C) \in T \lceil C \underline{p} U$$

Furthermore,

$$(y \lceil (B \cup C)) \lceil C = x$$

Hence,

$$x \in (T \lceil C \underline{p} U) \lceil C \qquad \text{(End of proof)}$$

<u>Property 2.7</u> Let $<T,A>$ and $<U,B>$ be two trace structures and C a set of symbols such that $A \cap B \subseteq C$. Then

$$(T \underline{p} U) \lceil C = (T \lceil C \underline{p} U) \lceil C$$

<u>Proof</u> Let $x \in (T \lceil C \underline{p} U) \lceil C$. We prove $x \in (T \underline{p} U) \lceil C$. The property then follows from Property 2.6. Let $y \in ((A \cap C) \cup B)^*$ be such that $y \lceil C = x \wedge y \in T \lceil C \underline{p} U$. Let $t \in T$ and $u \in U$ be such that $y \lceil (A \cap C) = t \lceil C \wedge y \lceil B = u$. Then

$$y \lceil A = y \lceil (A \cap C)$$

and

$$t \lceil ((A \cap C) \cup B) = t \lceil C$$

Hence,

$$y \lceil A = t \lceil ((A \cap C) \cup B)$$

Let $z \in (A \cup B)^*$ be such that

$$z \lceil A = t \wedge z \lceil ((A \cap C) \cup B) = y$$

(The existence of such a z follows from Property 2.2.) Then $z \lceil A \in T \wedge z \lceil B \in U$. Hence, $z \in T \underline{p} U$. Since $z \lceil C = x$, $x \in (T \underline{p} U) \lceil C$.   (End of proof)

By twice applying Property 2.6 and 2.7 respectively we find:

<u>Property 2.8</u> Let $<T,A>$ and $<U,B>$ be two trace structures and C a set of symbols. Then

$$\text{(i)} \qquad (T \underline{p} U) \lceil C \subseteq T \lceil C \underline{p} U \lceil C$$

(ii) if $A \cap B \subseteq C$ $(T \underline{p} U) \lceil C = T \lceil C \underline{p} U \lceil C$

In Properties 2.6, 2.7, and 2.8 the trace sets that are compared have equal alphabets. These properties, therefore, hold for trace structures as well.

A trace structure is called regular if its trace set is a regular set [5]. We mention without proof:

__Property 2.9__ The p-composite of two regular trace structures is a regular trace structure.

Let A be a set of symbols. We call $s \in A^*$ a *prefix* of trace $t \in A^*$ if ($\exists u: u \in A^*: t= su$) . Let T be a set of traces. The set of all prefixes of traces in T is denoted by PREF(T) . The set T is called *prefix-closed* if PREF(T) = T .

__Property 2.10__ Let T and U be two sets of traces. Then

(i) $T \subseteq PREF(T)$
(ii) $T \subseteq U \Rightarrow PREF(T) \subseteq PREF(U)$

Let <T,A> be a trace structure. Consider the relation $\sim$ defined on PREF(**T**) by s $\sim$ t if and only if

$$\{u \in A^* \mid su \in T\} = \{u \in A^* \mid tu \in T\}$$

Relation $\sim$ is an equivalence relation. Its classes, i.e. the elements of PREF(T)/$\sim$, are called the *states* of <T,A> . The state of which s is a member is denoted by [s] .

__Example 2.1__ Consider trace structure <T,A> with

T = $\{\varepsilon$, vp, vpvp, ... $\}$
A = $\{v,p\}$

Then

PREF(T) = $\{\varepsilon$, v, vp, vpv, vpvp, ... $\}$

i.e., the trace set of Example 1.1. The relation $\sim$ has two equivalence classes, viz. the elements of PREF(T) of even length and those of odd length. Trace structure <T,A> has, consequently, two states: [ε] and [v] .   (End of example)

__Property 2.11__ The p-composite of two trace structures with prefix-closed trace sets has a prefix-closed trace set.

__Proof__ Let <T,A> and <U,B > have prefix-closed trace sets, and let xy $\in$ T $\underline{p}$ U . Then $(x\lceil A)(y\lceil A) = (xy)\lceil A \in T$. Since T is prefix-closed, $x\lceil A \in T$. Likewise, $x\lceil B \in U$. Hence, $x \in T \underline{p} U$. (End of proof)

__Property 2.12__ Let <T,A> and <U,B> be two trace structures. Then

$$PREF(T \underline{p} U) \subseteq PREF(T) \underline{p} PREF(U)$$

__Proof__ According to Property 2.10(i) $T \subseteq PREF(T)$. With Property 2.5(i) we then find

$$T \underline{p} U \subseteq PREF(T) \underline{p} U$$

Likewise, we derive

PREF(T) $\underline{p}$ U $\subseteq$ PREF(T) $\underline{p}$ PREF(U)

And, hence,

T $\underline{p}$ U $\subseteq$ PREF(T) $\underline{p}$ PREF(U)

Applying Property 2.10(ii) yields

PREF(T $\underline{p}$ U) $\subseteq$ PREF(PREF(T) $\underline{p}$ PREF(U))

The result then follows from Property 2.11   (End of proof)

Property 2.13 Let <T,A> and <U,B> be two trace structures with A ∩ B = ∅ . Then

PREF(T $\underline{p}$ U) = PREF(T) $\underline{p}$ PREF(U)

Proof We prove PREF(T) $\underline{p}$ PREF(U) $\subseteq$ PREF(T $\underline{p}$ U) . Equality then follows from Property 2.12. Let x ∈ PREF(T) $\underline{p}$ PREF(U) . Then x⌈A ∈ PREF(T) ∧ x⌈B ∈ PREF(U) . Let y ∈ A* and z ∈ B* be such that (xy)⌈A ∈ T ∧ (xz)⌈B ∈ U . Then (xyz)⌈A ∈ T ∧ (xyz)⌈B ∈ U . Hence, xyz ∈ T $\underline{p}$ U , and, consequently, x ∈ PREF(T $\underline{p}$ U) (End of proof)

In Section 1 we said that the trace structure of a component should not reflect the component's internal structure, i.e. it should not reflect how it is composed of the trace structures of the subcomponents. Relations between subcomponents will be expressed as equalities between symbols in the alphabets of the trace structures of the subcomponents. To hide these relations between subcomponents, we introduce a second composition operation, q-composition, which is p-composition followed by the elimination of common symbols. The *q-composite* of two trace structures <T,A> and <U,B> , denoted by <T,A> $\underline{q}$ <U,B> , is the trace structure

<(T $\underline{p}$ U)⌈(A ÷ B) , A ÷ B>

(÷ denotes symmetric set difference, i.e. A ÷ B = (A ∪ B) \ (A ∩ B); symmetric set difference is associative.) This composition operator differs from the one in [7] in that the latter operator replaces eliminated symbols by "silent moves".

Property 2.14 Q-composition is symmetric.

Property 2.15 Let <T,A> , <U,B> , and <V,C> be three trace structures with A ∩ B ∩ C = ∅ . Then

(T $\underline{q}$ U) $\underline{q}$ V = T $\underline{q}$ (U $\underline{q}$ V)

Proof In this proof E denotes A ÷ B ÷ C and F   (A ÷ B) ∪ C .

    (T $\underline{q}$ U) $\underline{q}$ V
=   {definition of q-composition}
    ((T $\underline{p}$ U)⌈(A ÷ B) $\underline{p}$ V)⌈E
=   {E ⊆ F and Property 2.1(i)}
    (((T $\underline{p}$ U)⌈(A ÷ B) $\underline{p}$ V)⌈F)⌈E
=   {since A ∩ B ∩ C = ∅ , F = (A ÷ B) ∪ (C \ (A ∪ B))}
    (((T $\underline{p}$ U)⌈F $\underline{p}$ V)⌈F)⌈E
=   {(A ∪ B) ∩ C ⊆ (A ÷ B) ∪ C and Property 2.7}
    ((T $\underline{p}$ U $\underline{p}$ V)⌈F)⌈E
= (T $\underline{p}$ U $\underline{p}$ V)⌈E   (End of proof)

Whenever employing q-composition we will see to it that each symbol occurs in at most two alphabets of the constituting trace structures. Under this restriction q-composition is associative.

A number of properties of q-composition can immediately be derived from those of p-composition:

<u>Property 2.16</u>   Let $<T,A>$ and $<U,B>$ be two trace structures.

(i) $A \subseteq B \Rightarrow T \underline{q} U = \{t \in U \mid t{\restriction}A \in T\}{\restriction}(B \setminus A)$

(ii) $A = B \Rightarrow T \underline{q} U = \begin{cases} \varepsilon & \text{if } T \cap U \neq \emptyset \\ \emptyset & \text{if } T \cap U = \emptyset \end{cases}$

(iii) Let $<T',A>$ be a trace structure. Then

$$T \subseteq T' \Rightarrow T \underline{q} U \subseteq T' \underline{q} U$$

$$U \underline{q} (T \cup T') = (U \underline{q} T) \cup (U \underline{q} T')$$

$$U \underline{q} (T \cap T') \subseteq (U \underline{q} T) \cap (T \underline{q} T')$$

(iv)   The q-composite of two regular trace structures is a regular trace structure.

(v)   If T and U are prefix-closed $T \underline{q} U$ is prefix-closed.

(vi) $PREF(T \underline{q} U) \subseteq PREF(T) \underline{q} PREF(U)$

(vii) $A \cap B = \emptyset \Rightarrow PREF(T \underline{q} U) = PREF(T) \underline{q} PREF(U)$

The third formula of (iii) differs from the corresponding one for p-composition in that we have inclusion rather than equality. The following example shows the necessity of this difference:

$$A = B = \{a,b\}$$
$$T = \{ab\}$$
$$T' = \{ba\}$$
$$U = \{ab,ba\}$$

Then $T \cap T' = \emptyset$. Hence, $U \underline{q} (T \cap T') = \emptyset$. But $U \underline{q} T = U \underline{q} T' = \varepsilon$.

For disjoint alphabets q-composition is, just like p-composition, the shuffle operation.

In the remainder of this section we discuss some properties of two-symbol traces. There is a special trace structure called SYNC. Let k be a natural number, $k \geq 0$, and let b and c be two distinct symbols. $SYNC_k(b,c)$ is the trace structure with $\{b,c\}$ as its alphabet and

$$\{t \in \{b,c\}^* \mid 0 \leq \#_b t' - \#_c t' \leq k \text{ for every prefix } t' \text{ of } t\}$$

as its trace set. ($\#_b t'$ denotes the number of occurrences of b in t' .) Often we mean by $SYNC_k(b,c)$ just its trace set. $SYNC_k(b,c)$ is prefix-closed. It has $k + 1$ states, viz. $[b^i]$, $0 \leq i \leq k$. ($b^0 = \varepsilon$; $b^{i+1} = b^i b$)

Without proof, we mention:

<u>Property 2.17</u> Let a , b , and c be three distinct symbols. Then for $k \geq 1$ and $m \geq 1$

$$SYNC_k(a,b) \underline{q} SYNC_m(b,c) = SYNC_{k+m}(a,c)$$

Property 2.18 Let b and c be two distinct symbols and let $t \in \{b,c\}^*$ start with symbol b . Then

(i) $\#_b t - \#_{bb} t - \#_{cb} t - \#_c t + \#_{cc} t + \#_{bc} t = 1$

(ii) t ends on symbol b if and only if $\#_{bc} t = \#_{cb} t$

(iii) t ends on symbol c if and only if $\#_{bc} t = \#_{cb} t + 1$

Proof We prove (i) only. If t consists of a single b the property holds. We show that the property is invariant under the extension of t with one symbol. We call the extended sequence t' . There are four cases to be considered: one for each possible choice of the last two symbols of t' . If t ends on b and $t' = tb$

$$\#_b t' - \#_{bb} t' = \#_b t + 1 - (\#_{bb} t + 1) = \#_b t - \#_{bb} t$$

In this case the other terms are equal for t and t' . Also in the other three cases there are exactly two terms that are one greater for t' than for t , and the difference of these terms occurs in the property.   (End of proof)

Property 2.19 Let $t \in SYNC_k(b,c)$, $t \neq \epsilon$, and $k \geq 1$. Then

$$1 \leq \#_{bc} t - \#_{cb} t + \#_b t - \#_c t \leq k$$

Proof Let $t \in SYNC_k(b,c)$, $t \neq \epsilon$. Then

$$0 \leq \#_b t - \#_c t \leq k \qquad\qquad (2.1)$$

Properties 2.18(ii),(iii) yield

$$0 \leq \#_{bc} t - \#_{cb} t \leq 1 \qquad\qquad (2.2)$$

Since any prefix of t belongs to $SYNC_k(b,c)$, $\#_b t - \#_c t = 0$ implies that t ends on symbol c . Hence, using Property 2.18(iii)

$$\#_b t - \#_c t = 0 \Rightarrow \#_{bc} t - \#_{cb} t = 1 \qquad\qquad (2.3)$$

Likewise, using Property 2.18(ii),

$$\#_b t - \#_c t = k \Rightarrow \#_{bc} t - \#_{cb} t = 0 \qquad\qquad (2.4)$$

The property follows from (2.1) through (2.4).   (End of proof)

3. HIERARCHICAL COMPONENTS

In this section we discuss the program notation for the representation of components. We introduce this notation in three steps. First we consider components without subcomponents. Such a component C is represented by a text

$$\underline{com}\ C(A):\ S\ \underline{moc} \qquad\qquad (3.1)$$

S is a command. With a command S a trace structure $<TR(S)$, $AL(S)>$ is associated. A command can have one of five possible forms:

(i)   A symbol is a command. Command b has trace structure $<\{b\}$, $\{b\}>$.

(ii)   If S and T are commands then $S \mid T$ is a command.

$$TR(S \mid T) = TR(S) \cup TR(T)$$
$$AL(S \mid T) = AL(S) \cup AL(T)$$

(iii) If S and T are commands then S , T is a command.

$$TR(S , T) = TR(S) \underline{p} TR(T)$$
$$AL(S , T) = AL(S) \cup AL(T)$$

(iv) If S and T are commands then S ; T is a command.

$$TR(S ; T) = \{st \mid s \in TR(S) \wedge t \in TR(T)\}$$
$$AL(S ; T) = AL(S) \cup AL(T)$$

(v) If S is a command then S^* is a command.

$$TR(S^*) = TR(S)^*$$
$$AL(S^*) = AL(S)$$

($TR(S)^*$ denotes all finite concatenations of zero or more traces in $TR(S)$.)

With every component a trace structure is associated. Component C in (3.1) has trace structure

$$<PREF(TR(S)) , AL(S)>$$

We impose the syntactic restriction that A equals AL(S) .

Notice that a component has a prefix-closed trace set. In [4] PREF is applied to the trace sets of commands. This requires the introduction of a termination symbol to cater for sequential composition (the semicolon). A termination symbol at the end of a trace indicates that this is a complete trace, i.e. one that is not brought about by PREF . We will not introduce sequential composition of components and we, therefore, do not need a termination symbol.

Property 3.1 A component without subcomponents has a regular trace structure.

Example 3.1

$$\underline{com} \ sem_1(v,p): \quad (v ; p)^* \quad \underline{moc}$$

This is an example of a component without subcomponents. The trace structure of sem_1 is $SYNC_1(v,p)$. (End of example)

We introduce the rule that in commands the comma has the highest priority, followed by the semicolon, and then the vertical bar. The bar and the comma are commutative. The bar and the semicolon are associative. The semicolon distributes over the bar. The comma does in general not distribute over the bar, as the following example demonstrates.

$$TR((a;b) , (a \mid b)) = \{ab\} \underline{p} \{a,b\} = \emptyset$$

$$TR((a;b) , a \mid (a;b) , b) = (\{ab\} \underline{p} \{a\}) \cup (\{ab\} \underline{p} \{b\})$$
$$= \{ab\} \underline{p} \{ab\} = \{ab\}$$

According to Property 2.5(ii) the comma distributes over the bar if the two arguments of the bar have trace structures with equal alphabets.

We now turn to components that consist of subcomponents and relations. Such a component C is represented by a text

<u>com</u> C(A):

$$\underline{sub}\ s_0:\ C_0,\ \ldots\ ,\ s_{n-1}:\ C_{n-1}$$
$$a_0 = b_0,\ \ldots\ ,\ a_{m-1} = b_{m-1} \tag{3.2}$$

<u>moc</u>

Component C has n subcomponents s_i $(0 \le i < n)$. The n names s_i must be distinct. We say that subcomponent s_i is of type C_i , where C_i is a component with a given trace structure, i.e. a component that is textually represented elsewhere. To differentiate between A and the alphabets of the C_i we introduce compound symbols. A compound symbol is a symbol s.b , in which s is the name of a subcomponent and b a symbol in the alphabet of that subcomponent's type. Symbols that are not compound are called simple. If A is an alphabet of simple symbols then s.A denotes the alphabet of compound symbols obtained by changing each symbol b ∈ A into s.b . The trace set s.T is likewise obtained from T. The alphabet of a component contains simple symbols only. Each subcomponent has a trace structure. Let component C_i in (3.2) have trace structure $<T_i , A_i>$. The trace structure of subcomponent s_i is then

$$<s_i.T_i ,\ s_i.A_i>$$

We impose three syntactic restrictions on the equalities in (3.2):
. The a_j's and b_j's are symbols from alphabet B with

$$B = A \cup \bigcup_{i=0}^{n-1} s_i.A_i \tag{3.3}$$

. Symbols a_j and b_j $(0 \le j < n)$ belong to two different (of the n+1) alphabets.
. Each symbol in B occurs exactly once in the equalities.

We define the q-composite (in which symbols occurring in the same equality are considered to be the same symbol) of the trace structures of the subcomponents to be the trace structure of component C . Due to the third restriction each symbol in B \ A is either equal to a symbol in A or common to exactly two alphabets of the trace structures of the subcomponents. Thus (Property 2.15), this q-composition is associative and yields A as the alphabet:

$$TR(C) = s_0.TR(C_0)\ \underline{q}\ \cdots\ \underline{q}\ s_{n-1}.TR(C_{n-1})$$
$$AL(C) = A$$

By Property 2.16(v) the trace set of C is prefix-closed.

Example 3.2   Let sem_1 be given as in Example 3.1.

<u>com</u> sem_2(v,p):

$\underline{sub}$ s0,s1: sem_1

v = s0.v, s0.p = s1.v, s1.p = p

<u>moc</u>

(The text "s0,s1: sem_1" is short for "s0: sem_1, s1: sem_1".)

The trace structures of the two subcomponents are $SYNC_1(s0.v, s0.p)$ and $SYNC_1(s1.v, s1.p)$. The trace structure of sem_2 is, due to the equalities, the q-composite of $SYNC_1(v, s0.p)$ and $SYNC_1(s0.p, p)$, which is, by virtue of Property 2.17, $SYNC_2(v,p)$. (End of example)

The most general form a component can have is a combination of the two forms discussed, viz. a component with subcomponents, relations, and a command. Such a component C is represented by a text

<u>com</u> C(A):

    <u>sub</u> s_0: C_0, ... s_{n-1}: C_{n-1}

    $a_0 = b_0$, ... , $a_{m-1} = b_{m-1}$

    S

<u>moc</u>

The syntactic restrictions differ in two respects from those given earlier: each symbol in B, B as in (3.3), occurs at most once in the equalities, and the alphabet of S is $B \setminus (\{a_0, ... , a_{m-1}\} \cup \{b_0, ... , b_{m-1}\})$.
Component C has as its trace structure $<TR(C), AL(C)>$ with

$$TR(C) = PREF(TR(S)) \text{ } \underline{q} \text{ } s_0.TR(C_0) \text{ } \underline{q} \cdots \underline{q} \text{ } s_{n-1}.TR(C_{n-1}) \quad (3.4)$$

$$AL(C) = A$$

<u>Example 3.3</u> Let sem_1 be given as in Example 3.1.

    <u>com</u> $sem_4(v,p)$:

        <u>sub</u> s0,s1: sem_1

        s0.p = s1.v

        $(v; s0.v)^*$, $(s1.p; p)^*$

    <u>moc</u>

Using the fact that for trace structures with disjoint alphabets p- and q-composition are the same, and Properties 2.16(vii) and 2.17, we get that sem_4 has $SYNC_4(v,p)$ as its trace structure. (End of example)

Strictly speaking, this third form a component can have is superfluous. We can, if we wish, change any component into one that has either no subcomponents (and, consequently, no equalities) or no command. Component sem_4 of Example 3.3 would then become

    <u>com</u> $sem_4(v,p)$:

        <u>sub</u> s0,s1,t0,t1: sem_1

        s0.p = s1.v,

        t0.v = v, t0.p = s0.v,

        t1.v = s1.p, t1.p = p

    <u>moc</u>

We say that a component C has a component D as a composing

part if C has a subcomponent that either is of type D or has D
as a composing part. A component is called *recursive* if it has itself as a composing part. The way in which we have defined components excludes them to be recursive. We could have allowed recursion. But in that case one has to be more careful with formula
(3.4), which is then a recursive equation. It may have several solutions, of which the least nonempty one is defined to be the trace
set of C . This solution is not necessarily a regular set [10].
In these notes we restrict ourselves to nonrecursive components,
and, hence, to regular trace structures.

Property 3.2 A nonrecursive component has a regular trace structure.

Example 3.4 Let sem_1 be given as in Example 3.1 and let i > 1.

> com sem_i(v,p):
>
>> sub s: sem_{i-1}
>>
>> $((v \mid s.p) \; ; \; (p \mid s.v))^*$
>
> moc

(This is not a recursive component: sem_i does not have sem_i as a
composing part.)

We prove by mathematical induction that sem_i (i ≥ 1) has
$SYNC_i(v,p)$ as its trace structure. For i = 1 this is clear.
Assume for i > 1 that the trace structure of sem_{i-1} is
$SYNC_{i-1}(v,p)$. Hence, subcomponent s has trace structure
$SYNC_{i-1}(s.v , s.p)$. Let S denote the command of sem_i . Then

$$TR(sem_i) = PREF(TR(S)) \underline{q} SYNC_{i-1}(s.v , s.p)$$

According to Property 2.16(i)

$$TR(sem_i) =$$

$$\{t \in PREF(TR(S)) \mid t\lceil\{s.v , s.p\} \in SYNC_{i-1}(s.v , s.p)\}\lceil\{v,p\} \quad (3.5)$$

(i) Proof of $TR(sem_i) \subseteq SYNC_i(v,p)$:
From the structure of S we observe that for every trace $t' \in TR(S)$

$$\#_v t' + \#_{s.p} t' = \#_p t' + \#_{s.v} t'$$

and for every trace $t \in PREF(TR(S))$

$$\#_{s.v} t - \#_{s.p} t \leq \#_v t - \#_p t \leq \#_{s.v} t - \#_{s.p} t + 1$$

For every t in (3.5) $t\lceil\{s.v , s.p\} \in SYNC_{i-1}(s.v , s.p)$. Hence,

$$0 \leq \#_v t - \#_p t \leq i$$

Consequently, for every t in (3.5) $t \in SYNC_i(v,p)$. Since (3.5)
is prefix-closed, this holds for any prefix of t as well.

(ii) Proof of $SYNC_i(v,p) \subseteq TR(sem_i)$:
We may conclude this if there exists a function

$$f: SYNC_i(v,p) \to PREF(TR(S))$$

such that

$$f(t)\lceil\{v,p\} = t \; \wedge \; f(t)\lceil\{s.v , s.p\} \in SYNC_{i-1}(s.v , s.p)$$

We define such an f as follows:

$$f(\varepsilon) = \varepsilon$$
$$f(v) = v$$
$$f(ubc) = \begin{cases} f(ub) \; s.v \; c & \text{if } b = c = v \\ f(ub) \; s.p \; c & \text{if } b = c = p \\ f(ub) \; c & \text{if } b \neq c \end{cases}$$

Clearly, $f(t) \restriction \{v,p\} = t$. We show that

$$f(t) \restriction \{s.v , s.p\} \in SYNC_{i-1}(s.v , s.p)$$

It suffices to show that

$$0 \leq \#_{s.v} f(t) - \#_{s.p} f(t) \leq i - 1 \tag{3.6}$$

Property 2.19 yields for $t \in SYNC_i(v,p)$

$$0 \leq \#_{vp} t - \#_{pv} t + \#_v t - \#_p t - 1 \leq i - 1$$

Hence, by Property 2.18(i),

$$0 \leq \#_{vv} t - \#_{pp} t \leq i - 1$$

and, by the definition of f , (3.6).   (End of example)

Example 3.5

        <u>com</u> var(a0,a1,b0,b1):

            (a0 ; b0* | a1 ; b1*)*

        <u>moc</u>

This component has three states: $[\varepsilon]$, $[a0]$, and $[a1]$. It may
be interpreted as a one-bit variable. The symbols a0 and a1
then stand for "assignment of value 0 and 1 " respectively, and
the symbols b0 and b1 for "inspection of value 0 and 1 " respectively. Using this interpretation, the three states stand for
"variable uninitialized", "variable has value 0 ", and "variable
has value 1 ". Notice that the component is constructed in such a
way that it must be initialized before it can be inspected.

    Since the trace set of var is regular, there exists a finite-
state machine [5] accepting this trace set. Figure 1 shows the
state diagram of this machine.

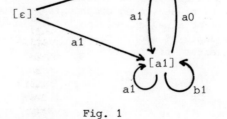

Fig. 1
Finite-state machine of component var

The initial state is $[\varepsilon]$. All states are final states. The sequences of labels encountered by walks from $[\varepsilon]$ through the graph are exactly the traces of component var . In [9] a circuit implementing this component is derived.   (End of example)

Example 3.6 Our final example is an i-bit buffer. For $i > 1$ it consists of a concatenation of two smaller buffers.

    <u>com</u> $queue_1$(a0,a1,b0,b1):

        (a0 ; b0 | a1 ; b1)*

    <u>moc</u>

For $i > 1$:

    <u>com</u> $queue_i$(a0,a1,b0,b1):

        <u>sub</u> q0: $queue_{i\ \underline{div}\ 2}$, q1: $queue_{i\ -\ i\ \underline{div}\ 2}$

        a0 = q0.a0, a1 = q0.a1,

        q0.b0 = q1.a0, q0.b1 = q1.a1,

        q1.b0 = b0, q1.b1 = b1

    <u>moc</u>

                   (End of example)

4. FINAL REMARKS

The trace theory presented can be extended with a composition operator expressing delay between the sending and the reception of signals. This allows concepts like "delay-insensitive" and "self-timed" to be formalized [9].

One of our motives to develop the material presented is our wish to translate program texts mechanically, by a "silicon compiler", into VLSI circuits. A method by which circuits may be derived from program texts can be found in [9].

The program notation to represent components needs further attention. A good extension would be the introduction of values and ports via which these values are communicated. In Examples 3.5 and 3.6 we have coded one-bit values. Comparing $queue_1$ with sem_1 we see that this coding has doubled the number of symbols and the length of the program text. The introduction of general values requires an abbreviation mechanism to prevent this textual growth.

We also need more theorems on trace structures. They must enhance our capability of designing components and arguing about them. The theorems should be aimed at leaving the realm of the individual traces. A nice example of such a theorem is Property 2.17. It expresses the net effect of the composite in terms of the net effects of the parts.

5. ACKNOWLEDGEMENTS

Acknowledgements are due to the members of the Eindhoven VLSI

Club. Much of the material presented took shape during the working sessions of this group.

6. REFERENCES

[1] Brock, J.D. & Ackerman, W.B. Scenarios: a model of non-determinate computation. In: Proc. Int. Coll. on Formalization of Programming Concepts. Lecture Notes in Computer Science 107, Springer-Verlag, 1981.

[2] Dijkstra, Edsger W. Cooperating sequential processes. In: Programming Languages (F. Genuys, ed.). Academic Press, 1968, pp. 43-112.

[3] Ginsburg, Seymour. The Mathematical Theory of Context-free Languages. McGraw-Hill, 1966.

[4] Hoare, C.A.R. Towards a theory of communicating sequential processes. Preliminary Report, Oxford University, 1978.

[5] Hopcroft, J.E. & Ullman, J.D. Formal Languages and their Relation to Automata. Addison-Wesley, 1969.

[6] Lauer, P.E. Synchronization of concurrent processes without globality assumptions. ACM Sigplan Notices 16,9, 1981, pp. 66-80.

[7] Milner, Robin. A Calculus of Communicating Systems. Lecture Notes in Computer Science 92, Springer-Verlag, 1980.

[8] Rem, Martin. The VLSI challenge: complexity bridling. In: VLSI 81 (J.P. Gray, ed.). Academic Press, 1981, pp. 65-73.

[9] Snepscheut, Jan L.A. van de. Deriving circuits from programs. These Proceedings.

[10] Udding, Jan Tijmen. On recursively defined sets of traces. Technical Note, Department of Mathematics and Computing Science, Eindhoven University of Technology, 1982.

Deriving Circuits from Programs

Jan L.A. van de Snepscheut

<u>Introduction</u>

[0] introduces a theory of traces and a notation for programs
consisting of a hierarchy of components. The trace structures of
components and programs are regular and, hence, one might construct
a finite state machine accepting a program's traces. The chip area
required to implement a finite state machine is, in the worst case,
proportional to the number of states of the trace structure, [1].
The number of states of the composite of some trace structures is,
in the worst case, the product of the numbers of states of the
composing trace structures. Hence, the chip area grows exponentially with the number of components in the program, which is not very
attractive. Another disadvantage of the traditional finite state
machines is that accepting traces is not very interesting: the
purpose of a machine is to receive inputs and to generate outputs.
In this paper the latter problem is solved by distinguishing, in the
trace structures, between input and output symbols. The former
problem is then solved by constructing a Mealy- or Moore-like finite
state machine per component and by introducing a communications
protocol between these machines. Hence, the circuit implementing
a program consists of two parts: one part for the finite state
machines and one part for their interconnections. In the worst case,
the chip areas required for these parts exhibit linear and quadratic
growth respectively with the number of   components in the program.
Before making a circuit, the trace structures of the program's
components are transformed such that their composition is a self-
timed composition. As we shall see, self-timedness admits a simple
implementation of finite state machines and imposes only few
restrictions on the layout.

<u>Warning</u> Not all properties mentioned in this paper are trivial;
yet, we do not present any proofs here.

<u>Author's address</u> Department of mathematics and computing science
Eindhoven University of Technology
The Netherlands

Composition with delay

[0] introduces a composition operation, called q-composition, that expresses a kind of synchronous communication between trace structures. This is sometimes referred to as "mutual inclusion" since, operationally seen, the trace structures involved perform the communication action simultaneously. Synchronous communication is attractive from a programming point of view. This paper discusses an implementation method that seems to be well suited to VLSI.

We introduce a composition operation, called r-composition, which is similar to q-composition but admits unbounded, finite, delay and overtaking. There is a direction in delay: a signal's sending precedes its reception. To express asymmetry, a trace structure's alphabet is partitioned into an input and an output alphabet. The result is called a directed trace structure. It is a triple, S say, $S = <\ tS,\ iS,\ oS\ >$, where iS is the input alphabet, oS is the output alphabet, iS and oS are disjoint, $tS \subseteq (iS \cup oS)^*$, and, hence, $<\ tS,\ iS \cup oS\ >$ is a trace structure. Symbols in iS are called input symbols or symbols of type input. Symbols in oS are called output symbols or symbols of type output.

Two directed trace structures S and T will be composed only when both $iS \cap iT$ and $oS \cap oT$ are empty, i.e. for any symbol they have in common, its type in S differs from its type in T . The q-composition of S and T is the q-composition of $<\ tS,\ iS \cup oS\ >$ and $<\ tT,\ iT \cup oT\ >$. The r-composite $S\ \underline{r}\ T$ of S and T is the directed trace structure $<\ tU,\ (iS \setminus oT) \cup (iT \setminus oS),\ (oS \setminus iT) \cup (oT \setminus iS)\ >$, where the trace set tU equals $\{u | \exists(s,\ t:\ s \in tS \wedge t \in tT :\ u(s\ \underline{r}\ t))\}$. The predicate $u(s\ \underline{r}\ t)$, pronounced as u is r-composable from s and t , is defined by

$$
\begin{aligned}
&u(s\ \underline{r}\ t) \\
&= (\ s = \varepsilon \qquad \wedge\ t = \varepsilon \qquad\quad \wedge\ u = \varepsilon) \\
&\vee\ (\ s = b\ s1 \qquad \wedge\ u = b\ u1 \qquad \wedge\ b \notin iT \cup oT \wedge u1(s1\ \underline{r}\ t)) \\
&\vee\ (\ t = b\ t1 \qquad \wedge\ u = b\ u1 \qquad \wedge\ b \notin iS \cup oS \wedge u1(s\ \underline{r}\ t1)) \\
&\vee\ (\ s = b\ s1 \qquad \wedge\ t = t0\ b\ t1 \wedge b \in oS \cap iT \wedge u\ (s1\ \underline{r}\ t0\ t1)) \\
&\vee\ (\ s = s0\ b\ s1 \wedge t = b\ t1 \qquad \wedge\ b \in iS \cap oT \wedge u\ (s0\ s1\ \underline{r}\ t1))\ .
\end{aligned}
$$

The difference between q- and r-composition is expressed in the last two lines. Looking operationally at the last line, for example, we see that T sends b to S where it is received after completing the sequence $s0$. If $s0 \neq \varepsilon$ there is a delay, and if $s0$ contains a symbol from $iS \cap oT$ there is overtaking. If, in the above definition, $s0$ and $t0$ are required to be empty, we have an alternative definition of q-composition.

Analogous to q-composition, r-composition of a number of directed trace structures is restricted such that any symbol occurs in at most two of them. As a result, r-composition is associative. (We say that a symbol occurs in a trace structure if it occurs in the trace structure's alphabet.) Notice that with r-composition we have left the realm of the regular trace structures and, hence, that of the finite state machines. This should not bother us now.

Using the two composition methods now available we define the
notion of delay-insensitive composition: the composition of a number
of trace structures is delay-insensitive if their q-composite and
r-composite are equal. As an example we consider

S = < {abd}, {a}, {b, d} > ,
T = < {ebf}, {b, e}, {f} > ,
U = < {ed}, {d}, {e} > ,
S $\underline{q}$ T = < {aedf, aefd, eadf, eafd}, {a, e}, {d, f} > ,
S $\underline{r}$ T = < {aedf, aefd, eadf, eafd, adef}, {a, e}, {d, f} > ,
S $\underline{q}$ T $\underline{q}$ U = < {af}, {a}, {f} > ,   and
S $\underline{r}$ T $\underline{r}$ U = < {af}, {a}, {f} > .

Observe that S $\underline{q}$ T $\neq$ S $\underline{r}$ T and S $\underline{q}$ T $\underline{q}$ U = S $\underline{r}$ T $\underline{r}$ U . Hence,
pairwise delay-insensitivity is not necessary to guarantee delay-
insensitivity of the composition of a number of trace structures.
It is, however, a sufficient condition. Observe also that
S $\underline{q}$ U = S $\underline{r}$ U , and U $\underline{q}$ T = U $\underline{r}$ T , yet S $\underline{q}$ T $\neq$ S $\underline{r}$ T . Hence,
delay-insensitivity is not transitive.

Interference

We have defined the trace structures described by programs
using q-composition, an operation resembling a telex-system where
sending and receiving occur simultaneously. If the composition is
delay-insensitive then one might, without affecting the result,
replace the telex by a mail-system with the possibility of unbounded,
finite, delay or overtaking. The communications in a VLSI circuit,
however, are more like a pneumatic post system with the additional
restriction that there is at most one item underway per tube. In
electrical systems this restriction is essential in order to prevent
a voltage transition from interfering with another one propagating
along the same wire, since interference might lead to the absorption
of one or both transitions or to the introduction of new ones.
Absorption of transitions equals an infinite delay and may cause a
grinding halt, whereas extra transitions may cause malfunctioning.
In the sequel we will refer to the bounding of the number of transitions on a wire as the exclusion of interference. As the
following example shows, delay-insensitivity does not exclude
interference.

S = <{ε, a, a b, a b b}, {a}, {b} > ,
T = <{ε, b, b b, b b c}, {b}, {c} > .

The composition of S and T is delay-insensitive, yet as many as
two b's can be sent by S before being received by T .
    A well-known technique of achieving non-interference is called
hand-shaking. With this technique a transmission is acknowledged
by a transmission in the opposite direction. These acknowledgements do not necessarily pair individual symbols; it is more common
to have the transmission of a symbol from a set of symbols acknowledged by the transmission of a symbol from another set. Such a set
of symbols is called a port. In the case of two communicating
components, both having one input and one output port, where the
input port of each component is connected with the other component's
output port, non-interference along those connections is guaranteed

by letting the transmissions via one connection and the other alternate. Observe the asymmetry between the two components caused by the fact that only one of them is allowed to do the first transmission. In the case of more ports various hand-shaking protocols may be envisaged, of which we will discuss two.

The input and output alphabets of a trace structure are partitioned into so-called (disjoint and non-empty) input and output ports respectively. The symbols in a port are called the port's values. Henceforth, it is assumed that the ports are the finest grain of interconnection, i.e. if two trace structures are composed they have for every port either all or none of the port's values in common. Consequently, ports are neither "split" nor "merged". The composition of a number of trace structures is defined to be free from interference if for every port P that is an output port of one of the composing trace structures, S say, and an input port of another one, T say,

$$\forall(s0, s1, t0, t1, u :$$
$$s0\ s1 \in tS \wedge t0\ t1 \in tT \wedge u(s0\ s1\ \underline{r}\ t0\ t1) :$$
$$0 \le \#_P s0 - \#_P t0 \le 1)$$

where $\#_P s$ denotes the number of occurrences of values of P in s.

The set of ports of a trace structure is partitioned into so-called (disjoint) signalling sets, where every signalling set is required to contain at least one input and one output port. If V is a signalling set and s a sequence of symbols, then s is called a weak signalling sequence w.r.t. V if s contains exactly one symbol of every port in V and no other symbols, and the the first symbol and the last symbol of s are of different type. If furthermore all symbols of equal type are adjacent then such a sequence is called a strong signalling sequence. A signalling sequence is called active if its first symbol is an output symbol, and passive otherwise. As an example, consider the signalling set V, $V = \{\{a\}, \{b\}, \{c\}, \{d\}\}$, with four ports, input symbols a and b, and output symbols c and d. The set of all strong and passive signalling sequences w.r.t. V is

{abcd, abdc, bacd, badc}

and the set of weak and passive signalling sequences w.r.t. V is

{abcd, abdc, bacd, badc, acbd, adbc, bcad, bdac} .

Finally we extend the definitions to trace structures: a trace structure T is called weak w.r.t. a signalling set V if every trace in $T \upharpoonright V$ is a prefix of a concatenation of weak signalling sequences w.r.t. V ; strong, active, and passive are defined similarly. Notice that in the case of a signalling set with two ports the strong and weak conditions coincide and correspond to the form of hand-shaking informally discussed above. The asymmetry between the hand-shaking components is expressed by their being active or passive. The notion of signalling sets was, in a restricted form, first introduced in [2].

We conclude this section with the definition of self-timed composition: a composition is called self-timed if it is both delay-insensitive and free from interference.

Some properties

Now we return to the components we want to implement. A component consists of a command, with trace structure T say, and n subcomponents, $n \geq 0$, with trace structures S_i say, $0 \leq i < n$. We define the composition of these $n + 1$ trace structures to be of class CL if the following conditions are satisfied:

- each S_i is weak and passive w.r.t. the signalling set consisting of all its input and output ports;
- the n trace structures S_i are such that their connections form no cyclic paths; (there is a connection from S_i to S_j if an output port of S_i equals an input port of S_j);
- $n = 0$ or T is strong and active w.r.t. the signalling set consisting of all ports that are not common to two of the n trace structures S_i.

For a discussion of class CL compositions see [3]; a proof of the following property can be found in [4].

Property 0

A composition of class CL is free from interference.
(End of Property 0)

Notice that a composition of class CL need not be delay-insensitive.

Making compositions delay-insensitive is easily achieved by the introduction of fresh symbols. Let S and T, $S = <tS, iS, oS>$, $T = <tT, iT, oT>$, be two trace structures, and let b and c be two fresh symbols. S' and T' are two trace structures,

$S' = < tS', iS \cup \{b\}, oS \cup \{c\} >$,
$T' = < tT', iT \cup \{c\}, oT \cup \{b\} >$,

where tS' and tT' are obtained from tS and tT respectively, by replacing in every trace the symbol d by the sequence bd for all d if $d \in oS \cap iT$, and by the sequence dc if $d \in iS \cap oT$. We then have

Property 1

$S \underline{q} T = S' \underline{q} T' = S' \underline{r} T'$.
(End of Property 1)

If all common symbols of S' and T' (including b and c) are grouped together in one input and one output port, then there is one common signalling set, V say, $V = \{(iS \cap oT) \cup \{b\}, (iT \cap oS) \cup \{c\}\}$.

Property 2

S' is strong and passive w.r.t. V.
T' is strong and active w.r.t. V.
The composition of S' and T' is free from interference.
(End of Property 2)

The above transformation of S and T into S' and T' is called the introduction of strong signalling; S' is called the passivated S, and T' the activated T.

We discuss one more transformation. Let S , S" , T , T" be four trace structures, where S" = < tS", iS, oS > , T" = < tT", iT, oT > , and tS" and tT" are obtained from tS and tT respectively by adding all traces obtainable by permuting in a trace consecutive symbols of equal type that are common to S and T . We then call S" and T" the pure versions of S and T . The pure versions of regular trace structures need not be regular. If, however, the length of all subsequences of common symbols of equal type is bounded, then regularity is preserved.

Property 3
    S r T = S" r T" .
(End of Property 3)

Property 4
    If the ports of S and T are the same as those of S" and T" then
        the composition of S and T is self-timed ≡
        the composition of S" and T" is self-timed .
(End of Property 4)

We conclude this section with an example. Consider the two trace structures S and T .
    S = < {abcdef}, {a, d, e}, {b, c, f} > ,
    T = < {bcghde}, {b, c, h}, {d, e, g} > .
Using fresh symbols i and j for the introduction of strong signalling yields S' and T' , the passivated S and activated T respectively:
    S' = < {aibicdjejf}, {a, d, e, i}, {b, c, f, j} > ,
    T' = < {ibicghdjej}, {b, c, h, j}, {d, e, g, i} > .
The pure versions of S and T are
    S" = < {abcdef, abcedf, acbdef, acbedf}, {a, d, e}, {b, c, f} >,
    T" = < {bcghde, bcghed, cbghde, cbghed}, {b, c, h}, {d, e, g} >.

Preparing for circuits
_____

A component is composed of a command and a number of subcomponents, each having its own trace structure. We assume that each
trace structure's alphabet has been partitioned into an input and
an output alphabet, these two alphabets into ports, and these ports
into signalling sets. Inspired by the previous section we propose
the following strategy to obtain a self-timed composition for a
component. If the composition is of class CL and it is delay-
insensitive, then, according to property 0, it is a self-timed
composition. Otherwise the trace structures are pairwise transformed by the introduction of strong signalling, which, according to
properties 1 and 2, does not affect the composite and leads to a
self-timed composition. Next we replace, in both cases, the trace
structures by their pure versions (with the same ports) which,
according to properties 3 and 4, affects neither the composite nor
the self-timedness. Finally, all prefixes of traces are added in
order to make the trace structures prefix-closed again.

The next step is the derivation of a number of communicating
finite state machines from a program. We do so by making a finite
state machine per component, i.e. a machine per command. A command
is a regular expression for which a finite state machine can be
derived. The transformations in the previous section were defined
on trace structures. Corresponding transformations can be defined
on finite state machines. For the introduction of strong signalling
this is trivial. In the case of making trace structures pure this
is not trivial but possible since the trace structures involved are
weak or strong w.r.t. the common signalling set, which bounds the
length of subsequences of common symbols of equal type.

The final step is the derivation of circuits from finite state
machines. Since each of the program's constituting expressions is
expected to be small, each of the corresponding circuits is expected
to be small. Therefore we allow ourselves the freedom of making
circuits under the assumption that every circuit is embedded in an
equipotential region [3]. We have to decide how the symbols from a
program are related to electrical signals. We may, for example,
represent a symbols's occurrence by a voltage level or by a voltage
transition. Voltage levels are, however, necessarily accompanied by
voltage transitions, since different levels are required for the
representation of a symbol's occurrence and non-occurrence. Furthermore, a level corresponds to a period of time and a transition
corresponds to a moment in time; the latter is closer to the notion
of a symbol's occurrence. Hence, we opt for voltage transitions.
This choice has one severe consequence. In our operational view of
r-composition, a symbol being sent is eventually received by a
different trace structure. When translating this to electrical
circuits we get: if a voltage transition is transmitted by a circuit
then, eventually, a different circuit will be ready to receive that
voltage transition. If we make no assumptions on the speed of
circuits and transmissions, a voltage transition may arrive at a
circuit before that circuit is ready to receive it. We deal with
this phenomenon by delaying voltage transitions at the receiving
circuit, where necessary, and by delaying subsequent voltage transitions at the sending circuit, where necessary. Notice that in the

case of synchronous systems both delays are regulated by the omni-
present clock. We will control these delays locally per circuit:
the delays at input terminals will be controlled by so-called privileges and the delays at output terminals by the signalling set
protocol.

Voltage transitions are alternately high-going and low-going
transitions. If all transitions are taken to represent a symbol's
occurrence we obtain what is often called 2-cycle signalling. In
order to economize on the amount of circuitry, however, 4-cycle
signalling is preferred. In that case only a high-going voltage
transition represents a symbol's occurrence. The low-going transitions are then irrelevant as far as the symbols are concerned. With
respect to interference of voltage transitions and to the delays
mentioned above, however, they are not irrelevant. Hence, a signalling sequence s leads to two voltage transition sequences s↑ and
s↓ , i.e. s in which each symbol is replaced by a high-going and
a low-going voltage transition respectively. In order to satisfy
the requirements of delay, s↑ must be completed before s↓ is
started and s↓ must be completed before t↑ is started, where t
is the next signalling sequence containing symbols from the same
signalling set as s . Of course, s↓ may overlap with any
sequence pertaining to a different signalling set. In order to
exclude interference it is not necessary that the order of the
transitions in s↓ and s↑ are equal; it suffices if they satisfy
the same requirements with respect to signalling, i.e. both s↓ and
s↑ are weak, strong, active, or passive.

With the above choices and consequences in mind we have to
construct a circuit per finite state machine. Such a circuit has an
input terminal per input symbol and an output terminal per output
symbol. The circuit receives voltage transitions at its input
terminals and generates them at its output terminals. The composition of components is realized by pairwise connecting, with a wire,
the output terminals and input terminals corresponding to the same
symbol. Due to the composition's self-timedness and the above
choices, there are no restrictions on the lengths of such wires.

Making circuits

In this section we restrict ourselves to trace structures with
only one signalling set and they are assumed to be passive w.r.t.
this set. We also assume that the trace set is not empty. According
to the previous section, the trace structure is pure and prefix-
closed.

The (unique) minimum state automaton corresponding to a regular
trace structure has at most one state from which no final state can
be reached. In the sequel we assume that state to have been removed.
Since the trace structures involved are prefix-closed, each of the
remaining states is a final state. These states correspond to the
equivalence classes of traces as defined in [0]. Among these states
are so-called signalling states, states that are equivalence
classes of those traces that are concatenations of signalling
sequences. All other states correspond to a proper prefix of such
a concatenation. A transition from one signalling state to another
corresponds to a signalling sequence and is called a signalling

state transition. We derive a circuit from the signalling states
and the signalling state transitions without considering the other
states.

We assume that the finite state machine is deterministic, i.e.
in any state the inputs (uniquely) determine both the next state
and the outputs. The former requirement can be met for any regular
trace structure by an appropriate machine construction. The latter
requirement, however, is independent of the machine construction and
imposes a restriction on the trace structures. In order to describe
this restriction we introduce the notion of a pattern and the
function f . An input pattern or an output pattern is a set of
symbols containing one symbol from every input port or output port
respectively, and no other symbols. For trace structure T ,
$T = < tT, iT, oT >$, the function $f([s], V, c)$ is, for any signalling state [s], set of input symbols V , and output symbol c ,
defined by

$f([s], V, c) = \exists(u: u \in tT \wedge u$ is a signalling sequence
                                 containing c :
                         $\forall(b: b \in iT \wedge b$ precedes c in u : $b \in V))$.
T is called deterministic if it satisfies, for every signalling
state [s], input pattern V , and for symbols b and c that are
values of the same output port,

$b = c \vee \neg f([s], V, b) \vee \neg f([s], V, c)$
and, for every signalling state [s], input pattern V , and output
pattern W ,

$\exists(b, c: b \in V \wedge c \in W: \neg f([s], V\backslash\{b\}, c) \vee \neg f([s], V, c))$.
The first requirement expresses that, given the inputs, the outputs
are determined; the second requirement expresses that, in order to
determine all outputs, all inputs are needed. The function f is
monotonic, i.e. if $V0 \subseteq V1$ then $f([s], V0, c) \Rightarrow f([s], V1, c)$.

The circuit of a command consists of a number of blocks, one
block per signalling state. Block [s] corresponds to signalling
state [s] and has a number of input and output terminals, viz.

one input terminal [s].b per input symbol b ,
one output terminal [s].c per output symbol c ,
one input terminal [s].N ,
one input terminal [s].P , and
one output terminal [s].Ω([t]) per signalling state [t] for
which a signalling state transition from [s] to [t] exists.

The blocks are independent of the trace structure's being weak or
strong. The way in which the command's circuit is constructed from
the blocks, however, differs for weak and strong trace structures.

Blocks can have a privilege. At any moment, one block has the
privilege, indicating that the circuit is in the corresponding
signalling state, or that it is making a signalling state transition
from that state. Presence or absence of the privilege is, per block
[s], recorded in a flip-flop [s].p . If a block, [s] say, has the
privilege, it inspects the inputs for high-going transitions and, if
appropriate, generates high-going output transitions, then waits for
low-going inputs, generates low-going outputs, and finally if all
inputs are low, which is indicated by a high voltage at input terminal [s].N , it inverts its privilege flip-flop and generates a
high-going voltage transition at terminal [s].Ω([t]) for that [t]
to which a signalling state transition has been completed. This

state [t] depends on the high-going inputs (and on [s], of course)
and, hence, is to be computed as those inputs are received, and to
be stored until the state transition is completed. One flip-flop
[s].q([t]) per [s].Q([t]) terminal is employed for this purpose.
Also the voltages to be generated for output symbols depend on the
inputs. They are to be kept high by the block only as long as the
inputs are high and, therefore, no storage elements are needed.
Finally, a block [s] not having the privilege generates low output
voltages only and waits until it receives the privilege via its
input terminal [s].P .

In order to describe the construction of electrical circuits,
we write boolean expressions and assignment statements. High and low
voltages correspond to the boolean values true and false respectively.
An assignment v:= e denotes that v is assigned such a voltage that
v = e holds. Existential and universal quantification are computed
with a number of OR and AND elements respectively. A flip-flop
ff has two input terminals, ff.R and ff.S , and two output
terminals, ff.Q and ff.$\overline{Q}$. If ff.R is assigned a high level
then ff.Q and ff.$\overline{Q}$ become low and high respectively. If ff.S
is assigned a high level then ff.Q becomes high and ff.$\overline{Q}$ low.
If both ff.R and ff.S are low, ff.Q and f.$\overline{Q}$ remain unchanged.

A block, [s] say, is constructed as follows. For each output
symbol c , c $\in$ oT , the level of its corresponding output terminal
[s].c is computed as follows

[s].c:= [s].p.Q $\wedge$ f([s], V, c)

where V is the set of input symbols whose corresponding input
terminals have a high voltage level, i.e.

[s].c:= $\exists$(u: su $\in$ tT $\wedge$ u is a signalling sequence
            containing c :
            [s].p.Q $\wedge\forall$(b: b $\in$ iT $\wedge$ b precedes c in u : [s].b)) .

The state transition flip-flops are driven similarly:

[s].q([t]).S:= $\exists$(u: su $\in$ tT $\wedge$ u is a signalling sequence $\wedge$
            [su] = [t] :
            [s].p.Q $\wedge\forall$(b: b $\in$ iT $\wedge$ b occurs in u : [s].b)).

Their other input is controlled by

[s].q([t]).R:= [s].p.$\overline{Q}$ $\wedge$ [s].q([t]).Q .

The privilege flip-flop [s].p is controlled by

[s].p.S:= [s].P

and

[s].p.R:= [s].N $\wedge$ $\exists$(t :: [s].q([t]).Q) .

The remaining output terminals are controlled by

[s].Q([t]):= [s].p.$\overline{Q}$ $\wedge$ [s].q([t]).Q .

We say that a block displays a voltage transition sequence u↑ if
the high-going voltage transitions at its terminals corresponding to
input and output symbols, when ordered in time, form the sequence
u↑ . If block [s] displays a voltage transition sequence u↑, then
su is a trace of T , i.e. su $\in$ tT . If u is a signalling
sequence then one q flip-flop will be set, viz. [s].q([su]) , and
after completing u↑ a sequence v↓ will be displayed. The
sequence v is a permutation of u and is not necessarily weak or
strong. Due to the monotonicity of f , u precedes v and the
first symbol of v is of type input.

The blocks thus obtained are combined to yield the entire circuit
implementing a component's command. The circuit has terminal b
per symbol b and no other terminals. The input terminals of the
circuit are connected to the corresponding input terminals of each
block:

$\quad$ [s].b:= b .

Controlling the circuit's outputs is more complex. As we have said
before, the low-going voltage transition sequence must be as weak
or strong as the high-going one. The low-going sequence displayed
by a block starts with an input. Hence, if there is only one input
port the low-going sequence satisfies the signalling condition. If
there is more than one input port, outputs must be delayed to become
low until all inputs are low in order to satisfy the weak or strong
conditions. To realize this delay, Muller C elements are used.
There are, then, two ways of computing the outputs: the immediate
output

$\quad\quad$ c:= $\exists$(s :: [s].c)

and the delayed output

$\quad\quad$ c:= $\exists$(b: b $\in$ iT : b) $\underline{C}$ $\exists$(s :: [s].c) ,

where x $\underline{C}$ y denotes the output of a Muller C element with inputs
x and y . There are three cases:

$\quad$ if there is only one input port, all outputs are immediate;
$\quad$ if there is more than one input port and the trace structure is
$\quad$ weak, arbitrarily one of the output ports is selected to have
$\quad$ delayed outputs for all symbols in that port, and all other
$\quad$ outputs are immediate;
$\quad$ if there is more than one input port and the trace structure is
$\quad$ strong, all outputs are delayed.

For every block [s] , [s].N indicates that all inputs are low,

$\quad$ [s].N:= ¬$\exists$(b: b $\in$ iT : b) .

The privilege passing is, for each state [t], controlled by

$\quad$ [t].P:= $\exists$(s :: [s].ϱ([t])) .

The initialization of the circuit must be such that all input and
output terminals have a low voltage level and that all flip-flops
have a low Q and a high $\overline{Q}$ output, with exception of the privilege
flip-flop of the initial state, [ε].p , whose $\mathbf{Q}$ output must be high
and $\overline{Q}$ low. This completes the circuit's construction.

$\quad$ With respect to the layout we recall the assumption that the
entire circuit implementing a command is embedded in an equipotential
or isochronic region. This assumption is essential in order to
express some requirements on relative speeds. The value assigned to
[s].p.R is computed from [s].N and $\exists$(t :: [s].q([t]).$\overline{Q}$) . It is
required that, per signalling sequence, the former becomes low
before the latter becomes high. This can be assured by making the
q flip-flops relatively slow. Both [s].q([t]).R and [s].ϱ([t])
and, hence, [t].p.S , are derived from [s].p.$\overline{Q}$ ∧ [s].q([t]).ϱ .
Since assigning a high level to [s].q([t]).R lowers [s].q([t]).$\mathbf{Q}$,
proper operation is again guaranteed by making the q flip-flops
relatively slow. A voltage transition arriving at an input terminal
of the circuit is sent to the corresponding input terminals of its
constituting blocks. The time differences between the moments at
which the voltage transitions arrive at those blocks must be less
than the time it takes a block to generate high outputs, receive low

inputs, generate low outputs, and then pass the privilege to the successor block. Furthermore, the high-going voltage transitions of the next signalling sequence are required to arrive after the currently privileged block has given up the privilege, but not necessarily after the successor block has received the privilege. Notice that, in the latter case, the conjunct [s].p.Ω in the computation of [s].c and [s].q([t]).S , represents the delay of input transitions until the circuit, more specifically block [s] , is ready for their reception. If the circuit is embedded in an isochronic region and the above timing restrictions are satisfied, the voltages at inputs of OR and AND elements computing the circuit's outputs, do not change in opposite directions. As a consequence, the circuit does not momentarily generate erroneous voltage transitions: the circuit is free from static and dynamic hazards.

We mention two optimizations that will be applied to the examples we will present. Privilege passing from a block to itself, i.e. along [s].Ω([s]), can be omitted if in that case also   inverting the privilege flip-flop is omitted. If a circuit consists of one block only, that block will always have the privilege and, hence, this need not be recorded (in a flip-flop). The circuits often contain OR and AND elements with either only one input or two inputs of which one is   constant. They can be removed but we have not done so in the examples. The circuits often compute boolean values described by an expression of the form E0 $\vee$ (E0 $\wedge$ E1). This expression may be reduced to E0 but, again, we have not done so in the examples.

Examples

As our first example we consider a full adder having three input ports, two output ports, and one signalling set w.r.t. which it is weak and passive. It is given that the full adder is used in a composition of class CL.

```
com full adder (in {a0, a1}, {b0, b1}, {c0, c1} ,
                out {d0, d1}, {e0, e1}) :
( a0, b0; d0, (c0; e0| c1; e1)
| b0, c0; d0, (a0; e0| a1; e1)
| c0, a0; d0, (b0; e0| b1; e1)
| a1, b1; d1, (c0; e0| c1; e1)
| b1, c1; d1, (a0; e0| a1; e1)
| c1, a1; d1, (b0; e0| b1; e1)
)*
moc
```

The adder is symmetric in its three input ports. The port {d0, d1} represents the carry-out and {e0, e1} the sum bit. The full adder circuit is shown in Figure 0. It consists of a single block only. The port {e0, e1} is chosen to have Muller C elements. With respect to correct operation this choice is arbitrary. With respect to speed, however, this choice is the preferred one: it allows for maximum concurrency in the undoing of carry-propagation in the low-going voltage transition sequence if full adders are cascaded.

We use the following notations when drawing circuits:

These pictures represent an inverter, satisfying $v := \neg u$, and an AND element, satisfying $u = v$ and $x := u \wedge v \wedge w$. An OR or C element is drawn similarly.

The other example we consider is the one-bit variable, which is also discussed in [0].

    <u>com</u> var (<u>in</u> {a0, a1}, <u>out</u> {b0, b1}) :
      (a0; $\overline{(b0)}$* |a1; $\overline{(b1)}$*)*
    <u>moc</u>

which is neither weak nor strong w.r.t. its signalling set. Hence, the variable is to be transformed by the introduction of strong signalling. Passivating it, using fresh symbols a and b , yields

    <u>com</u> var (<u>in</u> {a, a0, a1}, <u>out</u> {b, b0, b1}) :
      (a0; b; (a; b0)* |a1; b; (a; b1)*)*
    <u>moc</u> .

The trace structure has three signalling states and, hence, the circuit has three blocks. By optimization, two privilege passing flip-flops are omitted, thus leading to the circuit in Figure 1. The blocks [ε] , [a0 b] , and [a1 b] are drawn, in that order, from top to bottom. At the bottom the computation of N is drawn, and to the right of each block [s] the computation of [s].P   is drawn. Within each block we find, from left to right, the inputs, the outputs, the q flip-flops, N , and the p flip-flop.

Final remarks

    We have discussed the derivation of a circuit from a program. The resulting circuit consists of a number of smaller circuits, one per component, interconnected with wires of unrestricted length. These smaller circuits, in turn, consist of blocks embedded in isochronic regions. No global or local clock is used, not even a pausable clock.

    We have not discussed the construction of a circuit for a trace structure with more than one signalling set. Though equally possible it is more complex. Proceeding along the lines sketched above, we obtain circuits in which more than one block can be privileged, viz. any number from one through the number of signalling sets. Arbiters may be required in some of those blocks. Due to the self-timedness and the absence of clocks, the circuits do not suffer from the glitch phenomenon [3, 5]. The use of privileges for delaying inputs becomes more pronounced in this case.

One may wonder why we have chosen all encodings as "wasteful" as possible, viz. a wire per symbol, a p flip-flop per signalling state, and a q flip-flop per signalling state transition. This has been done in order to complicate the description not unnecessarily, to ease the avoidance of hazards, and to allow for generalization to trace structures with more than one signalling set.

Programs as introduced in [0] do not indicate the partitioning of their alphabets in inputs, outputs, ports, and signalling sets. The partitioning, however, influences the ease or even the possibility of checking whether a composition is of class CL and delay-insensitive, it influences the determinism of the trace structure, and it influences the size and speed of the resulting circuit. We doubt that a reasonable partitioning is derivable from a program not containing any of the above indications.

Acknowledgements

The Eindhoven VLSI club, especially Jan Tijmen Udding, is acknowledged for many contributions to this paper.

References

[0] M. Rem, J.L.A. van de Snepscheut, J.T. Udding (1983),
     Trace theory and the definition of hierarchical components,
     in this volume.

[1] R.W. Floyd, J.D. Ullman (1982),
     The compilation of regular expressions into integrated circuits,
     Journal of the ACM, vol. 29, no. 3, pp. 603-622.

[2] J.T. Udding (1982),
     On self-timed and delay-insensitive compositions,
     Internal memorandum JTU 1.

[3] C.L. Seitz (1980),
     System timing,
     in   C. Mead, L. Conway,
          Introduction to VLSI systems,
          Addison-Wesley, Reading, Mass., pp.218-262.

[4] J. Ebergen, J.L.A. van de Snepscheut (1981),
     A theorem on self-timed systems,
     Internal memorandum JAN 88a / JE 81/1a.

[5] Science and the citizen (1973),
     Scientific American, vol. 228, no. 4, pp. 43-44.

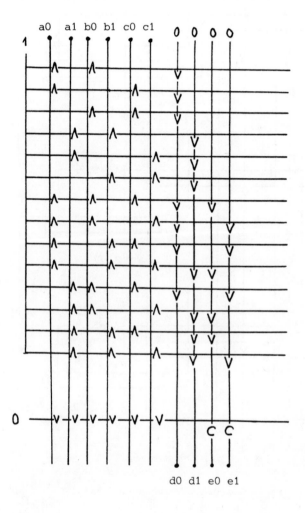

Figure 0   Circuit of component full adder .

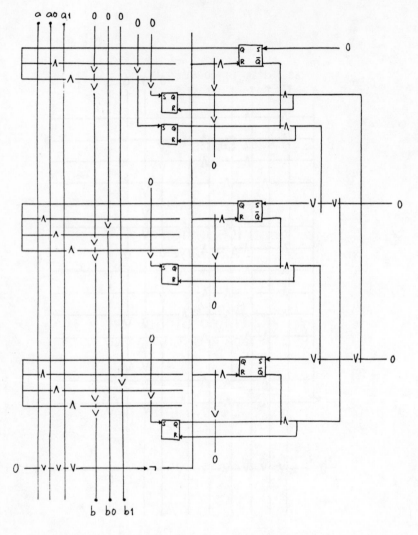

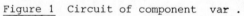

Figure 1   Circuit of component var .

System Building Blocks

Self-Timed IC Design with PPL's
A.B. Hayes

A Self-Timed Static RAM
E.H. Frank and R.F. Sproull

Design of the PSC: A Programmable Systolic Chip
A.L. Fisher, H.T. Kung, L.M. Monier, H. Walker and Y. Dohi

Self-Timed IC Design with PPL's

Alan B. Hayes

1. Introduction

PPL's [8] are a cellular IC design tool derived from PLA [9, 2] and SLA [6, 7] ideas. They allow IC designs to be created by placing logical symbols on a grid. Modules thus created may be interconnected by placing additional cells on the grid. A design may be checked with a simulator [5] which reads the file created in the process of laying out the design. When the designer is satisfied with a design he may finish the process of creating a die by placing cells from the PPL pad cell set on a larger grid around his design, and connecting power, ground and the input and output signals from the pads to the power, ground and IO points on his design with metal and/or poly wires. The wires also lay out on a grid that assures proper spacing between them. The cell symbols are then automatically replaced by the poly, metal and diffusion layout that realizes them and the resulting composite is fractured and dumped to tape in the format for the intended pattern generator.

The PPL design system is now being used to transfer portions of DDM2 [1], a completely self-timed data driven machine from SSI CMOS into LSI NMOS. In the process a technique for realizing self-timed stored state sequential circuits [3] that had been developed earlier and applied in the design of DDM2 has been transferred to PPL designs. This paper describes the technique as applied in DDM2 and as it has been transformed to work in PPL designs. The paper concludes with a brief description of some small self-timed PPL parts that have also been developed as part of the effort to integrate DDM2.

In what follows, the terminology and conventions used for self-timed systems are compatible with those used in [4], to which the reader is referred for background and additional detail.

2. Signaling Protocol and Construction Rules

The self timed state machines being described here are elements of larger self timed systems. They are linked to other elements of the system by REQUEST and ACKNOWLEDGE lines and associated data busses. Figure 1 shows a link and the single rail Four Cycle signalling protocol used in the circuits being described here.

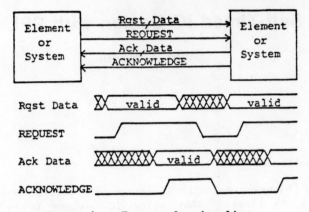

Figure 1:   Four cycle signaling.

As shown in Figure 1, a link, which is defined as the REQUEST and ACKNOWLEDGE lines plus the associated data busses, starts in the idle state with both REQUEST and ACKNOWLEDGE low.   From the idle state, the REQUEST goes high, indicating that Rqst Data is valid. This is followed by ACKNOWLEDGE going high which indicates both that the Ack Data is valid and that the Rqst Data has been accepted. REQUEST goes low to indicate that the Ack Data has been accepted and that the Requester is ready for the next cycle.   When the Acknowledger is ready for the next cycle, ACKNOWLEDGE will go low returning the link to the idle state.   Note that only the requester may initiate a cycle, that is, in the idle state only REQUEST may go high.

The state machines are constructed according to the following set of rules:

- In the initial state, all REQUESTS and ACKNOWLEDGES in the input and output set of the machine are low.

- The input set must contain at least one REQUEST.

This rule guarantees that the machine can start.

- If the input set contains more then one REQUEST, then external arbitration must be used to preclude concurrent REQUESTS.

- The output set must contain the associated ACKNOWLEDGE to every REQUEST in the input set.

- The input set may contain any number of ACKNOWLEDGES.

- The output set must contain the associated REQUEST to every ACKNOWLEDGE in the input set.

The machine is only capable of single-input-change operation (with respect to the REQUESTS and ACKNOWLEDGES). These rules and the signalling protocol are intended to insure that the machine will not receive multiple-input-changes.

Since the machine has no control over the occurrence of the initial low-to-high transition on REQUESTS in its input set, arbitration external to the state machine must be used to guarantee that only one of these links will be active at a time. The signalling protocol requires that a REQUEST that has gone high remain high until the associated ACKNOWLEDGE goes high. This and the rule that the associated ACKNOWLEDGE must be in the output set of the machine, allows the machine to preclude another transition on the REQUEST line until it is ready for that transition. Once the REQUEST goes low, the machine is similarly able to block the next transition on that or any other REQUEST line (because of the external arbiter) in the input set by holding the associated ACKNOWLEDGE high.

The rule that the associated REQUEST to every ACKNOWLEDGE in the input set of the machine must be in its output set, gives the machine the ability to control the number of REQUEST and ACKNOWLEDGE lines in its input set on which transitions may occur at any given time.

- On each state change, a transition will occur on one and only one of the REQUEST and ACKNOWLEDGE lines in the output set.

- There will be a one-to-one correspondence between transitions on the REQUEST and ACKNOWLEDGE lines in the input set and state changes.

The limitation of output set REQUEST and ACKNOWLEDGE transitions to exactly one per state change and the requirement that there be a state change for every input REQUEST and ACKNOWLEDGE transition, assures that the number of input set REQUEST and ACKNOWLEDGE lines on which a transition is possible at any time is also one, and that the machine, once started, will continue to run, at least until returning to the initial state or some other state in which all of its links are idle.

With the rules and signaling protocol outlined above, a self-timed state machine may be realized as shown in Figure 2.

The multiplexer selects, as a function of the current state of

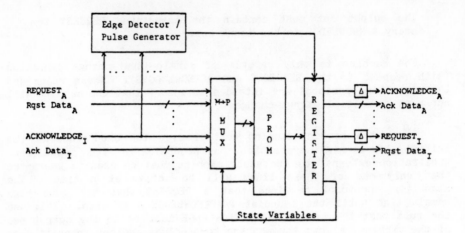

Figure 2:   Self timed state machine with state-selected input set.

the machine, that subset of the inputs to the machine which is necessary, along with the current state, to determine the next state of the machine. The multiplexer also assures that signals in the input set that are not guaranteed to be valid by the signaling protocol, and that might be changing value or be undefined, will not reach the PROM. The PROM holds the state table of the machine and generates the next state and outputs which are inputs to the register. State changes occur when the edge detector / pulse generator produces a pulse which it does in response to each transition on one of the input REQUEST or ACKNOWLEDGE lines. The pulse must be of sufficient width to satisfy the requirements of the register, and its trailing (trigger) edge must be delayed sufficiently to assure that the input signals have sufficient time to propagate through the multiplexer and PROM and to meet the setup time requirements of the register. Once the appropriate pulse width has been determined, the delay in the edge detector / pulse generator (Figure 3) is set to achieve it.

The delays on the output REQUEST and ACKNOWLEDGE lines are necessary in order to assure that the requirement that the associated data be valid before the rising edge of the REQUEST or ACKNOWLEDGE is met. In practice it is easier to design elements of the system to tolerate the skew that may arise from the variations in response time of the different register outputs. This can be done in the case of the state machine by increasing the pulse width by the maximum possible skew.

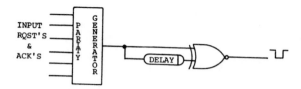

Figure 3:   Edge Detector / Pulse Generator realization.

3. Transferring the Model to PPL's

Calculating the appropriate pulse width and building a delay to realize it is quite easy in discrete systems. Different size registers do not require different delays as the necessary number of register chips are simply clocked in parallel. Registers are realized in the PPL cell set by placing a number of flip flops side by side as shown in figure 4. As the size of the register increases, the length of the clock lines increases, and the time necessary to drive them high and low increases. An additional complication arises from the necessity to provide the PPL flip flops with a two phase clock. To deal with these problems, a PPL circuit has been developed that generates a non overlapping two phase clock with phi 1 of sufficient duration to assure that the data has been loaded, which will adjust automatically to changes in the size of the register.

Figure 5 shows a gate representation of the clocking circuit for a 3 bit register. As in the discrete version of the edge detector / pulse generator, the REQUEST and ACKNOWLEDGE signals are inputs to a parity generator. The parity generator output is XORed with the output of a flip flop whose input it is. Phi 2 of the flip flop is tied high which converts it into a gated latch with phi 1 being the gate signal. The output of the XOR drives a two phase clock generator in which the cross connection is made through the phi 1 and phi 2 lines of the register flip flops. The phi 1 input to the flip flop that drives the XOR gate comes from the end of the phi 1 line through the register.

When an input to the parity generator changes, its output will change. The gated latch will hold the previous value of the parity generator output so the XOR gate output will become 1. This will force the output of the bottom NOR gate to 0. The clock lines of the flip flops are polysilicon and the loads on these lines are the gates of the pass transistors which connect the flip flops' internal latches to the input or output lines of the flip flop. This can be modeled as shown in figure 6 where the poly has been replaced by resistors and the transistor gates have been replaced by capacitors to ground. From this it can be seen that when the output of the

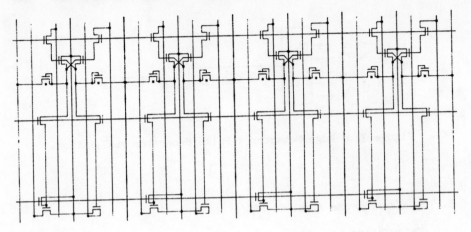

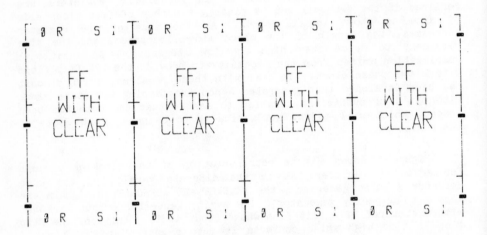

Figure 4:   Symbolic and schematic representation of a PPL
register realization.

bottom NOR gate goes to zero, that the signal will propagate from
right to left on the phi 2 line and then to the input of the top NOR
gate. With both its inputs 0, the top NOR gate's output will go to
1.   This will propagate from right to left on the phi 1 line,
loading the register with the data on its inputs.   When the signal
reaches the flip flop driving the XOR gate, it will gate the value
from the parity generator output into the latch which will cause the
XOR gate output to go to 0.   This will cause the output of the top
NOR gate to go to 0 which will propagate through the phi 1 line to
the input of the bottom NOR gate.   The output of the bottom NOR gate
will now go to 1 which will propagate through the phi 2 line,

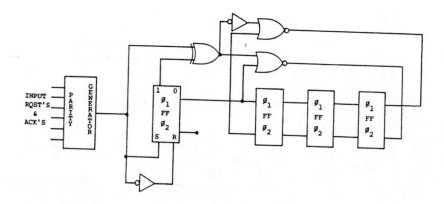

Figure 5:   Clocking circuit for a 3 bit register.

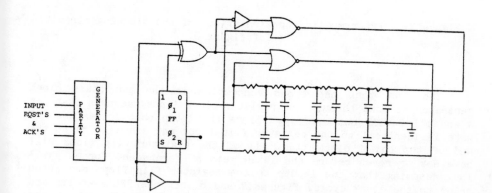

Figure 6:   Clock circuit with RC model for phi 1 and phi 2 clock
lines.

enabling the new state of the register to its outputs and completing
the clock cycle.

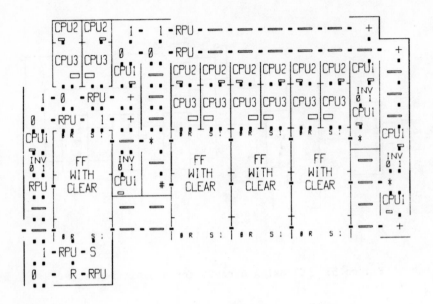

Figure 7:   PPL symbolic representations of the clocking circuit
of figure 5.

The fact that the cross connection in the two phase clock
generator is done through the register clock lines, assures that phi
1 and phi 2 will not overlap. Since the latch that drives the XOR
gate is identical to the register flip flops and since it is at the
end of the phi 1 clock line, all of the register flip flops will
have set or reset before the latch sets or resets and forces phi 1
low, assuming that the inputs to the register flip flops are setup
prior to the clock cycle. Figures 7 and 8 show the PPL symbolic and
schematic representation of the circuit of figure 5, minus the
parity generator. The circuit can be modified for any register size
by adding the appropriate number of flip flops.

This solves a portion of the timing problem for the PPL version
of the circuit of 2 in that it assures the generation of the
appropriate duration non-overlapping 2 phase clock for the state and
output register regardless of the size register that is required for
a particular state machine. The problem that remains is to assure
that the inputs to the register are set up at the time that the
clock pulse occurs.

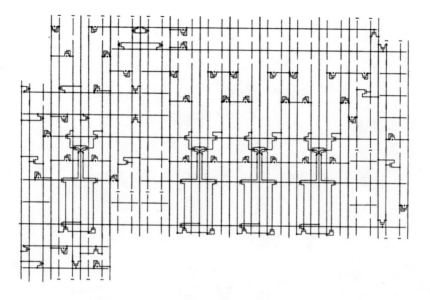

Figure 8: PPL schematic representations of the clocking circuit of figure 5.

Figure 9 shows the state diagram of 6 state machine. The signals with arrows in front of them are the REQUEST or ACKNOWLEDGE signals into the machine whose transition (in the direction implied by the arrow) results in the indicated state transition, or are the REQUEST or ACKNOWLEDGE signals out of the machine whose value is changed as a result of the state transition. The PPL realization of the state machine is shown in figure 10. From left to right, the first 3 bits of the register are the state variables (A=0, B=1, C=2, etc.), followed by CS, Cycle Rqst, D3 Rqst, D2 Rqst, D1 Rqst, Path, and ECD Ack. There is one row for each of the state transitions in the machine. The first row under the register, for example, corresponds to the transition from state A to state A on ECD Rqst going high, setting ECD Ack. The first two rows define the state transitions that occur in response to transitions of ECD Rqst. The next nine rows are the state transitions triggered by Cycle Ack, and the last 6 are the 2 state transitions each triggered by D2 Ack, D3 Ack and D1 Ack.

Note that the signal (REQUEST or ACKNOWLEDGE) that triggers the state transition is at the extreme right of the row that defines the

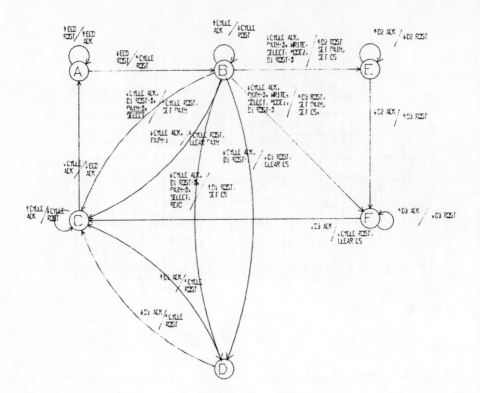

Figure 9:   State diagram of a 6 state machine.

transition, and that the pullup (RPU) for the row is as close as possible to the right end. All of the other inputs to the row will be valid prior to the transition of the trigger signal. This is guaranteed by the state machine construction rules and the 4 cycle signaling protocol rules. The rows are polysilicon and the principle capacitive loads on them are the gates of the transistors that are the row outputs (S, R, R/2, and S/2). The rows then are analogous to the clock lines in the register, and when the transition occurs on the trigger signal releasing the final pulldown on the row, the signal will propagate from right to left.

The parity generator which drives the clock generator is located in the bottom left corner of the state machine. Rather then connecting the input REQUEST and ACKNOWLEDGE signals directly to it as would be done in a discrete realization of the state machine, the signals are connected through the rows that define the state transitions that they trigger. The S/2 or R/2 in the column to left of the register sets or resets a latch and drives an input to the parity generator. As the S/2 or R/2 is to the left of all of the other outputs of the row, all of the other row outputs will have

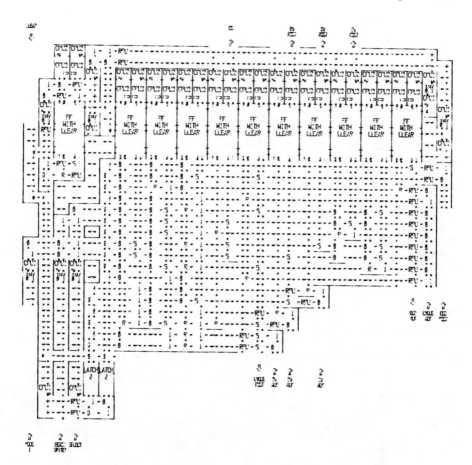

Figure 10: PPL realization of the machine of Figure 9.

been activated when the S/2 or R/2 output activates. The latch on the parity generator input line is necessary to hold the correct state of the line during the time when none of the rows driving the input are active. Normally one latch and parity generator input would be required for each REQUEST or ACKNOWLEDGE line into the state machine. In this case it is possible to OR ECD Rqst and Cycle Ack into a single input as cycles on these signals are non-overlapping, that is, ECD Ack goes from 0 to 1 to 0 while Cycle Rqst remains 0 and Cycle Rqst then goes from 0 to 1 to 0 while ECD Rqst remains 0. Similarly, D1 Ack, D2 Ack and D3 Ack are non-overlapping and may share a single latch and parity generator input.

The column wires connected to the flip flop inputs are pulled up, when phi 1 is low, by a 6 micron wide by 48 micron long depletion transistor (CPU3), while the column wires into the latch and parity generator are pulled up by the 6 by 12 micron depletion transistors internal to the latch. The device pulling down the column wires, when they are active, (S, R, S/2, R/2) is a 12 micron wide by 6 micron long enhancement transistor. The column wires are metal and are sufficiently short to be regarded as equipotential regions [4] loaded with a lumped capacitance equal to the sum of the capacitances from the gates of transistors being driven by the wire, the capacitance of transistor sources or drains connected to the column, and the metal to substrate capacitance of the wire itself. Of these, by far the largest contribution is from transistor gates. The flip flop input columns, when phi 1 is low, are driving no gates whereas the columns connected to the latch and parity generator are driving the gates of the transistor internal to the latch plus the transistor of the 1 or 0 cell in the parity generator. The combination of the lower capacitive and resistive loads on the columns that are inputs to the register flip flops and the fact that the device driving these lines low will be activated before the device driving the columns into the latch and parity generator, assures that the inputs to the state machine register will be setup prior to the initiation of the clock cycle.

This completes the description of the technique for realizing self-timed state machines in PPL's.

4. Other Self-Timed PPL Parts

Most of the integration effort to date has involved controllers that involve little other then state machines. As a result, few other self-timed PPL parts have been developed so far. Figure 11 shows an eight bit latch that was designed as part of one of the controllers. Its principle of operation is quite similar to that of the state machine clocking circuit. Figure 12 is a schematic of a single bit of the register. It is simply an RS latch with pass transiistors on its inputs. The latch is constructed from two inverter cells and two CPU2 column pullups. These are used rather then one of the PPL latch cells as an 8:1 ratio device is needed since it will be driven through pass transistors.

A register is constructed by connecting together the desired number of bits and adding the control circuit shown on the right in Figure 11. The control circuit contains a modified register bit - there is no pass transistor on the reset input, and some additional logic. When ST REG RQST is 1 and the modified register bit is 0, the input to the row of pass transistors on the reset sit of the register bits will go high. This signal will propagate from right to left on the row enabling the pass transistors. At the left end of the row ohmic contacts (#) carry the signal down to the row of pass transistors on the set side of the register bits. The signal will propagate back along this row finally enabling the pass transistor on the set input to the modified register bit, allowing the latch to be set. This will cause the signal enabling the pass transistors to go low. When the low has propagated across the reset

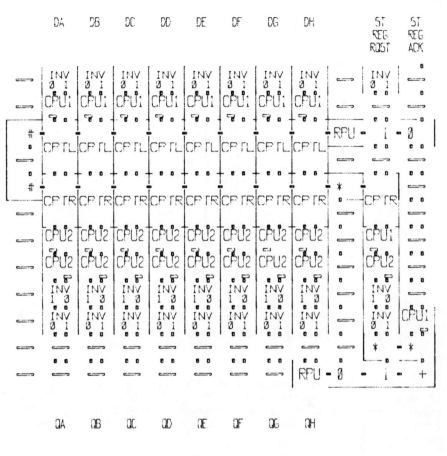

Figure 11:   8 bit self-timed latch.

pass transistors and back across the set pass transistors disabling them all, ST REG ACK, which is the AND of the control latch being 1 and the pass transistor enable line being 0, will become one. When ST REG RQST goes to 0, the control bit will be reset immediately, as there is no pass transistor on the reset input, and ST REG ACK will go to 0.

Figure 13 shows a self-timed ripple carry counter stage. When RQST IN goes to 1, one of the two right hand latches will be set as a function of the value in the left hand latch (COUNT). IF COUNT is 1, the RQST OUT latch will be set. This latch being set will propagate a request to the next cell and will reset COUNT to 0. If COUNT is 0 when RQST IN goes to 1, the ACK latch will be set. This will set COUNT to 1 which will then result in the ACK line, which is the OR of the ACK outputs of all of the counter stages becoming one. When RQST IN goes to 0, it will reset whichever of ACK and RQST OUT

Figure 12:   Self-timed latch cell.

is set resulting, ultimately in the ACK signal going to 0. Figure
14 shows a 3 bit counter. Note the additional logic on the MSB
stage which results in the counter wrapping around from 7 to 0 and
generating an ACK rather then hanging up trying to set the
nonexistent next stage.

 Ripple carry counters are an example of where self-timed design
has a substantial advantage over synchronous design. A synchronous
design would have to adjust the clock period for the worst possible
delay, whereas a self-timed system will be able to proceed as soon
as the counter has actually completed. In half the cases, the ACK
will be generated by the first stage, in a quarter by the second, in
an eighth by the third etc. Yielding the formula:

$$\text{Average Delay} = (1/2)*1+(1/4)*2+(1/8)*3+(1/16)*4+ \dots$$

$$= 1/2 + 2/4 + 3/8 + 4/16 + \dots$$

$$= 1/2 + 1/4 + 1/8 + 1/16 + \dots$$
$$+ 1/4 + 1/8 + 1/16 + \dots$$
$$+ 1/8 + 1/16 + \dots$$
$$+ 1/16 + \dots$$
$$+ \dots.$$

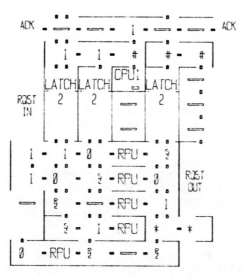

Figure 13:   Self-timed ripple carry counter stage.

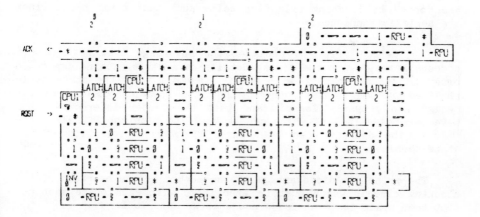

Figure 14:   Self-timed 3 bit ripple carry counter.

In the limit as the number of stages approaches infinity this becomes:

$$\text{Average delay} = 1 + 1/2 + 1/4 + 1/8 + \cdots$$
$$= 2,$$

which is appreciably better then the result for a synchronous ripple carry counter as the number of stages approaches infinity.

5. Conclusions

A method has been described for realizing self timed state machines in integrated circuits designed using PPL's. The method consists of 2 parts:

- a clocking circuit which will generate a non-overlapping two phase clock cycle for an arbitrary size register, where the duration of the phi 1 phase of the cycle is automatically adjusted to the register size, and

- a layout discipline for the folded PLA holding the state table which guarantees that the inputs to the state register will be valid at the time that the clock cycle occurs.

The method depends on certain properties of the NMOS PPL cell set, i.e. that row and clock wires are polysilicon, and that registers are formed by locating flip flop cells such that their clock lines are serially connected.

The method offers a designer the advantage that he need not concern himself with the timing details of a state machine design in order to assure that it will work. Assuming that the state table realized by the PLA is correct, that the rows and columns of the design are properly loaded, and that the proper interconnections have been made (all of which can be verified with the PPL simulator [5].), the designer can be assured of correct operation of the state machine.

The principle disadvantage of the method is the overhead of the clocking circuit which must be associated with each state machine. (But then space isn't an issue now that we have VLSI.)

In addition to the state machine two other small self-timed PPL parts were described. All of these parts could, of course, be realized in a custom NMOS design. The use of the PPL design system is not necessary to doing self-timed or any other kind of IC design, but it offers the designer significant advantages, including that it:

- allows design at a symbolic level,

- requires no geometric design rule checks, and

- is supported by a simulator that will check the placement and loading rules associated with PPL designs.

6. Acknowledgement

The work reported in this paper was supported in part by General Instrument Corp. under contract number 5-20326 with the University of Utah.

REFERENCES

[1]   A. L. Davis.
      The Architecture and System Method of DDM-1: A Recursively-
          Structured Data Driven Machine.
      In (editor), Proceedings of the Fifth Annual Symposium on
          Computer Architecture. , 1978.

[2]   D. L. Greer.
      An Associative Logic Matrix.
      IEEE Journal of Solid State Circuits SC-11(5):177-190,
          October, 1976.

[3]   A. B. Hayes.
      Stored State Asynchronous Sequential Circuits.
      IEEE Trans. Comput C-30(8):596-600, August, 1981.

[4]   C. Mead and L. Conway.
      Introduction to VLSI Systems.
      Addison-Wesley, Reading, Mass, 1980, chapter 7,System Timing.
      Chapter by C. L. Seitz III.

[5]   B. E. Nelson.
      ASYLIM: A Simulation and Placement Checking System for Path-
          Programmable Logic Integrated Circuits.
      Master's thesis, Department of Computer Science, University of
          Utah, October, 1982.

[6]   S. S. Patil.
      An Asynchronous Logic Array.
      Project MAC Technical Memo TM-62, MIT, May, 1975.

[7]   G. W. Preston.
      A Programmable Matrix Logic Array, The General Logic Circuit.
      Received from J. Robert Jump at Rice University.

[8]   Smith, K. F.,Carter, T. M., Fisher, D. A., Nelson, B. E.,
      Hayes, A. B .
      The Path Programable Logic (PPL) User's Manual.
      Technical Report 82-070, Department of Computer Science,
          University of Utah, July, 1982.

[9]   R. A. Wood.
      A High Density Programmable Logic Array Chip.
      IEEE Trans. Comput C-28():, September, 1979.

A Self-Timed Static RAM

Edward H. Frank
Carnegie-Mellon University
Department of Computer Science
Pittsburgh, PA 15213

Robert F. Sproull
Sutherland, Sproull, and Associates, Inc.
Pittsburgh, PA 15213

ABSTRACT

This paper presents the design of a self-timed static RAM. Although the memory array uses a conventional six-transistor static cell, extra circuitry is associated with each column and each row of the memory to make the memory self-timed. This circuitry detects several conditions: address line precharge complete, word line driven, bit line precharge complete, read complete, and write complete. To increase the read speed, each column of the memory uses an unclocked sense amplifier. The memory design includes a controller that implements a four-phase request/acknowledge interface. Although the memory is intended for use as part of a single-chip processor and not as a stand-alone chip, we have fabricated a 64 by 64 bit test chip and measured its performance.

1 INTRODUCTION

Although building memories using multi-project chip (MPC) fabrication is not always advisable, sometimes the exploration of a new architecture forces us to design memories that can be placed on the same chip as an experimental processor. While MPC memories compete poorly with commercial memory products that are denser, faster, and more reliable, on-chip memory can supply far greater memory bandwidth than off-chip memory can. Since several recent architecture proposals require large bandwidths to relatively small memories, we have seen on-chip memory emerge in MPC projects. For example, the RISC chip has 138 thirty-two bit registers on the

---

This Research was sponsored by the Defense Advanced Research Projects Agency (DOD), ARPA order No. 3597, monitored by the Air Force Avoinics Laboratory under Contact F33615-81-K-1539. Edward Frank is also supported by a John and Fannie Hertz Foundation Fellowship

chip [3], and the Programmable Systolic Chip has more than 4K bits of dynamic RAM for microcode and data memory [2]. Morover, as the technology available to the MPC designer improves, thereby making it possible to fabricate more sophisticated memories using MPC, it is likely that we will see even more projects with on-chip memory.

In this paper we describe a static RAM designed to be part of a single-chip processor being constructed at CMU [5]. The decision to make the memory self-timed follows directly from the decision to use self-timing in the processor as a whole. Even though the self-timed memory has a slower cycle time than its non-self-timed counter-part, system performance will not suffer unduly because processor and memory ac-tivities are overlapped. In other words, the processor is roughly as slow as the memory.

Although the memory is based on the familiar six-transistor static memory cell, the fact that the memory is self-timed makes our design quite novel. In particular, the memory accepts requests only when *all* of the appropriate internal signals are precharged and signals an acknowledgment only when *all* of the appropriate signals have changed and the valid data have been read or written.

In the remaining sections of this paper we present the design of the memory cir-cuits, the design of the memory controller, and the performance of a 64 word by 64 bit test chip.

2 THE MEMORY ARRAY

Figure 2-1 shows the components of the memory array. In the upper middle of the figure is a single six-transistor memory cell. Dual-rail bit lines run down along a column of cells to read/write circuitry associated with each column: precharging tran-sistors, a sense amplifier, and a write driver. Word lines run horizontally along a row of cells to word line drivers and address decoders associated with each row. Address lines run down from the address decoders to address line drivers. All critical transistor dimensions are given as L/W in units of lambda.

The basic idea of a self-timed read cycle is to start with all word lines low and all bit lines high, and then allow a single selected word line to come high, which in turn results in one of the bit lines in each column to come low, thus indicating the state of the selected memory cell. When every column has detected a bit line low, the cycle is complete. In the sections below, the memory activities are explained in greater detail, making frequent reference to Figure 2-1.

2.1 Precharging

Before a memory cycle may begin, the address lines and bit lines must be precharged and sensing circuitry must indicate that the memory array is in an idle state. The signal PRECHARGE ENABLE activates pullups on the address and bit lines to begin precharging. The signal ADDRESS PRECHARGE DONE, generated by the address line drivers, will become high when all address lines have been successfully precharged high. Moreover, after all address lines are precharged, all word lines will be driven low, a state that is sensed by the ~WORD LINE DRIVEN signal. The comple-

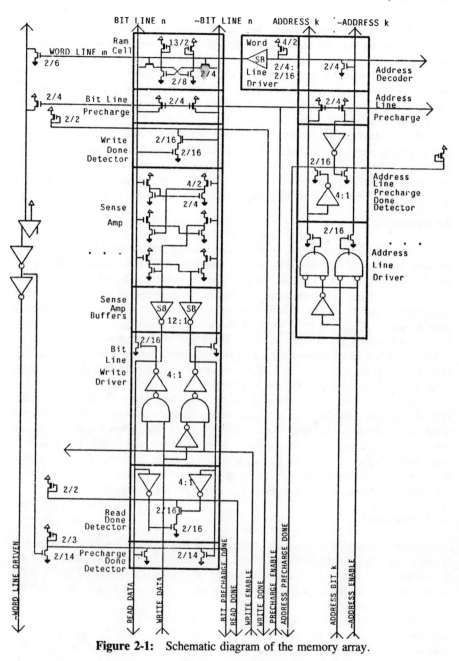

Figure 2-1: Schematic diagram of the memory array.

tion of bit line precharging is detected by the signal BIT PRECHARGE DONE, which detects that all bit lines are high and also detects that all word lines are low. Note that the bit line precharge detector examines the output of the sense amplifier, so that the memory waits for any delay through the sense amplifier before beginning a read or write cycle. Furthermore note that because the metal bit lines are essentially equipotential [7] there is no difficulty with the precharge drivers and the precharge detectors being at the same end of the bit line.

When ADDRESS PRECHARGE DONE and BIT PRECHARGE DONE indicate that precharging is complete, PRECHARGE ENABLE may be deactivated. However, since all precharge pullups are depletion-mode transistors, the entire memory array remains precharged indefinitely. Moreover, the sense amplifier is designed to require bit lines to be precharged all the way to VDD rather than VDD less a threshold drop.

2.2 Addressing

Each row, or word, of the memory is addressed with a word line driven by a decoder, which is in turn driven by address line drivers. The address line drivers are enabled by a single address enable line that is gated with the appropriate address bit to drive the appropriate pulldown transistor. The gating is necessary so that the addressing structures can be "cleared" during precharging, i.e., the memory is returned to a state in which all address lines are high and no word line is low. When ~ADDRESS ENABLE is activated, one address line in each pair will be driven low, which will result in one word line being driven high. Because all address lines are returned to a high state by precharging after the cycle, there are never any "glitches" on the word lines: either no word line is high or the single addressed word line is high.

An unusual part of a memory row is the word line driven detector, which consists of a NOR gate formed from pulldown transistors at the ends of the rows. When a word line is driven, the output of the detector will be low; when no word lines are driven, the output will rise. This detector is not used for read cycles, but is used in write cycles and as an indication that all word lines have become low during precharging. Note that this detector must be at the opposite end of the word line from the word line driver because poly-silicon wires of this length are not equipotential.

2.3 A Read Cycle

After precharging is complete, a read cycle begins when ~ADDRESS ENABLE is invoked, thereby driving low one of the two address lines in each dual-rail pair. Somewhat later, exactly one address decoder line rises and is amplified to drive the corresponding word line. Each memory cell in the row begins to pull down one of the two complementary bit lines.

The signals on the bit lines of a column are amplified by a differential sense amplifier patterned after one used by MOSTEK in a 256K ROM [6]; our design uses somewhat larger voltage swings than the original and has depletion-mode pullups. The buffers on the sense amplifier outputs use a ratio of 12:1 because the output threshold of the amplifier is lower than that for conventional logic circuits. (The "follower" nature of the connection from the bit lines to the amplifier stages is responsible for the lower threshold.) Although the amplifier is differential, when both bit lines are high both outputs are high and thus provide an indication that the bit lines are precharged.

The primary advantage of this amplifier is that it is unclocked. Although clocked sense amplifiers such as those used in dynamic RAM's are more sensitive and faster (see, for example, [4]), they require an appropriately-timed clock signal, which is not feasible in a self-timed environment. The objective of a self-timed memory is to allow the activities of each word line and each bit to proceed at their own rate and to announce completion of a read cycle only after all bits have been successfully read. Viewed in this way, the clock signal for the sense amplifier would have to be the "read complete" signal; but deriving that signal requires sensing all bit lines, thus yielding a circularity that cannot be broken.

The unclocked nature of the sense amplifier also avoids other problems often found in extremely sensitive dynamic RAM sense amplifiers. For example, there is no problem of a sense amplifier "misreading" a bit due to unequal precharging of the bit lines. The static design of the memory cells and the fact that the sense amplifier is unclocked guarantee that the amplifier outputs will settle to the correct value. If unequal precharging of the bit lines is severe, one of the complementary outputs of the sense amplifer remains low, and the precharge done detector never announces that the bit lines are precharged. The memory simply stops; it will not start another cycle, and will of course never signal the completion of the cycle. Such grossly unequal precharging is not likely to be a problem in our design because the precharge transistors are depletion devices that pull the bit lines all the way to VDD and because the sense amplifier is not extremely sensitive.

The complementary outputs of the sense amplifier are used by the read done detector, a NOR gate distributed among the read circuits for all bits, which detects when one of each pair of bit lines has come low. When this condition occurs, the READ DONE signal rises, indicating that valid data is available for all bits. The read cycle is purely self-timed; the cycle is declared complete only when valid data are available for the entire word. Note that the pullup for the read done detector (as well as for the write done detector and the bit precharge done detector) is at the opposite end of the memory from the output of the gate. In this way we can be certain that the output goes high only when all of the pulldowns have turned off as the long diffusion wire that connects the pulldowns is not equipotential.

2.4 A Write Cycle

Like a read cycle, a write cycle begins by activating ~ADDRESS ENABLE; the selected word line is eventually driven high. The WRITE ENABLE signal is also activated, which is gated to control the write driver, two large pulldown transistors. Each memory cell in the word writes the appropriate data.

The completion of a write cycle is detected by two mechanisms. First, the write done detector, a NOR gate distributed among the write circuitry for all bits, allows the WRITE DONE signal to rise when one bit line in each column has been driven low. This detector operates on the bit lines directly, rather than on the sense-amplified versions of the bit lines, in order that the threshold of the write done detection circuitry be as close as possible to that of the memory cell write circuitry.

In addition to detecting that the bit lines have been driven appropriately, it is necessary to detect that the selected word line has been driven high, indicating that the memory cells in the word have been enabled for writing. The ~WORD LINE DRIVEN signal, derived from a large NOR of all word lines, detects this condition. We conclude that a write cycle is complete when WRITE DONE has come high and ~WORD LINE DRIVEN has come low.

The determination that the write cycle is complete is actually indirect. It is impractical to detect directly the completion of the writing event, namely the setting of the flip-flop in each memory cell of the word. Instead, we detect that the signals driving the memory cells have completed their transitions and then *assume* that the memory cells have actually written the data by the time write done and word line driven detectors have signalled completion. This assumption is based on simulations which show that the time to flip a memory cell (a few nanoseconds) is more than an order of magnitude less than the time to detect and drive the done signals (several tens of nanoseconds). However, it is remotely conceivable that threshold variations across the chip might allow a done signal to precede writing a memory cell.

2.5 Speedup

The large NOR gates in the done-detection circuitry are slow and are responsible for a large fraction of the total memory cycle time. We have taken two steps to speed up these circuits. First, the word line driven detector is arranged to be sensitive to high-to-low transitions so that a sense amplifier, identical to the one used on memory bit lines, can be used to derive the done signal. Second, the 64 by 64 bit memory is split in half (i.e., into two 32-bit words) so that the by-bit NOR gates have only 32 pulldowns rather than 64. Because of this split, two signals are generated for each condition, e.g., READ DONE (LEFT) and READ DONE (RIGHT). The paired signals are finally joined in the memory controller.

3 THE MEMORY CONTROLLER

The purpose of the memory controller is to simplify interfacing to the memory; a schematic is shown in Figure 3-1. The controller provides an interface to the memory user that consists of three signals: REQUEST, READ/~WRITE, and ACKNOWLEDGE. The REQUEST and ACKNOWLEDGE signals implement a standard four-phase signalling. The READ/~WRITE signal is high to indicate that the requested cycle should be a read and low to indicate a write. The sequence relationship of the READ/~WRITE signal is identical to that of WRITE DATA BIT signals. It might have been cleaner to provide separate READ REQUEST and WRITE REQUEST signals to the controller. However, since the memory design assumes that an equipotential or bundling constraint be applied to the data read and write signals, it seems acceptable to apply the same constraint to this controller signal.

The principal component of the controller is a C-element (or "last-of") that remembers that the memory has been precharged or that a cycle has started. A cycle may start when a request arrives and all of the precharge-detection conditions are met (~WORD LINE DRIVEN, ~READ DONE, BIT PRECHARGE DONE, and ADDRESS PRECHARGE DONE). When the cycle begins, ADDRESS ENABLE is activated, as is WRITE ENABLE if a write cycle is requested.

Acknowledgement is generated using the done conditions discussed in the previous section. For a read cycle, READ DONE signals completion; for a write cycle, both WRITE DONE and WORD LINE DRIVEN are required.

When REQUEST is removed *and* when all of the precharge detectors have signalled that the precharge conditions no longer hold, the last-of flips and the memory begins precharging again by driving PRECHARGE ENABLE. The sequencing of the signals in the memory and the controller is shown in Figures 3-2*a* and 3-2*b*. Actual timing measurements are presented in the next section.

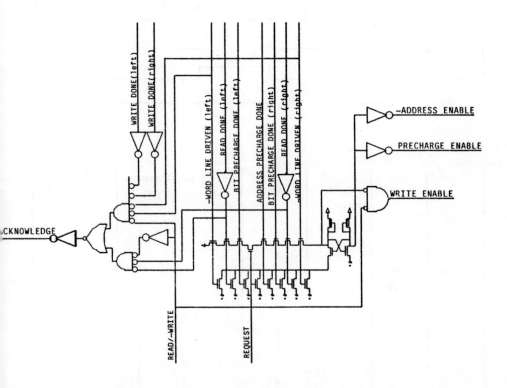

Figure 3-1:   Schematic diagram of the memory controller.

4 IMPLEMENTATION AND PERFORMANCE

A 64 by 64 bit test implementation of the memory design has been fabricated using the MOSIS implementation system with lambda equal to 2 microns. In addition to pins for data, REQUEST, ACKNOWLEDGE, and READ/~WRITE signals, the test chip also has pins for directly accessing some of the done detectors and for forcing the memory controller to either drive PRECHARGE ENABLE or ADDRESS ENABLE, and, if

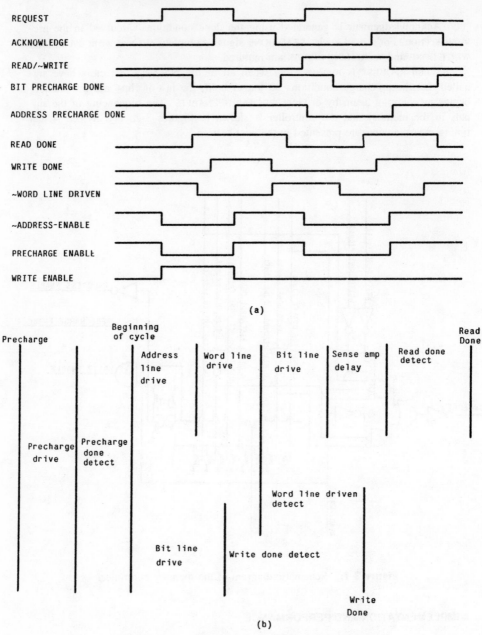

Figure 3-2: Sequencing of signals in the memory.

appropriate, WRITE ENABLE. A checkplot of the chip is shown in Figure 4-1. The penalty in chip area for the self-timed design is not large. Only 5.2% of the total chip area (.89 mm^2 of 17.2 mm^2) is devoted to circuitry to detect the various *done* conditions; more details are given in Table 4-1.

Table 4-1: Area overhead of the self-timed circuitry.

Section	Total Area (mm^2)	% in detection circuitry
Column driver/Sense	.029	23.7
64 bit column	.244	2.9
Address Drivers	.010	31.6
64 bit column	.092	3.7
Row Drivers/Decoders	.022	29.9
64 bit row	.237	2.8
Memory controller	.070	
Total (64 by 64 array and controller)	17.2	5.2

The performance of the test chip is acceptable for our application; details are given in Table 4-2. The times given are estimates of on-chip times obtained by measuring times at the pins and subtracting a pad-driver delay measured independently; only two chips were available to make these measurements. Write access time is shorter than read access time because bit lines are driven by more powerful drivers in the write circuit than in the memory cell and because there are fewer events in series in a write cycle. From the numbers in Table 4-2 it is apparent that the detection circuitry for the self-timing is responsible for a large fraction of the overall memory cycle time. In fact simulations and measurements have shown that almost all of the time in a memory cycle can be attributed to charging and discharging the enable signals, which are run in poly-silicon, and the done signals, which are run in diffusion. If two-level metal had been available to us, then this problem would have been less severe as this signals could have been run in metal. However, it is also clear that we should have paid more attention to the design of the driving and detection circuitry. Indeed, we concur with the conclusion of Cherry and Roylance, who, in their paper *A One Transistor RAM for MPC Projects* [1] state that "there is much more to a ... RAM than sense amplifiers."

5 CONCLUSIONS

The design we have described demonstrates that MPC fabrication of self-timed memories is feasible, thus allowing us to build a variety of systems with these memories as integral on-chip components. The obvious disadvantage of our design is that it is somewhat slow. While the performance of the memory could probably be improved substantially by more sophisticated circuit design, the current performance is adequate for our application. The strength of this design lies in its simplicity and in the self-timed interface to it that allows it to be applied reliably in a number of self-timed systems.

Table 4-2:   Summary of the performance of the test chip.

Power dissipation (at VDD = 5V)	.95 watts
Read Cycle	
1. REQUEST to ACKNOWLEDGE	245ns
2. REQUEST to valid data at sense amp output	145ns
3. Falling of REQUEST to falling of ACKNOWLEDGE	195ns
4. Falling of REQUEST to data precharged	160ns
5. REQUEST to ~WORD LINE DRIVEN	135ns
Self-timed read cycle (1+3)	440ns
Read cycle time without self-timing (2+4)	305ns
Write Cycle	
6. REQUEST to ACKNOWLEDGE	220ns
7. Falling of REQUEST to falling of ACKNOWLEDGE	175ns
8. REQUEST to ~WORD LINE DRIVEN	135ns
Self-timed write cycle (6+7)	395ns

REFERENCES

[1]   Cherry, J., and Roylance, G.
       A one transistor RAM for MPC projects.
       In *Proc. of the 2nd Caltech Conf. on VLSI*, pages 329-341. Caltech, Pasadena
           CA, Jan, 1981.

[2]   Dohi, Y., Fisher, A., Kung, H.T., Monier, L., and Walker, H.
       The design of the PSC: A programmable systolic chip.
       In *Proc. of the 3rd Caltech Conf. on VLSI*. March, 1983.

[3]   Fitzpatrick, D.T., Foderaro, J.K., et. al.
       A RISCy approach to VLSI.
       VLSI Design II(4):14-20, Fourth Quarter, 1981.

[4]   Foss, R.C., and Harland, R.
       Peripheral circuits for one-transistor cell MOS RAMS.
       IEEE Journal of Solid-State Circuits SC-10:255-261, October, 1975

[5]   Frank, E.H.
       The Fast-1: A data-driven multiprocessor for logic simulation.
       Thesis proposal VLSI Memo 122, Carnegie-Mellon University, Computer
           Science Department, Pittsburgh, PA, October, 1982.

[6]   Scherpenberg, F.A., and Sheppard, D.
       Asynchronous circuits accelerate access to 256K read-only memory.
       Electronics :141-145, June, 1982.

[7]   Seitz, C.
       *Chapter 7, System Timing, in Introduction to VLSI systems by C. Mead, and
           L. Conway.*
       Addison-Wesley, Reading, Mass, 1980.

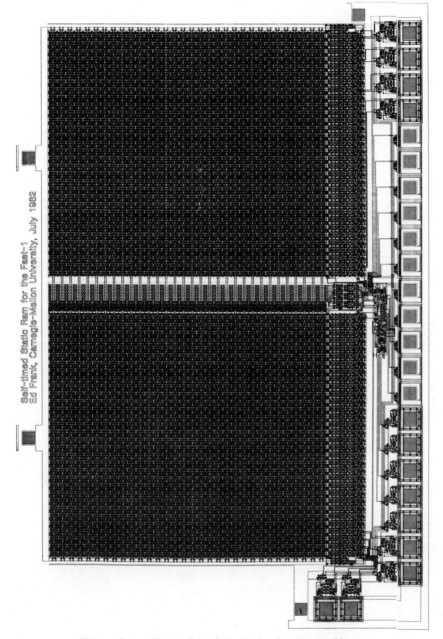

Figure 4-1:   Check plot of the 64 by 64 bit test chip.

Design of the PSC: A Programmable Systolic Chip*

Allan L. Fisher, H.T. Kung, Louis M. Monier, and Hank Walker
Department of Computer Science, Carnegie-Mellon University
Pittsburgh, Pennsylvania 15213

Yasunori Dohi**
Department of Computer Engineering, Yokohama National
University
Tokiwadai, Hodogaya-ku, Yokohama, 240 Japan

Abstract—The *programmable systolic chip* (PSC) is a high performance, special-purpose, single-chip microprocessor intended to be used in groups of tens or hundreds for the efficient implementation of a broad variety of systolic arrays. For implementing these systolic arrays, the PSC is expected to be at least an order of magnitude more efficient than conventional microprocessors. The development of the PSC design, from initial concept to a silicon layout, took slightly less than a year. This project represents an integration of many disciplines including applications, algorithms, architecture, microprocessor design, and chip layout. This paper describes the goals of the project, the design process, major design features and current status.

1. Introduction

Many *systolic algorithms* have recently been proposed as solutions to computationally demanding problems in signal and image processing and other areas (see, e.g., [12]). By exploiting the regularity and parallelism inherent to given problems and by employing high degrees of parallelism and pipelining, systolic algorithms achieve high performance with regular communication structures and low I/O requirements. Typically, systolic systems are large regular arrays of simple cells called *systolic cells*, and their implementations are relatively inexpensive. The development cost of a systolic system can be further reduced if it can be amortized over a large number of units typical of a general-purpose processor. One way to achieve this is to provide a single programmable systolic cell, many copies of which can be connected and programmed to implement many systolic algorithms. The CMU *programmable systolic chip* (PSC), a prototype single-chip microprocessor now being fabricated in nMOS, is aimed at

*This research was supported in part by the Defense Advanced Research Projects Agency (DoD), ARPA Order No. 3597, monitored by the Air Force Avionics Laboratory under Contract F33615-81-K-1539. Allan L. Fisher was supported in part by a National Science Foundation fellowship and in part by an IBM fellowship.

**Work performed while visiting CMU.

exploring the design space of systolic processors of this flexibility. As illustrated in Figure 1, PSCs will be used as the basic cells of a wide variety of systolic arrays.

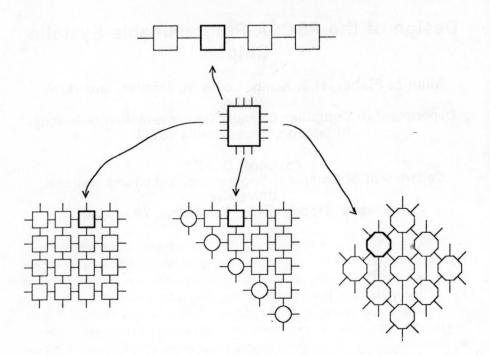

Figure 1:   PSC: A building-block chip for a variety of systolic arrays.

When a systolic array is in operation, data continuously flow from cell to cell in a regular manner and each cell performs some simple functions on the passing data. This nature of computation imposes new requirements for programmable processors which are quite different from those found in a conventional microprocessor. It is these requirements which the PSC attempts to meet.

Section 2 of this paper recounts the thinking that led to the PSC architecture. Section 3 discusses the process of designing the PSC, and Section 4 briefly discusses some of the applications envisioned for the PSC. Section 5 describes the current status of the project, along with future plans, and some reflections and concluding remarks are given in the last section. Appendix I describes the PSC memory design, a major design effort that made the PSC on-chip writable control store possible.

2. Rationale and Architecture Features

The major goals established at the outset of the PSC project were to design a processor that would be flexible enough to implement many different systolic algorithms, cost-effective enough to establish the practical usefulness of the PSC concept, novel enough to gain new architectural insights and simple enough to be implemented in available LSI technology by non-expert circuit designers in a reasonable amount of time. A secondary goal was to use and test a variety of design tools, with an eye toward gaining insight into VLSI design methodology.

In order to ensure sufficient flexibility to cover a broad range of applications and algorithms, a set of target applications was chosen. An obvious candidate was the field of signal and image processing; many systolic algorithms for such problems have already been designed (see, e.g., [10; 13]). Another choice was error detection and correction coding, Reed-Solomon coding (see, e.g., [14; 17]) in particular. In addition, sorting of large files was chosen as a representative of data-processing applications. The performance goal of the PSC was that PSC-based systolic arrays (using tens or hundreds of PSCs) would achieve orders of magnitude speed-ups over other existing approaches for *all* these target applications. To reach this goal the following design decisions were made:

- *Single-chip processor.* Keeping a processor cell on a chip has two advantages. First, it allows the functional blocks of the processor to operate together without paying the time penalty of off-chip communication. Second, it allows systolic arrays to be constructed with minimum chip count. Moreover, since each chip contains only one cell, reconfigurability of cell interconnections can easily be achieved by custom inter-chip wiring on PC boards. At current and near-future circuit densities, extra silicon area can more profitably be devoted to increasing the functionality and memory space of a programmable cell than to replicating such cells on a single chip.

- *On-chip parallelism.* Partition of the processor's function into units which can operate in parallel enhances performance.

- *Microprogrammed control.* In order to provide flexibility in programming new applications and to promote parallelism within the chip, the processor uses a horizontal microprogrammed control structure.

- *On-chip dynamic RAM.* The high density of the dynamic RAM makes it possible to implement an on-chip writable control store and a register file to support flexible programming with minimum off-chip I/O.

- *Eight-bit data.* In order to keep the chip size down and hence keep yield reasonable, a modest word size of eight bits was chosen. The yield problem is particularly important to the project, since the intended applications require a large number of working chips. For instance, the decoder implementation of Reed-Solomon codes will use more than one hundred PSCs. In order to support arithmetic modulo 257, however, eight-bit words are augmented to nine bits in appropriate places. Facilities are also provided for multiple-precision computation.

- *On-chip bit-parallel multiplier-accumulator.* Multiplication within a single machine cycle is necessary to provide good performance in signal processing and coding applications.

- *High-bandwidth inter-chip and on-chip communication.* A principal feature of systolic arrays is the continuous flow of data between cells. Efficient implementation of such arrays therefore requires several wide I/O ports for inter-chip communication and high bandwidth data paths for on-chip communication. Provision is also made for the transmission of pipelined *systolic control* signals, which are crucial for controlling systolic arrays at runtime.

Note that none of the conventional, commercially available microprocessor components fulfill the stated goals. Unlike the PSC, conventional microprocessors do not have fast, on-chip multiplier circuits which are crucial for high-speed signal and image processing, they do not have enough I/O bandwidth to pass data from cell to cell with a speed sufficient to balance the computation speed, they lack significant internal parallelism, and they usually do not have on-chip RAM for program memory. Existing signal-processing microprocessors, while possessing multiplication hardware, also usually lack one or more of these features.

The design decisions summarized above resulted in the following architectural features for the PSC, which are also sketched in Figure 2:

- 3 eight-bit data input ports and 3 eight-bit data output ports,
- 3 one-bit control input ports and 3 one-bit control output ports,
- eight-bit ALU with support for multiple precision and modulo 257 arithmetic,
- multiplier-accumulator (MAC) with eight-bit operands and 16-bit accumulator,
- 64-word by 60-bit writable control store,
- 64-word by 9-bit register file,
- three 9-bit on-chip buses, and
- stack-based microsequencer.

A simple two-phase clock is used for the whole chip, except for the dynamic RAM, which uses a slight variant of this scheme. During φ_1, the buses are precharged, the functional blocks (ALU, MAC, memories) compute, and the next microinstruction is fetched and transmitted to the functional blocks. During φ_2, the results of the previous operations are transmitted by the buses, the functional blocks are precharged, and the microinstruction address is updated. Off-chip communication takes place during φ_1, interleaved with on-chip bus transactions.

Each of the three buses can be written by one of eight sources and read by any of ten destinations. The choice of three buses was a compromise between several considerations: small size of microcode, layout area for the buses, and high degree of internal parallelism. The limitation imposed by this number of buses seems to be small, and is alleviated by the ability of each of the functional blocks to hold its inputs over more than one cycle.

The microsequencer, as sketched in Figure 3, is made up of a microprogram address

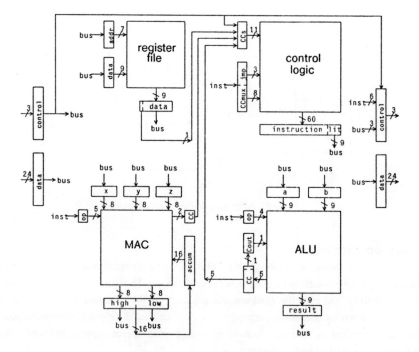

Figure 2: PSC block diagram

register, a four-word return address stack, two condition code multiplexers, an adder, and a small PLA for control bit decoding. It is able to fetch instructions in sequence, branch to any of four locations depending on any two of eleven condition bits (including looping in place), branch to a literal address, and jump to and return from subroutines. The four-way branch feature is quite useful in making fast decisions based on the many status bits available.

The logic design of the MAC is similar to that of the corresponding TRW chip [19]. It is a bit-parallel multiplier-accumulator with precharged operand inputs. Test results indicate that it multiplies in approximately 200 ns. The ALU is similar to the one used in the OM2 [15], with a smaller instruction set. It uses a precharged carry chain, uses a small PLA to generate 15 bits of control from a 4-bit opcode, and has provisions for finite-field and multiple precision arithmetic. The other major design is the memory, which is rather ambitious and is described in Appendix I. The writable control store is loaded through a shift register, which can also be used to separately test the WCS and the data paths.

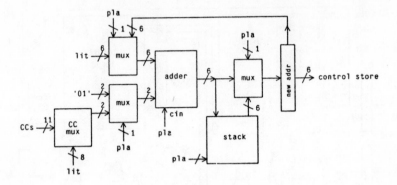

Figure 3: Microsequencer block diagram

3. Design Process

The project began in October, 1981. Our very first ambition was to define a systolic processor "general enough" to be used in many applications. The structure of the processor was still to be defined, and even the word size was debated: should we have a very small processor (1, 2 or 4 bits) and squeeze many such cells on a chip, or should we use a larger word size? It soon became evident that we were looking for a single-chip microcomputer, but with features unknown to any product on the market.

Some time in November, the decision was made to design and implement a chip. An eight-bit word size seemed to be the best compromise: the chip had a good chance to be useful while still reasonable to design with our facilities. The chip was intended to be general-purpose, so we had to find an architecture that would serve many different applications. To this end, we selected a small set of target applications, very different in nature and covering several important domains, as already mentioned in the preceding section. For these applications, previously known systolic algorithms were reviewed, and some new systolic algorithms such as one for computing polynomial GCDs needed in the Reed-Solomon codes were developed [3]. The requirements of these algorithms led to an initial design.

The initial design of the PSC was specified in ISPS [1]. The choice of ISPS as our hardware description language was a natural one, since the language was initially developed at CMU and is well-supported in our local environment. The PMS program, part of the ISPS system, was useful in constructing simulations of entire arrays from the description of a single cell and the arrays' connectivity. Microcode for several systolic algorithms was written and tested as an aid to design evaluation. This process was iterated several times as experiments and measurements resulted in changes to the specification. The ISP language and a table-driven microassembler were invaluable in this effort. In addition to the machine-level simulator (which at one point was recoded in C for extra speed), we made use of a high-level simulator for systolic arrays, written in Lisp, as a first step in the coding of a few rather complicated systolic algorithms.

Concurrently, design and layout of several individual pieces of the chip were begun. In particular, the multiplier-accumulator (MAC) and the memory were the first units to receive attention. This somewhat "bottom-up" approach to the layout of the chip was prompted mainly by the fact that the architecture was not yet fully defined. This certainly had a negative impact on the final layout of the chip—a fairly large area is given over to routing—but it allowed the layout to be completed earlier than would otherwise have been possible. Since we viewed the chip as a prototype, this seemed a reasonable price to pay. The piecemeal approach has the added advantage that subparts can be tested and characterized individually, aiding in the design and testing of the whole.

The ISP specification of the design was finalized in early May, and from that time layout of the chip was the major activity. Most of the layout was done with Caesar [16], except for two small PLAs produced by a PLA generator. Both of these programs were supplied by the VLSI CAD group at Berkeley, who also supplied Cifplot, a checkplotting program. Design rules, including the MOSIS buried contact rules, were checked by a locally written DRC [7]. We also made heavy use of ACE [6], a high performance edge-driven circuit extractor written at CMU. The output of the extractor was checked by hand for connectivity, and all of the parts of the chip were simulated at the logic level using simulators written at MIT [20]. Critical parts of the design were simulated at the circuit level with Spice [5].

4. Applications

The PSC can be used as the basic systolic cell for many systolic systems in many application areas. In this section we give a flavor of how it is applied in a few target applications, and discuss its cost-effectiveness with respect to these applications. Most of these applications have been microcoded; this code has been simulated by ISPS [1] simulators and used to evaluate architectural specifications of the chip. We estimate that a commercial nMOS implementation of the PSC could operate with a cycle time of 200 ns; our first implementation should run within a factor of two or three of this speed. We shall assume the 200 ns period in our performance estimates. In this section, we will only describe each application briefly. For detailed information, the reader is referred to a forthcoming PSC paper on applications.

4.1. Digital Filtering in Signal and Image Processing

Many digital signal and image processing applications require high-speed filtering capabilities. Mathematically, a filtering problem with $h + k$ taps is defined as follows:

given the weights $\{w_1, w_2, ..., w_h\}$, $\{r_1, r_2, ..., r_k\}$, the initial values $\{y_0, y_{-1}, ..., y_{-k+1}\}$ and the input data $\{x_1, ..., x_n\}$,

compute the output sequence $\{y_1, y_2, ..., y_{n+1-h}\}$ defined by

$$y_i = \sum_{j=1}^{h} w_j x_{i+j-1} + \sum_{j=1}^{k} r_j y_{i-j}$$

If the $\{r_i\}$ are all zero, then the problem is called a finite impulse response (FIR) problem, and otherwise an infinite impulse response (IIR) problem. It is known that both types of filtering can be performed by systolic arrays [9; 10; 12]. Figure 4 illustrates a systolic array for FIR filtering for $h = 3$.

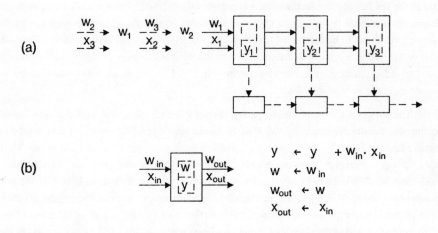

Figure 4: (a) Systolic FIR filtering array and (b) its cell definition.

Based on this scheme, digital filters (FIR or IIR) with eight-bit data and weights and m taps can be computed with a linear systolic array composed of m PSCs, taking one sample each 200 ns. Thus with 40 PSCs, a 40 tap filter can be computed at a rate of 400 million operations per second (MOPS), counting each inner product step (eight-bit multiply, 16-bit add) as two operations. This is equivalent to 600 MOPS for pure eight-bit arithmetic.

For applications requiring more accuracy, a filter with 16-bit data and eight-bit coefficients and m taps can be computed with m PSCs, taking one sample each 1.2 μs. Thus with 40 PSCs, a 40 tap filter can be computed at a rate of 67 MOPS, counting each inner product step as two operations. This is equivalent to 200 MOPS for eight-bit arithmetic.

4.2. Error-Correcting Codes

Among various error-correcting codes, Reed-Solomon codes are most widely used today for deep space communications, where burst errors occur frequently [14; 17]. A popular Reed-Solomon code, which for example has been adopted by the European Space Agency, is a scheme in which a codeword consists of 224 message symbols followed by 32 parity symbols. This code can correct up to 16 symbol errors per codeword through a decoding process.

Before a message is transmitted, it is first encoded. It is well known that encoding a message, i.e., obtaining parity symbols from the given message symbols, is equivalent to a polynomial division. It can therefore be carried out with the systolic division array described in [11]. Since in this case the divisor—the *generator polynomial* in the terminology of error-correcting codes—is monic, no cell in the systolic array needs to perform a numerical division. Each cell is basically the same as used in the systolic filtering array discussed above, except that integer arithmetic modulo 257 is used.

Decoding, which is much more complex than encoding, is usually done in four steps. We will not describe the steps in detail here; the reader is referred to the texts on error-correcting codes cited earlier. We simply point out that each of the steps—syndrome computation, solution of the key equation, error location, and error evaluation—corresponds to some polynomial computation over the finite field $GF(257)$ that can be effectively carried out by systolic arrays.

All of the steps mentioned above can be implemented with appropriate systolic arrays made up of the same PSCs, with different microcode for each array. We estimate that using a linear array of 112 PSCs, Reed-Solomon decoding can be performed with a throughput of 8 million bits per second. Encoding is much easier; it requires only about 16 PSCs to achieve the same throughput. As far as we know, the fastest existing Reed-Solomon decoder with the same mathematical characteristics uses about 500 chips but achieves a throughput of only 1 million bits per second.

4.3. Fingerprint Computation

It has recently been proposed [8; 18] that a randomly chosen irreducible small polynomial $K(x)$ be used to "fingerprint" a long character string or file by computing the residue of that string, viewed as a large polynomial, modulo $K(x)$. One can show that if the fingerprints of two given files agree then, with very high probability, the two files are identical, and on the other hand if the fingerprints do not agree then the files must be different. Since comparing fingerprints is much easier than comparing entire files, the fingerprint method represents a significant savings in computing time for applications where large files must be compared, provided that fingerprints of files can be stored and used several times. For example, we can check whether or not a file stored on a disk has been tampered with by computing its fingerprint and comparing it with the old fingerprint of the original file. (Of course we assume here that the old fingerprint, which presumably is a very small file, can be safely stored.)

Assume that each given file has n bytes and the randomly chosen irreducible polynomial over $GF(257)$ is of degree d. Using the same analysis as in [18], one can prove that the probability of a false match, i.e., two different files having the same fingerprint, satisfies the following relation:

$$\text{probability of mismatch} \leq \frac{n}{257^d}$$

For example, for files of no more than 10^{10} bytes, we can make the probability of mismatch smaller than 10^{-12} by choosing $d=10$. Fingerprints can be computed by systolic arrays, since the problem is equivalent to polynomial division. Thus for this case, with $d=10$ PSCs connected as a linear systolic array, a file can be reduced to its fingerprint at the rate of 8 million bits per second.

Note that it is possible to implement fingerprint computations over $GF(2)$, as originally suggested in [18]. We suspect that it is more cost-effective to implement them over $GF(257)$, as we do with the PSC, since in this case a parallel multiplier that works on bytes rather than bits can be used. A detailed study on this matter is yet to be performed.

4.4. Disk Sort/Merge

To demonstrate the breadth of PSC applications, it was decided in the early specification stage in the fall of 1981 that the PSC should also be useful for non-numerical processing. Disk sort/merge was chosen as the target application in this area. It has been estimated that 20% of all computer time is spent on sort/merge applications, much of it on merging from disks [2]. Many banks sort hundreds of megabytes of data daily, each sort taking hours. Thus any nontrivial speed-up we can get for the sort/merge process will be significant in practice. A systolic array consisting of about 17 PSCs, augmented with an appropriate amount of buffer memory, can achieve an order of magnitude of speed-up over conventional minicomputers for sort/merge.

The systolic array implements a binary tree selection sort-merge, in which one PSC is assigned to each level of the tree and all of the nodes at that level are held in the associated buffer memory. At each step in the algorithm, one value advances at each level of the tree. This systolic sorter has been simulated using the ISPS description of the PSC. A systolic cycle consists of 10 microinstructions, which is about 10×200 ns $= 2$ μs. The throughput roughly matches a typical disk transfer rate. The total amount of buffer memory needed depends on the disk size and the maximal number of merging passes we want to tolerate. For example, with 16 Mbytes of memory a 1 Gbyte disk can be sorted in two passes; with only 3 Mbytes, sorting would take three passes.

5. Status and Future Plans

The layout of the PSC was completed in early October and submitted to MOSIS [4] for fabrication. The chip has about 25,000 transistors and 74 bonding pads, and is packaged in an 84-contact ceramic leadless chip carrier. Using Mead-Conway design rules [15] with $\lambda = 2$ microns, the chip measures 7 mm by 9 mm. Spice simulations indicate that the chip should operate at a clock rate between 2 and 4 MHz. We received one batch of chips in December, and discovered that most parts of the chip worked, although some portions were non-functional due to small layout errors. The yield of chips without apparent production flaws was only about 10%. The layout errors have been corrected, and we expect to receive new chips in March. Tests of the working portions of the chips indicate that they should operate with a cycle time of less than 400 ns, in the range of 2.5-3 MHz. We estimate that a commercial nMOS implementation of the PSC could operate with a cycle time of 200 ns or less.

Once the design of the PSC is validated, we plan to use it to build some systolic arrays, both as demonstrations of the PSC and in furtherance of other research. This jump from experimental design to limited production raises serious issues of yield, packaging, and testing. As a result, we plan to carry out these demonstrations in two stages, the first of which will dodge most of the hard questions.

The first prototype systolic array will be a linear array consisting of about ten cells. This arrangement can be used to implement a large number of systolic algorithms [12]. The small size of the machine means that we can count on getting the right number of working chips by getting ten times that many packaged chips from MOSIS and discarding those that don't work.

Similarly, one or two circuit boards can be wire-wrapped and populated with PSCs in expensive and area-inefficient zero-insertion-force sockets. We expect a machine of this type to be operational in the summer of 1983.

The successor to the ten-processor array will contain approximately 100 cells. This could be used, for example, as a high-speed Reed-Solomon decoder. For this larger number of processors, testing of packaged chips at 10% yield will be prohibitively expensive. We expect to have to resort to testing wafers and having only working chips packaged. Circuit board density will also be an issue; it may be advisable to redesign the chip to fit it into a package with smaller pinout, or to use wave soldering or some other "exotic" means of circuit board attachment. All of these issues, of course, have been encountered and dealt with in industrial settings; their novelty here stems from our academic setting and our expectation that "limited production" runs will become a normal mode of operation in the growing area of special-purpose VLSI system implementation.

6. Concluding Remarks

Very few large IC designs have been attempted as yet by small research teams in university environments. For the PSC project there was basically no "proven" design methodology to follow. We conducted the PSC design in a more or less intuitive and pragmatic manner; this meant that we had a lot of flexibility to make quick design decisions. We find that this rather informal style of design has in general served the project well. Of course, better designs, in several respects, could have resulted if we had used a more rigid approach. For example, immediately after the ISP architecture was finalized, first-order decisions about the timing and the floor plan for the whole chip could have been made, and then all the functional blocks could have been designed accordingly. Following this discipline would have probably saved a fair amount of chip area. On the other hand, casting decisions in stone early on could have impeded the research goals of the project. In particular, significant design changes, already difficult because of the strong interaction between application, architecture, and chip layout, would have been impossible. We wonder if we would have done much better by following these more rigid design disciplines.

Constraints on time and manpower posed a real problem. The pace of research in the area of special-purpose processors suggests that projects of long duration will be partially obsolete by the time they are completed; availability of researchers and the desire to keep the PSC project manageable dictated that not too many people be involved. Although the design project involved more than four people and lasted for about one year, it was less than a four man-year project. Because of these limitations, it was usually impossible to evaluate more than a few alternatives at any stage in the design. The current PSC architecture cannot be claimed to be optimal in any sense; all we can say is that it is intuitively reasonable and has been evaluated with several programs for the target applications. Our feeling is that a fast-paced project followed by a second iteration will produce better results than a project that moves more slowly, because of the information available from all stages of the first try.

One way to lessen the effort required by such a project, as well as to improve the final result, is to have improved design tools. One tool that we particularly missed having was a router. The layout editor that we used was not designed with global routing as an important objective, and the overall routing on the chip turned out to be the most tedious part of the design, as well as the most difficult to modify. Although a very clever global router would have been best, even a very simple channel router would have been quite helpful.

Another area where the tools at hand were deficient was in wirelist analysis. An electrical rules checker would have been helpful in finding inverter ratio errors and similar problems. It would also have been useful to be able to check the connectivity of the wirelist by some means other than by hand. One approach to this problem, currently being investigated at CMU, is comparison of wirelists derived from schematics with those extracted from the layout.

We found that even though the PSC is of fairly modest complexity (25,000 transistors and 250,000 CIF rectangles), our layout analysis tools began to take an inconveniently large amount of resources to run. Five megabyte files and 60 CPU-minute jobs on a VAX-11/780 were common. Since most of the changes that we made to the layout were incremental, incremental artwork analysis could be very helpful for large designs.

Improved design tools are also important at a higher level. For example, one of the best ways to evaluate architectural alternatives is to test them against application code. However, such code is difficult and time-consuming to generate in the face of ever-changing and incomplete hardware specifications. Design iterations involving changes in microcode could be facilitated by flexible microcompilers capable of generating near-optimal code.

Acknowledgments

The PSC project has profited greatly from the assistance of John Zsarnay (microcoding several applications), Monica Lam (high-level Lisp simulator of systolic arrays), and Onat Menzilcioglu (hardware multiplier, jointly with Dohi).

I. Appendix: The PSC Memory Design

The PSC contains a 64 word by 60 bit writable control store (WCS) and a 64 word by 9 bit register file [21]. Both of these are implemented with a three-transistor dynamic RAM cell design. The WCS has a shift register on its input/output. This is normally used for loading the microcode, but is also used for testing the WCS and the rest of the PSC. The WCS takes its address input from the microsequencer and puts its data out on the microinstruction bus (MIB). The register file takes its address from a system bus and puts its data out on another bus. The register file does not contain a shift register.

I.1. Circuit Design

The original die size goal for the PSC was 6.5 mm square with $\lambda = 2$ microns. Static RAM implementations for the WCS and register file were too large to meet this goal. A 4-transistor dynamic RAM was also too large [22]. A 3-transistor dynamic RAM was the only acceptable alternative, despite its disadvantages, such as the need for refreshing, and a more complex circuit design.

The basic read cycle of the RAM is that first the bit lines are precharged to V_{dd}. A WORD is then read selected and begins discharging bit lines if a high voltage is stored in the corresponding cell. The bit lines are read by a sense amplifier that drives the output shift/load register. Address input to the decoder is overlapped with precharge.

On a write cycle, cells that are to have a low voltage stored in them have their bit lines pulled down after precharge. At the same time, a word is write selected, storing the bit line values.

Figure 5 is a schematic of the salient features of the WCS circuit design. The register file does not contain a shift register and has smaller output buffers. The address drivers are super-buffers with an input latch. The decoder is a NOR array. The row drivers are bootstrapped pass transistors for both read and write select. The corresponding clock signal goes high in order to drive a select line high, and then low to discharge it. The latch signal tracks the poly delay of the select lines, so that the address remains stable until the select line is completely discharged. Unselected row lines are held low by pulldown resistors.

The RAM cell transistors are all minimum size to reduce area. This results in slow readout. An unaided RAM cell would typically take 120-150ns to discharge a bit line. In order to meet a maximum cycle time goal of 200ns, a sense amplifier must be used to speed readout.

The sense amplifier used is a cross-coupled pulldown pair. A shield resistor hides the bit line capacitance, so that only a low-capacitance node must be discharged. The reference voltage comes from a voltage divider that is in reality a very low ratio inverter. The pulldown of the inverter is connected to the bit line. The reference voltage swings from 4.3V to 4.6V as the bit line swings from 5V to 4V. This creates a 0.6-0.7V signal to sense. The sense amp discharges the node with the lower voltage when PHI1BAR goes high.

The WCS has a pass transistor that hides the large output buffer capacitance from the sense amp until sensing is over, increasing sensitivity. This transistor also provides isolation during shift register operation. The register file buffer is small enough that it does not need this isolation. The register file buffer does not contain shift register connections.

The WCS write logic is connected to the buffer output for read-write refresh cycles and data loading from the shift register. The register file write logic is connected to either the buffer output for refresh cycles, or the data bus for write cycles.

The precharge logic is a bootstrapped pullup driver. This logic is not enabled until after the select lines are low, as in the case of the address drivers. This prevents precharge from destroying a just-written low value. Precharge occurs during PHI2. The WCS and register file clocks are generated from off-chip PHI1 and PHI2 signals, which are separate from the logic PHI1 and PHI2 clocks. This enables refresh to be hidden from the logic by halting the logic clocks during a refresh cycle. During the refresh cycle, the address input comes from a refresh counter, which is incremented on each refresh cycle.

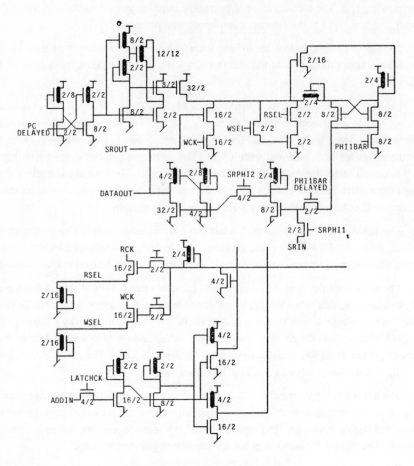

Figure 5: WCS Circuit Schematic.

I.2. Layout Design

The RAM cell is 12.5 x 18 lambda. The buried contact rules are more aggressive than the MOSIS rules.

The global layout of the WCS has two features of interest. The row decoder/driver is located in the middle of the array to minimize poly select line delay. The sense amplifiers/shift registers are twice the pitch of one column, so they are stacked on top of one another. This results in significant area being consumed by the peripheral logic. The shift register runs from the right through the odd bit pairs and then back through the even bit pairs.

I.3. Problems

The PSC was designed without the aid of advance floorplanning. The RAMs were designed for minimum area rather than for convenient pitch. A 64 x 60 array was selected for the control store. This results in a 2130μ x 3939μ block size. It was later discovered that a shorter and wider block was more desirable. This could be implemented with a 32 x 120 array with output multiplexing. Unfortunately, this is not possible with the 3-transistor RAM cell design without using read/write cycles for all writes. This is because an entire row must be write-enabled at one time, so odd and even columns must both be written with valid data. Other RAM designs do not suffer from this problem.

In the case of the register file, the pitch of a 128 x 9 array was unacceptable; a 64 x 18 design was required. For the same reason as above, this could only be done using another RAM design. Due to lack of manpower, the register file was simply reduced to a 64 x 9 design for the first version of the PSC.

References

[1]   Barbacci, M.R. Instruction Set Processor Specifications (ISPS): The Notation and Its Application. *IEEE Transactions on Computers* C-30(1):24-40, January, 1981.

[2]   Bayer, R. and Haerder, T. Preplanning of Disk Merges. *Computing* 21(1):1-16, 1978.

[3]   Brent, R.P. and Kung, H.T. *Systolic VLSI Arrays for Polynomial GCD Computation.* Technical Report, Carnegie-Mellon University, Computer Science Department, May, 1982.

[4]   Cohen, D. and Tyree, V. Quality Control from the Silicon Broker's Perspective. *VLSI Design* III(4):24-30, July/August, 1982.

[5]   Dowell, R., Newton, A.R. and Pederson, D.O. SPICE VAX User's Guide.
        Department of Electrical Engineering and Computer Science, U. C. Berkeley.

[6]   Gupta, A. ACE: A Circuit Extractor.
        Submitted for publication.

[7]   Haken, D. *A Geometric Design Rule Checker.* VLSI Doc. V053, Carnegie-Mellon University, Computer Science Department, June, 1980.

[8]   Karp, R.M. and Rabin, M.O. *Efficient Randomized Pattern-Matching Algorithms.* Technical Report TR-31-81, Center for Research in Computing Technology, Harvard University, 1981.

[9]   Kung, H.T. Let's Design Algorithms for VLSI Systems. In *Proceedings of Conference on Very Large Scale Integration: Architecture, Design, Fabrication,* pages 65-90. California Institute of Technology, January, 1979. Also available as a CMU Computer Science Department technical report, September 1979.

[10]   Kung, H.T. Special-Purpose Devices for Signal and Image Processing: An Opportunity in VLSI. In *Proceedings of the SPIE, Vol. 241, Real-Time Signal Processing III*, pages 76-84. The Society of Photo-Optical Instrumentation Engineers, July, 1980.

[11]   Kung, H.T. Use of VLSI in Algebraic Computation: Some Suggestions. In Wang, P.S. (editor), *Proceedings of the 1981 ACM Symposium on Symbolic and Algebraic Computation*, pages 218-222. ACM SIGSAM, August, 1981.

[12]   Kung, H.T. Why Systolic Architectures? *Computer Magazine* 15(1):37-46, January, 1982.

[13]   Kung, H.T. and Leiserson, C.E. Systolic Arrays (for VLSI). In Duff, I. S. and Stewart, G. W. (editors), *Sparse Matrix Proceedings 1978*, pages 256-282. Society for Industrial and Applied Mathematics, 1979. A slightly different version appears in *Introduction to VLSI Systems* by C. A. Mead and L. A. Conway, Addison-Wesley, 1980, Section 8.3.

[14]   MacWilliams, F.J. and Sloane, N.J.A. *The Theory of Error-Correcting Codes.* North-Holland, Amsterdam, Holland, 1977.

[15]   Mead, C.A. and Conway, L.A. *Introduction to VLSI Systems.* Addison-Wesley, Reading, Massachusetts, 1980.

[16]   Ousterhout, J. Caesar: An Interactive Editor for VLSI Circuits. *VLSI Design* (Fourth Quarter):34-38, 1981.

[17]   Peterson, W.W. and Weldon, E.J., Jr. *Error-Correcting Codes.* MIT Press, Cambridge, Massachusetts, 1972.

[18]   Rabin, M.O. *Fingerprinting by Random Polynomials.* Technical Report, Center for Research in Computing Technology, Harvard University, 1981.

[19]   Schirm IV, L. Multiplier-Accumulator Application Notes.
        TRW LSI Products, January, 1980.

[20]   Terman, C.J. User's Guide to NET, PRESIM, and RNL/NL.
        Laboratory for Computer Science, MIT, July, 1982.

[21]   Walker, H. *The Control Store and Register File Design of the Programmable Systolic Chip.* Technical Report, Carnegie-Mellon University, Computer Science Department, January, 1983.

[22]   Walker, H. *A 4-Kbit 4-transistor Dynamic RAM.* Technical Report, Carnegie-Mellon University, Computer Science Department, January, 1983.

Special-Purpose Chip Architectures

The VLSI Design of a Reed-Solomon Encoder Using Berlekamp's Bit-Serial Multiplier Algorithm
T.K. Truong, L.J. Deutsch, I.S. Reed, I.S. Hsu, K. Wang and C.S. Yeh

A VLSI Chess Legal Move Generator
J. Schaeffer, P.A.D. Powell and J. Jonkman

New VLSI Architectures with Reduced Hardware
J. Ja'Ja' and R.M. Owens

The VLSI Design of a Reed-Solomon Encoder Using Berlekamp's Bit-Serial Multiplier Algorithm

T.K. Truong and L.J. Deutsch
Communications Systems Research Section
Jet Propulsion Laboratory

I.S. Reed, I.S. Hsu, K. Wang, and C.S. Yeh
University of Southern California

E.R. Berlekamp has developed for the Jet Propulsion Laboratory a bit-serial multiplication algorithm for the encoding of Reed-Solomon (RS) codes, using a dual basis over a Galois field. The conventional RS-encoder for long codes often requires look-up tables to perform the multiplication of two field elements. Berlekamp's algorithm requires only shifting and exclusive-OR operations. It is shown in this paper that the new dual-basis (255, 223) RS-encoder can be realized readily on a single VLSI chip with NMOS technology.

I. INTRODUCTION

One of the major obstacles in the exploration of deep space is being able to communicate at an acceptable quality with a spacecraft that is several astronomical units from the Earth. Such deep space probes must be as light and as small as possible in order to make good use of their limited fuel and power resources. For this

303

reason, the receiving antennas on board a probe have been kept rather small. Voyager 2, for example, has only a five meter dish for reception of uplink data from the Earth. Also, the transmitters on such spacecraft are of relatively low power.

The link from Earth to deep space takes advantage of high power ground-based transmitters and huge transmitting antennas. Also, the amount of data that needs to be sent to a spacecraft such as Voyager is very small so that low data rates may be used. This allows the amount of transmitted energy per bit of information to be very large - more than enough for the spacecraft to receive good quality data with its small antenna.

The downlink telemetry, i.e. the data sent from the spacecraft to the Earth, is a different story entirely. In this case, great amounts of data need to be transmitted with a low power transmitter and a small antenna. Even though the ground-based receiving antennas are large and the receivers are of very low noise, there is still a considerable degradation in the quality of the data. For this reason, all deep space probes use some form of channel coding.

Channel coding is a form of information processing that adds redundancy to the data that is to be transmitted so that if some of it is lost or received in error the original data can still be extracted. There are many different types of channel codes. Some familiar examples are repetition codes, in which each bit of data is merely repeated a number of times, and Hamming codes, in which a series of parity checks on the data is actually sent rather than the data itself. The problem with the repetition codes is that they are very wasteful in throughput. Even the Hamming codes increase the number of transmitted bits per second too much for deep space work. More information on the various types of codes can be found in [1].

The coding that will be used by Voyager 2 at Uranus and Neptune encounters is known as a concatenated coding system. There are actually two codes involved with the output of the first, or "outer", encoder being passed to the input of the second, or "inner", encoder. The inner code is a convolutional code. Bits are sent through a seven-bit shift register. Two sets of taps from this register are used to generate two independent parity checks on its contents each time a new bit is shifted in. These parity checks are output from the encoder. Since the register length is seven and two bits are output for each input bit, the code is known as a (7, 1/2) convolutional code. The errors produced by using this kind of code in a noisy channel tend to occur in clumps called "bursts."

The outer code is a Reed-Solomon code. Reed-Solomon codes are a form of block code. This means that a fixed number of bits from the input, called a block or a frame, is converted to a different fixed number of bits at the output. In the case of Voyager, each set of 1,784 input bits is converted to 2040 output bits. This is done in such a way that the first 1,784 of the output bits are the same as the input bits. A code that has this property is called "systematic." Systematic codes are useful in communication applications since the decoding process need not be performed when the extra correction power of the code is not needed. Reed-Solomon codes are particularly good at correcting the kind of burst errors that are produced by the inner convolutional code. More details on concatenated coding may be found in [2], [3] and [4].

Reed-Solomon codes are an example of non-binary coding. The input sequence can be thought of as a polynomial over an extension field of $GF(2)$, the field of binary numbers. A number in this extension field will be referred to as a Reed-Solomon symbol. Suppose this field is $GF(2^m)$ and let the input to the encoder be

$$I(x) = \sum_{i=2t}^{n-1} b_i x^i$$

where each b_i is an element of $GF(2^m)$ and is represented by a string of m bits. The integer t will determine the number of errors that the code can correct. The output of the encoder is a second polynomial, called a Reed-Solomon codeword,

$$C(x) = \sum_{i=0}^{n-1} c_i x^i$$

where each c_i is in $GF(2^m)$, $c_i = b_i$ for $i=2t, 2t+1, \ldots n-1$, and $C(x)$ is divisible by a particular polynomial $g(x)$ whose coefficients are also in $GF(2^m)$. The polynomial $g(x)$ is called the "generating polynomial" of the Reed-Solomon code. It has the form

$$g(x) = \prod_{i=b}^{b+2t-1} (x - \gamma^i) = \sum_{i=0}^{2t} g_i x^i \ . \tag{1}$$

The encoder determines $C(x)$ by dividing $I(x)$ by $g(x)$ to get

$$I(x) = q(x)g(x) + r(x)$$

where $q(x)$ and $r(x)$ are polynomials over $GF(2^m)$. Then

$$\begin{aligned}
C(x) &= q(x)g(x) \\
&= I(x) - r(x) \qquad\qquad (2) \\
&= I(x) + r(x)
\end{aligned}$$

where the last equality is true since addition and subtraction are the same operation over extensions of the binary field. It is evident from equation (2) that the Reed-Solomon encoder need only determine the remainder polynomial $r(x)$ and add it to the input.

The Reed-Solomon code discussed above would be described by following set of parameters:

m = number of bits per symbol

$n = 2^m - 1$ = length of a codeword in symbols

t = maximum number of symbol errors that can be corrected (if t or fewer symbols arrive at the decoder in error, then the original codeword can still be reconstructed)

$2t$ = number of check symbols that are appended to the input to form the Reed-Solomon codeword.

$k = n - 2t$ = number of input information symbols in a codeword.

In the case of the Voyager code, which is the one that was implemented in VLSI, $m = 8$, $n = 255$, $t = 16$, and $k = 223$. This code is usually referred to as the (255, 223) Reed-Solomon code and has been adopted as a guideline for deep space communication by

both the National Aeronautics and Space Administration (NASA) and

the European Space Agency (ESA) [5,6].

Circuits for performing the necessary polynomial division are

well known. Examples of such circuits can be found in [7] and one

possibility is illustrated in Figure 1. The blocks R_i and Q are

delays of one Reed-Solomon symbol time. Initially, all these

registers are set to zero and both switches (controlled by the

signal SL) are set to position A. The input Reed-Solomon symbols

c_{n-1}, c_{n-2},...c_{2t} are fed into the circuit and also output on

the $C(x)$ line. After the last of the input symbols have been

loaded, the switches are moved to position B and the polynomial $r(x)$

is read out of the internal register.

The number of distinct multipliers needed to implement the

circuit of Figure 1 can be reduced almost in half by careful

selection of the generator polynomial $g(x)$. If b in (1) is chosen

so that

$$2b + 2t - 1 = 2^m - 1 \qquad (3)$$

then the coefficients of $g(x)$ are symmetric - i.e. $g(x)$ becomes a

self reciprocal polynomial that satisfies $g_o = g_{2t}$, $g_1 =$

g_{2t-1},... . For the NASA/ESA code, b was chosen to be 112 so

that $g(x)$ does indeed have this property.

The complexity of the design of a Reed-Solomon encoder results

from the computation of the products zg_i. The traditional

parallel multiplication schemes involve either too much hardware or large look-up tables. Both of these methods would prohibit the fabrication of a single chip encoder in today's technology. A single chip implementation is desirable since it would mean reduced weight and power on a deep space probe.

Recently, Berlekamp [8] developed a bit-serial multiplication algorithm that has the features necessary to solve this problem. A detailed mathematical tutorial on the algorithm may also be found in [9]. In this paper, it is shown how the Berlekamp multiplier has been used to design a single chip (255, 223) encoder on a single NMOS chip.

There has been much interest recently in the design of such Reed-Solomon encoders. In [10] Liu and Lee describe a conceptual design for a multi-chip encoder using conventional multipliers. A modular approach was also taken by Kung in [11]. Kung extends the idea to Reed-Solomon decoding as well. Berlekamp has actually built a working encoder using the new algorithm in discrete parts. Our design, however, is the first to exploit bit-serial finite field arithmetic to produce a single chip encoder.

II. Berlekamp's Bit-Serial Multiplier Algorithm Over $GF(2^m)$

In order to understand Berlekamp's multiplier algorithm some mathematical preliminaries are needed. Toward this end the mathematical concepts of the trace and a complementary (or dual) basis are introduced. For more details and proofs see [8], [12] and [13].

Definition 1: The trace of an element β belonging to $GF(p^m)$, the Galois field of p^m elements, is defined as follows:

$$Tr(\beta) = \sum_{k=0}^{m-1} \beta^{p^k}$$

In particular, for p = 2,

$$Tr(\beta) = \sum_{k=0}^{m-1} (\beta)^{2^k}$$

The trace has the following properties

(1) $[Tr(\beta)]^p = \beta + \beta^p + \ldots + \beta^{p^{m-1}} = Tr(\beta)$, where $\beta \in GF(p^m)$.

This implies that $Tr(\beta) \in GF(p)$; i.e., the trace is in the ground field $GF(p)$.

(2) $Tr(\beta + r) = Tr(\beta) + Tr(r)$, where β, $r \in GF(p^m)$.

(3) $Tr(c\beta) = cTr(\beta)$, where $c \in GF(p)$.

(4) $Tr(1) \equiv m \pmod{p}$.

Definition 2: A basis $\{\mu_j\}$ in $GF(p^m)$ is a set of m linearly independent elements in $GF(p^m)$.

Definition 3: Two bases $\{\mu_j\}$ and $\{\lambda_\kappa\}$ are said to be complementary of the dual of one another if

$$Tr(\mu_j \lambda_k) = \begin{cases} 1, & \text{if } j = k \\ \\ 0, & j \neq k \end{cases}$$

Lemma: If α is a root of an irreducible polynomial of degree m in $GF(p^m)$, then $\{\alpha^k\}$ for $0 \leq k \leq m - 1$ is a basis of $GF(p^m)$. The basis $\{\alpha^k\}$ for $0 \leq k \leq m - 1$ is called the natural basis of $GF(p^m)$.

Theorem 1 :   Every basis has a complementary basis.

Corollary 1:   Suppose the bases $\left\{ \mu_j \right\}$ and $\left\{ \lambda_k \right\}$ are complementary. Then a field element z can be expressed in the basis $\left\{ \lambda_k \right\}$ by the expansion

$$z = \sum_{k=0}^{m-1} z_k \lambda_k = \sum_{k=0}^{m-1} \text{Tr}(z\mu_k) \, \lambda_k$$

Proof: Let $z = z_0 \lambda_0 + z_1 \lambda_1 + \ldots + z_{m-1} \lambda_{m-1}$. Multiply both sides by α^k and take the trace. Then by Def. 3 and the properties of the trace,

$$\text{Tr}(z\alpha^k) = \text{Tr} \left(\sum_{i=0}^{m-1} z_i (\lambda_i \mu_k) \right) = z_k \quad \text{Q.E.D.}$$

The following corollary is an immediate consequence of Corollary 1.

Corollary 2:   The product $w = zy$ of two field elements in $GF(p^m)$ can be expressed in the dual basis by the expansion

$$w = \sum_{k=0}^{m-1} \text{Tr}(zy\mu_k) \, \lambda_k$$

where $\text{Tr}(zy\mu_k)$ is the kth coefficient of the dual basis for the product of two field elements.

These two corollaries provide a theoretical basis for the new RS-encoder algorithm.

III. A Simple Example of Berlekamp's Algorithm Applied to an RS-Encoder

This section follows the treatment in Ref. 9. This example is included to illustrate how Berlekamp's new bit-serial multiplier algorithm can be used to realize the RS-encoder structure presented in Fig. 1.

Consider a (15, 11) RS code over $GF(2^4)$. For this code, $m = 4$, $n = 15$, $t = 2$, and $n - 2t = 11$ information symbols. Let α be a root of the primitive irreducible polynomial $f(x) = x^4 + x + 1$ over $GF(2)$ so α satisfies $\alpha^{15} = 1$. An element z in $GF(2^4)$ is representable by 0 or α^j for some j, $0 \leq j \leq 14$. The element z can be represented also by a polynomial in over $GF(2)$. This is the representation of $GF(2^4)$ in the natural basis $\left\{ \alpha^k \right\}$ for $0 \leq k \leq 3$. That is, $z = u_o + u_1 \alpha + u_2 \alpha^2 + u_3 \alpha^3$, where $u_k \in GF(2)$ for $0 \leq k \leq 3$.

In Table 1, the first column is the index or logarithm of an element in base α. The logarithm of the zero element is denoted by an asterisk. Column 2 shows the 4-tuples of the coefficients of the elements expressed as polynomials.

The trace of the element z in $GF(2^4)$ is found by Def. 1 and the properties of the trace to be

$$TR(z) = u_0 Tr(1) + u_1 Tr(\alpha) + u_2 Tr(\alpha^2) + u_3 Tr(\alpha^3)$$

where $Tr(1) \equiv 4 \pmod{2} = 0$, $Tr(\alpha) = Tr(\alpha^2) = \alpha + \alpha^2 + \alpha^4 + \alpha^8 = 0$ and $Tr(\alpha^3) = \alpha^3 + \alpha^6 + \alpha^9 + \alpha^{12} = 1$. Thus $Tr(z) = u_3$. The trace element α^k in $GF(2^4)$ is listed in column 3 of Table 1.

By Def. 2 any set of four linearly independent elements can be used as a basis for the field $GF(2^4)$. To find the dual basis of the basis $\left\{\alpha^j\right\}$ in $GF(2^4)$ let a field element z be expressed in dual basis $\left\{\lambda_0, \lambda_1, \lambda_2, \lambda_3\right\}$. From Corollary 1 the coefficients of z are $z_k = Tr(z\alpha^k)$ for $0 \le k \le 3$. Thus $z_0 = Tr(z)$, $z_1 = Tr(z\alpha)$, $z_2 = Tr(z\alpha^2)$ and $z_3 = Tr(z\alpha^3)$. Let $z = \alpha^i$ for some i, $0 \le i \le 14$. Thus a coefficient z_k, for $0 \le k \le 3$, of an element z in the dual space can be obtained by cyclically shifting the trace column in Table 1 upward by k positions where the first row is excluded. These appropriately shifted columns of coefficients are shown in Table 1 as the last four columns. In Table 1 the elements of the dual basis, $\lambda_0, \lambda_1, \lambda_2, \lambda_3$, are underlined. Evidently $\lambda_0 = \alpha^{14}$, $\lambda_1 = \alpha^2$, $\lambda_2 = \alpha$ and $\lambda_3 = 1$ are the four elements of the dual basis.

In order to make the generator polynomial g(x) symmetric b must satisfy the equation $2b + 2t - 2 = 2^m - 1$. Thus b = 6 for this code. The Υ in Eq. (1) can be any primitive element in $GF(2^4)$. It will be shown in Section IV that Υ can be chosen to simplify the binary mapping matrix. In this example let $\Upsilon = \alpha$. Thus the generator polynomial is given by

$$g(x) = \prod_{j=6}^{9} (x - \alpha^j) = \sum_{i=0}^{4} g_i x^i$$

where $g_0 = g_4 = 1$, $g_1 = g_3 = \alpha^3$ and $g_2 = \alpha$.

Let g_i be expressed in the basis $\left\{1, \alpha, \alpha^2, \alpha^3\right\}$. Let z, a field element be expressed in the dual basis; i.e., $z = z_0\lambda_0 + z_1\lambda_1 + z_2\lambda_2 + z_3\lambda_3$. In Fig. 1 the products zg_i for $0 \le i \le 3$ need to be computed.

Since $g_3 = g_1$, it is necessary to compute only zg_0, zg_1 and zg_2. Let the products zg_i for $0 \leq i \leq 2$ be represented in the dual basis. By Corollary 2 zg_i can be expressed in the dual basis as

$$z \begin{bmatrix} g_0 \\ g_1 \\ g_2 \end{bmatrix} = \sum_{k=0}^{3} \begin{bmatrix} T_0^{(k)}(z) \\ T_1^{(k)}(z) \\ T_2^{(k)}(z) \end{bmatrix} \lambda_k \tag{4}$$

where $T_i^{(k)}\alpha(z) = \text{Tr}(zg_i\alpha^k)$ is the kth coefficient (or kth bit) of zg_i for $0 \leq i \leq 2$ and $0 \leq k \leq 3$.

The present problem is to express $T_i^{(k)}(z)$ recursively in terms of $T_i^{(k-1)}(z)$ for $1 \leq k \leq 3$. Initially for $k = 0$,

$$\begin{bmatrix} T_0^{(0)}(z) \\ T_1^{(0)}(z) \\ T_2^{(0)}(z) \end{bmatrix} = \begin{bmatrix} \text{Tr}(zg_0) \\ \text{Tr}(zg_1) \\ \text{Tr}(zg_2) \end{bmatrix} = \begin{bmatrix} \text{Tr}(z\alpha^0) \\ \text{Tr}(z\alpha^3) \\ \text{Tr}(z\alpha) \end{bmatrix} = \begin{bmatrix} z_0 \\ z_3 \\ z_1 \end{bmatrix} \tag{5}$$

where $\text{TR}(z\alpha^j) = \text{TR}((z_0\lambda_0 + z_1\lambda_1 + z_2\lambda_2 + z_3\lambda_3)\alpha^j) = z_j$ for $0 \leq j \leq 3$. Equation (5) can be expressed in a matrix form as follows:

$$\begin{bmatrix} T_0^{(0)}(z) \\ T_1^{(0)}(z) \\ T_2^{(0)}(z) \end{bmatrix} = \begin{bmatrix} 1 & 0 & 0 & 0 \\ 0 & 0 & 0 & 1 \\ 0 & 1 & 0 & 0 \end{bmatrix} \begin{bmatrix} z_0 \\ z_1 \\ z_2 \\ z_3 \end{bmatrix} \tag{6}$$

The above matrix is the 3 x 4 binary mapping matrix of the problem.

To compute $T_i^{(k)}(z)$ for $k > 0$, observe that $T_i^{(k)}(z) = \text{Tr}((\alpha z)g_i\alpha^{k-1}) = T_i^{(k-1)}(\alpha z)$. Hence $T_i^{(k)}(z)$ is obtained from $T_i^{(k-1)}(z)$ by replacing z by $y = \alpha z$. Let $\alpha z = y = y_0\lambda_0 + y_1\lambda_1 + y_2\lambda_2 + y_3\lambda_3$, where $y_m = \text{Tr}(y\alpha^m) = \text{Tr}(z\alpha^{m+1})$ for $0 \le m \le 3$. Then $T_i^{(k)}$ is obtained from $T_i^{(k-1)}$ by replacing z_0 by $y_0 = \text{Tr}(z\alpha) = z_1$, z_1 by $y_1 = \text{Tr}(z\alpha^2) = z_2$, z_2 by $y_2 = \text{Tr}(z\alpha^3) = z_3$ and $y_3 = \text{Tr}(z\alpha^4) = \text{Tr}(z(\alpha + 1)) = z_0 + z_1$.

To recapitulate $zg_i = T_i^{(0)}\lambda_0 + T_i^{(1)}\lambda_1 + T_i^{(2)}\lambda_2 + T_i^{(3)}\lambda_3$, where $0 \le i \le 3$ and $z = z_0\lambda_0 + z_1\lambda_1 + z_2\lambda_2 + z_3\lambda_3$, can be computed by Berlekamp's bit-serial multiplier algorithm as follows:

(1) Initially for $k = 0$, compute $T_0^{(0)}(z)$, $T_1^{(0)}(z)$ and $T_2^{(0)}(z)$ by Eq. (10). Also $T_3^{(0)}(z) = T_1^{(0)}(z)$.

(2) For $k = 1, 2, 3$, compute $T_i^{(k)}(z)$ by

$$T_i^{(k)}(z) = T_i^{(k-1)}(y)$$

where $0 \le i \le 3$ and $y = \alpha z = y_0\lambda_0 + y_1\lambda_1 + y_2\lambda_2 + y_3\lambda_3$ with $y_0 = z_1$, $y_1 = z_2$, $y_2 = z_3$ and $y_3 = z_0 + z_1 = T_f$, where $T_f = z_0 + z_1$ is the feedback term of the algorithm.

The above example illustrates Berlekamp's bit-serial multiplier algorithm. This algorithm developed in Ref. 8 requires shifting and XOR operations only. Berlekamp's dual basis RS-encoder is well-suited to a pipeline structure which can be implemented in VLSI design. The same procedure extends to the design of a (255, 223) RS-encoder over $GF(2^8)$.

IV.   A VLSI Architecture of a (255, 223) RS-Encoder with Dual-Basis
       Multiplier

Since $g_3 = g_1$ an architecture is designed to implement

(255, 223) RS-encoder using Berlekamp's multiplier algorithm. The

circuit is a direct mapping from an encoder using Berlekamp's

bit-serial algorithm as developed in the previous sections to an

architectural design. This architecture can be realized quite

readily on a single NMOS VLSI chip.

Let $GF(2^8)$ be generated by α, where α is a root of the

primitive irreducible polynomial $f(x) = x^8 + x^7 + x^2 + x + 1$ over

$GF(2)$. The natural basis of this field is $\{1, \alpha, \alpha^2, \alpha^3, \alpha^4, \alpha^5,$

$\alpha^6, \alpha^7\}$. From Corollary 1 the coefficients of a field element j

can be obtained from $z_k = Tr(\alpha^{j+k})$ for $0 < k < 7$, where $\alpha^j =$

$z_0\lambda_0 + \ldots + z_7\lambda_7$. It can be shown that the dual basis $\{\lambda_0, \lambda_1,$

$\ldots, \lambda_7\}$ of the natural basis is the ordered set $\{\alpha^{99}, \alpha^{197}, \alpha^{203},$

$\alpha^{202}, \alpha^{201}, \alpha^{200}, \alpha^{199}, \alpha^{100}\}$.

It was mentioned previously that γ in Eq. (1) can be chosen to

simplify the binary mapping matrix. Two binary matrices, one for

the primitive element $\gamma = \alpha^{11}$ and the other for $\gamma = \alpha$, were

computed. It was found that the binary mapping matrix for $\gamma =$

α^{11} had a smaller number of 1's. Hence this binary mapping

matrix was used in the design. For this case the generator

polynomial $g(x)$ of the RS-encoder over $GF(2^8)$ was given by

$$g(x) = \prod_{j=112}^{143} (x - \alpha^{11j}) = \sum_{i=0}^{32} g_i x^i \qquad (7)$$

where $g_0 = g_{32} = 1$, $g_1 = g_{31} = \alpha^{249}$, $g_2 = g_{30} = \alpha^{59}$, $g_3 = g_{29} = \alpha^{66}$, $g_4 = g_{28} = \alpha^4$, $g_5 = g_{27} = \alpha^{43}$, $g_6 = g_{26} = \alpha^{126}$, $g_7 = g_{25} = \alpha^{251}$, $g_8 = g_{24} = \alpha^{97}$, $g_9 = g_{23} = \alpha^{30}$, $g_{10} = g_{22} = \alpha^3$, $g_{11} = g_{21} = \alpha^{213}$, $g_{12} = g_{20} = \alpha^{50}$, $g_{13} = g_{19} = \alpha^{66}$, $g_{14} = g_{18} = \alpha^{170}$, $g_{15} = g_{17} = \alpha^5$, and $g_{16} = \alpha^{24}$.

The binary mapping matrix for the coefficients of the generator polynomial in Eq. (7) is computed and shown in Ref. 14. The feedback term T_f in Berlekamp's algorithm is found in this case to be:

$$T_f = \text{Tr}(\alpha^8 z) = \text{Tr}\left((\alpha^7 + \alpha^2 + \alpha + 1) \, z\right) = z_0 + z_1 + z_2 + z_7 \quad (8)$$

In the following, a VLSI chip architecture is designed to realize a (255, 223) RS-encoder using the above parameters and Berlekamp's algorithms. An overall block diagram of this chip is shown in Fig. 2. In Fig. 2 VDD and GND are power pins. CLK is a clock signal, which in general is a periodic square wave. The information symbols are fed into the chip from the data-in pin DIN bit-by-bit. Similarly the encoded codeword is transmitted out of the chip from the data-out pin DOUT sequentially. The control signal LM (load mode) is set to 1 (logic 1) when the information symbols are loaded into the chip. Otherwise, LM is set to 0.

The input data and LM signals are synchronized by the CLK signal, while the operations of the circuit and output data signal are synchronized by two nonoverlapping clock signals $\phi 1$ and $\phi 2$. To save space, dynamic registers are used in this design. A logic diagram of a 1-bit dynamic register with reset is shown in Fig. 3.

The timing diagram of CLK, $\phi 1$, $\phi 2$, LM, DIN and DOUT signals are shown in Fig. 4. The delay of DOUT with respect to DIN is due to the input and output flip-flops.

Figure 5 shows the block diagram of a (255, 223) RS-encoder over $GF(2^8)$ using Berlekamp's bit-serial multiplier algorithm. The circuit is divided into five units. The circuits in each unit are discussed in the following:

(1) Product Unit: The Product Unit is used to compute T_f, T_{32}, ..., T_0. This circuit is realized by a Programmable Logic Array (PLA) circuit [15]. Since $T_0 = T_{32}$, $T_1 = T_{31}$, ..., $T_{15} = T_{17}$, only T_f, T_{31}, ..., T_{17} and T_{16} are actually implemented in the PLA circuit. T_0, ..., T_{15} are connected directly to T_{32}, ..., T_{17}, respectively.

(2) Remainder Unit: The Remainder Unit is used to store the coefficients of the remainder during the division process. In Fig. 5, S_i for $0 \leq i \leq 30$ are 8-bit shift registers with reset. The addition in the circuit is a modulo 2 addition or Exclusive-OR operation. While c_{32} is being fed to the circuit, c_{31} is being computed and transmitted sequentially from the circuit. Simultaneously c_i is computed and then loaded into S_i for $0 \leq i \leq 30$. Then c_{30}, ..., c_0 are transmitted out of the encoder bit-by-bit.

(3) Quotient Unit: In Fig. 5, Q and R represent a 7-bit shift register with reset and an 8-bit shift register with reset and parallel load, respectively. R and Q store the currently operating coefficient and the next

coefficient of the quotient polynomial, respectively. A

logic diagram of register R is shown in Fig. 6.   The

value of z_i is loaded into R_i every eight clock

cycles, where $0 \leq i \leq 7$.   Immediately after all 223

information symbols are fed into the circuit, the control

signal SL changes to logic 0.   Thenceforth the contents

of Q and R are zero so that the check symbols in the

Remainder Unit sustain their values.

(4) I/O Unit: This unit handles the input/output operations.

In Fig. 5 both F_0 and F_1 are flip-flops. A pass

transistor controlled by $\phi 1$ is inserted before F_1 for

the purpose of synchronization. Control signal SL selects

whether a bit of an information symbol or a check symbol

is to be transmitted.

(5) Control Unit: The Control Unit generates the necessary

control signals. This unit is further divided into 3

portions, as shown in Fig. 7. The two-phase clock

generator circuit in Ref. 15 is used to convert a clock

signal into two nonoverlapping clock signals $\phi 1$ and $\phi 2$.

In Fig. 8 a logic diagram of the circuit for generating

control signals START and SL is shown. Control signal

START resets all registers and the divide-by-8 counter

before the encoding process begins. Control signal LD is

simply generated by a divide-by-8 counter to load the

z_i's into the R_i's in parallel.

Since a codeword contains 255 symbols, the computation of a

complete encoded codeword requires 255 "symbol cycles." A symbol cycle is the time interval for executing a complete cycle of Berlekamp's algorithm. Since a symbol has 8 bits, a symbol cycle contains 8 "bit cycles." A bit cycle is the time interval for executing one step in Berlekamp's algorithm. In this design a bit cycle requires a period of the clock cycle.

The layout design of this (255, 223) RS-encoder is shown in Fig. 9. Before the design of the layout each circuit was simulated on a general-purpose computer by using SPICE (a transistor-level circuit simulation program) (Ref. 16). The total circuit requires about 3000 transistors, while a similar discrete design requires 30 CMOS IC chips. This RS-encoder design will be fabricated and tested in the near future.

V.   Concluding Remarks

A VLSI structure is developed for a Reed-Solomon encoder using Berlekamp's bit-serial multiplier algorithm. This structure is both regular and simple.

The circuit in Fig. 2 can be modified easily to encode an RS code with a different field representation and different parameters other than those used in Section IV. Table 2 shows the primary modifications needed in the circuit to change a given parameter.

Table 1.   Representations of elements over $GF(2^4)$
generated by $\alpha^4 = \alpha + 1$

Power j	Elements in natural basis $\alpha^3\alpha^2\alpha^1\alpha^0$	$Tr(\alpha^j)$	Elements in dual basis $z_0 z_1 z_2 z_3$
*	0000	0	0000
0	0001	0	0001 λ_3
1	0010	0	0010 λ_2
2	0100	0	0100 λ_1
3	1000	1	1001
4	0011	0	0011
5	0110	0	0110
6	1100	1	1101
7	1011	1	1010
8	0101	0	0101
9	1010	1	1011
10	0111	0	0111
11	1110	1	1111
12	1111	1	1110
13	1101	1	1100
14	1001	1	1000 λ_0

Table 2. The primary modifications of the encoder circuit in Fig. 2 needed to change a parameter

Parameter to be changed	The value used for the circuit in Fig. 2	New value	The circuits of Fig. 2 that require modification
1. Generator polynomial	Eq. (8)	$g(x)$	The PLA of the Product Unit needs to be changed
2. The finite field used	$GF(2^8)$	$GF(2^m)$	All registers are m-bit resistors, except Q is a (m-1)-bit register. A divide-by-m counter is used. (The generator polynomial may not be changed.)
3. Error-correcting capability	16	t	2t-2 shift registers are required in the Remainder Unit. (The generator polynomial is also changed.)
4. Number of information symbols	223	k	None is changed, since k is implicitly contained in the control signal LM.

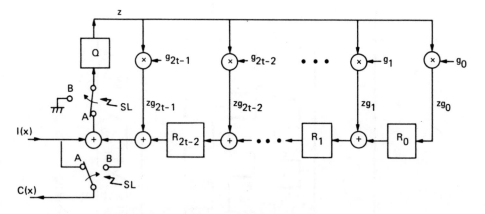

Figure 1. A structure of a t-error correcting RS-encoder

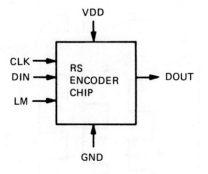

Figure 2. Symbolic diagram of a RS encoder chip

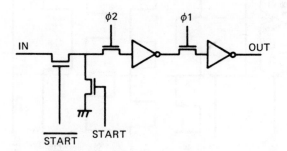

Figure 3. Logic diagram of a 1-bit dynamic register with reset

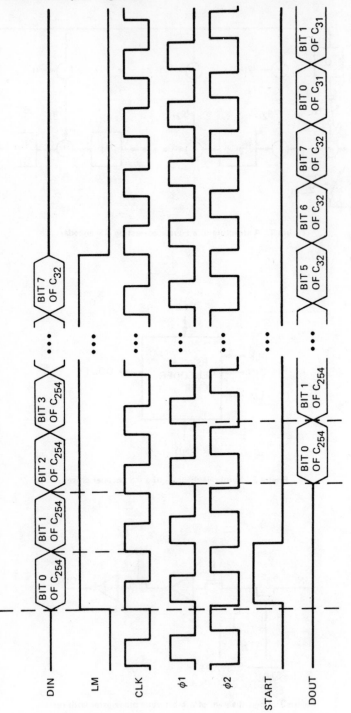

Figure 4. The signals of DIN, LM, CLK, $\phi 1$, $\phi 2$, START, and DOUT

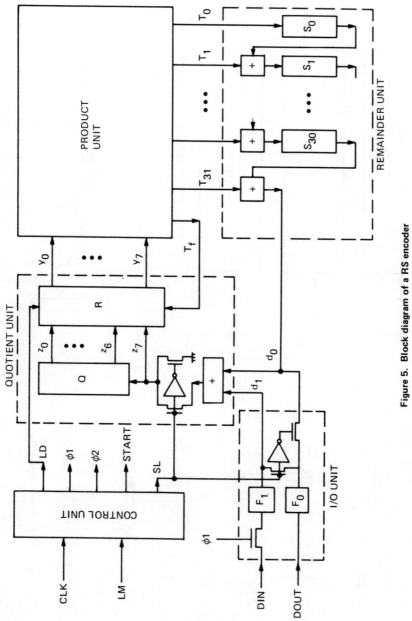

Figure 5. Block diagram of a RS encoder

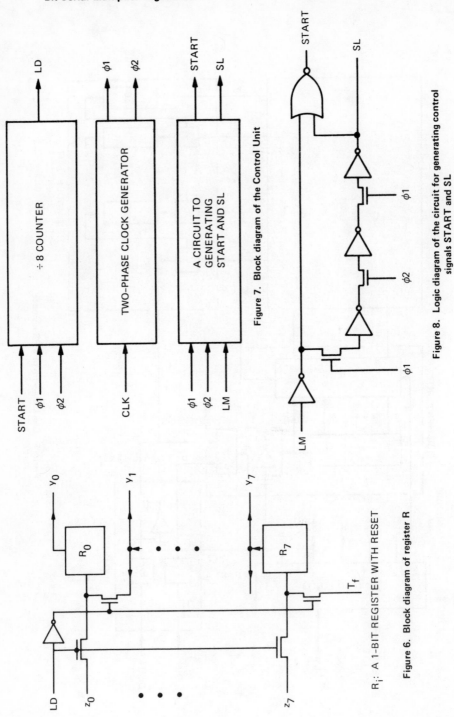

Figure 7. Block diagram of the Control Unit

Figure 8. Logic diagram of the circuit for generating control
signals START and SL

R_i: A 1-BIT REGISTER WITH RESET

Figure 6. Block diagram of register R

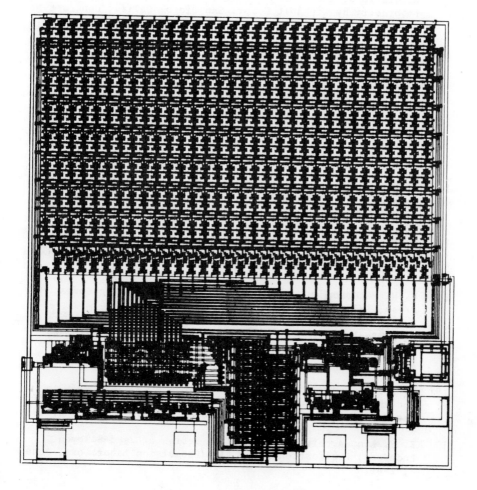

Figure 9. Layout of the (255, 223) RS-encoder chip

REFERENCES

[1]   R. J. McEliece, The Theory of Information and Coding,
      Addison-Wesley Publishing Company, Reading, Mass., 1977.

[2]   J. P. Odenwalder, "Concatenated Reed-Solomon/Viterbi Channel
      Coding for Advanced Planetary Missions: Analysis,
      Simulations, and Tests," submitted to the Jet Propulsion
      Laboratory by Linkabit Corp., San Diego, Calif., Contract No.
      953866, December 1974.

[3]   R. L. Miller, L. J. Deutsch, and S. A. Butman, On the Error
      Statistics of Viterbi Decoding and the Performance of
      Concatenated Codes, Publication 81-9, Jet Propulsion
      Laboratory, Pasadena, Calif., September 1, 1981.

[4]   K. Y. Liu and J. J. Lee, An Experimental Study of the
      Concatenated Reed-Solomon/Viterbi Channel Coding System and
      Its Impact on Space Communications, Publication 81-58, Jet
      Propulsion Laboratory, Pasadena, Calif., August 15, 1981.

[5]   R. R. Stephens and M. F. Pellet, "Joint NASA/ESA Telemetry
      Channel Coding Guideline: Issue 1," NASA/ESA Working Group
      (NEWG) Publication, January 1982.

[6]   H. F. A. Reefs and A. R. Best, "Concatenated Coding on a
      Spacecraft-to-Ground Telemetry Channel Performance," Proc.
      ICC 81, Denver, 1981.

[7]   W. W. Peterson and E. J. Weldon, Error-Correcting Codes, MIT
      Press, 1972.

[8]   E. R. Berlekamp, "Bit Serial Reed-Solomon Encoders," IEEE
      Trans. on Information Theory, Volume IT-28 No. 6, November
      1982.

[9]   M. Perlman and J. J. Lee, Reed-Solomon Encoders -
      Conventional vs Berlekamp's Architecture, Publication 82-71,
      Jet Propulsion Laboratory, Pasadena, Calif., December 1, 1982.

[10] K. Y. Liu, Architecture for VLSI design of Reed-Solomon
      Encoders, Publication 81-87, Jet Propulsion Laboratory, June
      8, 1981.

[11]

[12] P. J. MacWilliams and N. J. A. Sloane, The Theory of
      Error-Correcting Codes, North-Holland Publishing Company,
      1978.

[13] I. N. Herstein, Topics in Algebra, Blaisdell, 1964.

[14]   T. K. Truong, L. J. Deutsch, and I. S. Reed, The Design of a
       Single Chip Reed-Solomon Encoder, Publication 82-84, Jet
       Propulsion Laboratory, Pasadena, Calif., November 15, 1982.

[15]   C. Mead and L. Conway, Introduction to VLSI Systems,
       Addison-Wesley Publishing Company, Calif., 1980.

[16]   L. W. Negal and D. O. Pederson, "SPICE - Simulation Program
       with Integrated Circuit Emphasis, Memorandum No. ERL-M382,
       Electronics Research Laboratory, University of California,
       Berkeley, April 12, 1973.

A VLSI Chess Legal Move Generator

Jonathan Schaeffer, P.A.D. Powell and Jim Jonkman
VLSI Research Group, Institute for Computer Research
University of Waterloo, Waterloo, Ontario, Canada

ABSTRACT

Constructing a chess legal move generator (LMG) illustrates the design and evaluation of a variety of algorithms and the problems encountered implementing them in VLSI. Software-based algorithms for LMG and their possible hardware implementations are examined. Several new approaches exploiting parallelism and systolic structure are developed. The implementations of several algorithms are compared: the space, time, complexity, and feasibility tradeoffs provide interesting insights into the development of custom-designed VLSI circuits for non-numeric applications.

1. Introduction

Programming a computer to play chess has been an area of active research in computer science since the late 1950s. The first software approaches attempted to simulate human thought processes in computers. A (perhaps) surprising result of this research was that some very fast programs with limited "chess intelligence" could out-perform other programs with very sophisticated "chess intelligent" software. In the late-'70s, with the availability of faster, cheaper hardware, computer chess became less a software problem and more a hardware problem, as people tried to build machines to analyze chess moves at greater and greater speeds. The purpose of analyzing chess moves is to determine the optimal move by "looking ahead" as many moves as possible. The faster the hardware, the more chess positions can be analyzed per second and the farther ahead one can look. Potentially, this will improve the quality of moves that the program selects to play.

In November 1981 the chess program Prodigy, co-authored by Jonathan Schaeffer and Howard Johnson from the University of Waterloo, competed in the North American Computer Chess Championships. The program was written in B [1] and ran on a Honeywell 66/80, executing 1-1.5 million instructions per second. The program was very "chess-intelligent", but on a stand-alone machine, Prodigy could analyze only about 50-60 positions per second. Also competing in the tournament was Belle, a special-purpose hardware/software "chess machine" designed by Ken Thompson and Joe Condon of Bell Telephone Laboratories [2]. The resulting machine, using limited software, calculates 160,000 chess positions per second! As a result, Belle is the current world computer chess champion and is ranked a Master by the United States Chess Federation[1]. Belle is capable of looking ahead 8-9 full ply (one move by one player, or one half move) exhaustively for a typical

---

[1] Although fewer than 1% of all serious chess players ever achieve the title of Master, Belle still has a considerable way to go to be able to compete at the world championship level.

middlegame chess position (i.e. selects its next move based on the results of considering every possible move sequence from the current position, 8-9 half moves deep). With advancing VLSI technology, computation speeds will soon allow the 10 ply barrier to be surpassed; a chess machine performing at this level might surpass even the very best of the human players.

A typical chess program has three major components: legal move generation, evaluation, and look-ahead tree searching. The legal move generator finds all moves which can legally be made from a given position. The look-ahead tree searcher's job is to use the legal move generator to find moves and the evaluator to assess the "goodness" of those moves. By calling itself recursively, the tree searcher builds a (large) tree of possible move sequences and examines it carefully to chose the optimal move.

The algorithms used by the evaluation and tree-searching components, which are software approaches to simulate "thinking", are what determines the "personality" of the chess program. By contrast, legal move generation is a utility in all computer chess programs, whose function is fixed by the rules of the game. Since this routine is used extensively, comprising perhaps 50% of the program's total execution time, and is a simple, well-defined, and computation-bound algorithm, it is a natural candidate for a hardware implementation. In addition, there are relatively few legal moves possible for any given position on a chess board. The limit seems to be about 100 moves for one player [3], with an average of 40 moves per position.

Most legal move generators (LMG), whether hardware or software, do not actually generate true legal moves. In fact, we are discussing a "pseudo" legal move generator, in that moving into or out of check is not considered. By generating pseudo legal moves, and then allowing the tree searching algorithms to eliminate the illegal ones (usually discovered by having one's king captured!), the problems in designing the LMG are considerably simplified[2]. Nor will we consider castling and en passant moves; they are infrequent and complicate any software/hardware design. When these "exceptions" occur, they can readily be handled by the tree-searching algorithms.

The LMG must maintain a current position on a chessboard and allow the user to modify it (put pieces on and remove them from squares). The user can request the legal moves for the current position for either white or black. The minimum requirement of the LMG is to output a list of legal moves defined as from-square/to-square pairs. As shown in Figure 1.1, standard chess notation refers to squares by their column (a-h) and row (1-8) coordinates. Interpreting the co-ordinate as a number in the range 0-63, a from/to pair would consist of a pair of 6 bit binary integers. We will follow the convention of encoding the column as the most significant 3 bits and the row as the least significant 3 bits of the move.

In this paper, we discuss the design and implementation of a VLSI chess legal move generator. In the first sections, we provide motivation for the VLSI approach by examining alternative solutions in hardware and software. Next, the feasibility of implementing these algorithms in VLSI is examined and shown to be unsatisfactory. Three new algorithms are introduced which are candidates for a VLSI

---

[2] In practice, few moves involve checks, and thus this process usually returns the correct set of legal moves.

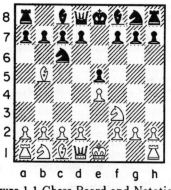

Figure 1.1 Chess Board and Notation

implementation. Finally, the implementation of the chosen algorithm is described. The paper attempts to provide an insight into the design and development process, and to indicate how an algorithm can be transformed into apparently different implementations, while still performing the same function.

2. Software Approaches

The common software approaches to the LMG problem fall into one of three categories: brute-force, incremental, and bit-vector. The basic idea of the brute-force approach is to

(1) find all pieces for the side to move,

(2) for each piece found, explore each direction that it can move, and

(3) for any square that the piece can move to, record a legal move.

The algorithm is very simple, but it is also very slow. Figure 2.1 illustrates a sample coding of the algorithm.

The incremental approach takes advantage of the observation that approximately 80% of the legal moves remain unchanged from move to move during a game. When a move is made, we update the list of legal moves by deleting those which are no longer legal and adding those which are now possible. This approach was implemented successfully in Prodigy, but the authors concluded that since programming complexity increased substantially, the average computation time per legal move generated was greater than for an optimized brute-force approach.

The bit-vector algorithm [4] is again more complex than the brute force algorithm, but in this case the complexity is handsomely repaid by the gains in speed. Properties of a position may be represented as bit vectors, with each bit corresponding to a square on the board. For example, *piece* bit-vectors could be used to indicate occupied squares; the *piece* vectors for the white and black pieces in Figure 2.2 would have the (hexadecimal) values $1430 and $4A00 respectively using a row-major convention. To obtain the bit-vector representing the legal moves for the white knight on square c3,

(1) generate the white *piece* vector. Its bitwise complement, the *possible* vector, indicates the squares the white pieces can move to. In the example, the *possible* vector is $ebcf.

```
# Pseudo-code for a typical brute force LMG.
# Pawn moves are ignored.
# Colour      - colour of the side to move
# Board       - 64 element matrix specifying whether a square
#               is EMPTY or occupied by a WHITE or BLACK piece.
for square := each of the 64 squares do
    if Board[square].colour == Colour then
        # We have found a piece of the right colour
        for each of the directions that the piece can move do
            nextsquare := square;
            while piece can still move in this direction do
                nextsquare := nextsquare in this direction;
                if nextsquare is off the board then
                    break;
                if Board[nextsquare].colour == Colour then
                    # We have found a piece of the same colour
                    break;
                if Board[ nextsquare ].colour <> EMPTY then
                    # capture move found record legal move;
                    break;
                endif;
                # nextsquare is empty
                record legal move;
            endwhile;
        endfor;
    endif;
endfor;
```

Figure 2.1 Brute Force Algorithm

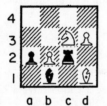

White piece	0001	0100	0011	0000
Black piece	0100	1010	0000	0000
White possible	1110	1011	1100	1111
Knight reach	0101	1000	0000	1000
Knight move	0100	1000	0000	1000
Knight capture	0100	1000	0000	0000

Figure 2.2 Board Properties Represented as Bit Strings

(2) generate the *reach* vector (representing the squares the knight can possibly move to) by using the current location of the knight as an index into a 64 entry table of precomputed *reach* vectors. Here, the *reach* vector has the value $5808.

(3) form the *move* vector (representing the set of all legal moves for the knight) by the bit-wise AND of the *reach* and *possible* vectors. In the example, the *move* vector has the value $4808, signifying legal moves to b1, a2, and a4.

(4) the *capture* vector can formed by the bit-wise AND of the *move* vector and the opponent's *piece* vector. This gives us the sub-set of legal moves which involve captures. In the example, the *capture* vector has value $4800, indicating that the moves to a2 and b1 are captures.

This approach is extremely fast and simple to implement for king, knight and pawn moves (even en passant). Unfortunately, queen, rook, and bishop moves do not fall out as simply, although again, a bit representation is helpful. Generating the *reach* vectors for these pieces is more difficult because they depend on the board configuration (i.e. the locations of the blocking pieces). The legality of a move to a square is "context sensitive" in the sense that information about the particular to-square is insufficient to determine legality; one must know the occupancy of all intervening squares from the from-square.

3. Software-to-Hardware Mapping

The main purpose for implementing legal move generation in hardware is to obtain substantially faster performance for a reasonable cost. The cost of implementation can be measured in many ways: hardware dollar costs, design costs in terms of man hours of effort, debugging costs, etc. These costs differ for various hardware alternatives.

Translating an algorithm from a higher level language to microcode is conceptually the simplest implementation, assuming that the *host* processor used is microcode-based, that the microcode can be altered, and that the modifications can be made without seriously hurting the processor's performance. In fact, the sophistication and wariness of the manufacturers of powerful microcode-based processors usually make this a difficult (if not impossible) approach. In addition, the resulting microcode is often not portable, even between different models of the same family of machines. All three of the software approaches would be arduous to convert without the benefits of a high-level microcode generation language.

Another approach is to implement the algorithm on a microprocessor which can be accessed either as an external device or as a co-processor. Given the speeds of current microprocessors, it is clear that a single microprocessor would not be substantially faster than a large mainframe type computer, or even a "super-minicomputer". However, by using a group of 32 microprocessors (one for each piece) which are broadcast the current board position simultaneously, substantial parallelism can be attained.

The bit-slice methodology is usually used to implement a dedicated peripheral or co-processor. The bit-slice hardware consists of a data path and associated control circuits, allowing the data path architecture to be tailored to a particular algorithm, and the execution to be optimized for both the algorithm and the data path architecture. The availability of general-purpose ALU chips and programmable ROMs makes this a fairly simple piece of hardware to design and construct [5]. However, the coding and debugging of the bit-slice program can be as difficult as using the microcode method. The bit-vector algorithm is a good candidate for this type of hardware implementation.

A random-logic implementation would not be significantly faster than the bit-slice implementation described above. However, it is perhaps worthwhile to construct a VLSI circuit to implement the data paths part of the bit-slice design, as the barrel shifter, registers, ALU, and other data path functions would be quite easy to implement on a chip.

4. Systolic Approach

The above algorithms and their hardware implementations have been constrained by the sequential nature of their structure. This is true even if software is mapped to hardware and then parallelism is introduced. What about using a non-sequential or systolic [6] approach?

Each non-border square of the chessboard has 8 immediate neighbours. Consider a piece which is to move horizontally from left to right, as illustrated in Figure 4.1. We observe that

(1) The set of horizontal legal moves depends only on the other pieces in the same row, not on any feature of the other rows (*row independence*), and

(2) along a given row at most one legal move is possible to each square. That is, only the piece of the proper colour nearest left of a square can move rightwards to that square (*move uniqueness*).

Row independence and *move uniqueness* can be extended to leftward, vertical and diagonal directions. For vertical moves, each of the columns is independent, and for the diagonal moves, the diagonals are independent of each other.

Assume that we are generating the legal moves for the white pieces moving horizontally left to right. Imagine a row of the chess board as a sequence of squares, each sending a message to its right neighbour (Figure 4.2). Each square uses the incoming message, the piece (if any) on it, and a globally available direction (horizontal, vertical, diagonal, etc.) to generate an output message and detect legal moves, using the following method:

(1) The *move capabilities* of a piece on a square are determined by the colour which is to move and the direction of motion. If the square is unoccupied, its move capability is C_NONE meaning that no move can originate from it; if it is occupied by an opposite-coloured piece, then its capabilities are C_THEIRS. If a piece is the right colour, but cannot move in the direction being considered, its square has capability C_OURS; kings, which can only move one square, confer the capabilities C_ONE, while queens, rooks and bishops confer the capability C_MANY.

Figure 4.1 Row Independence and Move Uniqueness

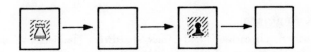

Figure 4.2 Row Communication

(2) To generate horizontal rightward moves, we start at the left boundary square. If it is occupied by a black piece (capability C_THEIRS), or by a white piece which cannot make a move in the rightward direction (capability C_OURS), or is empty (capability C_NONE), we send the output message M_NONE to the adjacent right square, indicating a legal move cannot originate from the boundary square. If a king (moving one square, thus conferring capability C_ONE) occupies the square, we send M_ONE to the adjacent right square; if it is occupied by a queen or rook (capability C_MANY), we send M_MANY, meaning that we can keep moving until blocked by an occupied square or the other boundary.

(3) Now consider the next square and its possible input messages and occupants. If it is empty (capability C_NONE) and its input message is M_NONE, no piece can move to it from the boundary square, nor can a legal move originate from it, so a M_NONE message is sent to the square's right neighbour. If the input message is M_ONE or M_MANY and the square is empty (capability C_NONE), then a legal move to the square is possible. Here the input message M_ONE will result in an output message of M_NONE (because no further move in this direction is possible), while the input message M_MANY will yield the output message M_MANY (because additional moves to subsequent neighbouring squares are possible). If the square is occupied by a white piece (capability C_OURS or C_ONE or C_MANY) another piece cannot move to it and the M_NONE, M_ONE, or M_MANY message generated by the square's own occupant would be output instead. If the square is occupied by a black piece (capability C_THEIRS) and the input message is M_NONE, the output message will also be M_NONE. However, if the input message is M_ONE or M_MANY, then there is a possible legal move - capture of the black piece. The output message will be M_NONE, as the capturing piece cannot move through the square.

This method may be applied inductively to the remaining squares to obtain all the rightward legal moves along the row. The leftward, vertical and diagonal directions may be treated in the same manner. If the boundary square input message is initialized to M_NONE, then the algorithm becomes independent of the square's location, and depends only on the occupant, the connectivity of the squares, the side to move, and the direction of motion.

Knights can move along "supra-diagonals" (Figure 4.3) in each of 8 directions. By using a row composed of the "supra-diagonals" of the board, we can generate legal moves for knights as we do for kings. If we separate regular from knight move generation, a square occupied by a knight will have C_OURS move capability for regular move generation and C_ONE for knight move generation.

Pawns are highly non-symmetric pieces, for they can move in four possible ways, depending on their position and whether they are to capture or merely advance. The squares receive as input the side to move, the direction (vertical, diagonal), and the type of move (normal or knight). If we enhance the direction information and introduce a new "piece type" for pawns, we can handle legal moves in the following manner.

A pawn on its home square is called a P2 pawn (i.e. can move forward 2 squares), while those elsewhere are P1 pawns. For forward moves (vertical up for white or vertical down for black), a square occupied by a P1 or P2 pawn has move

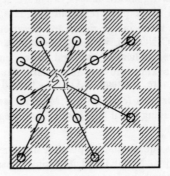

Figure 4.3 Supra-diagonals

capability C_PONE or C_PTWO, and will generate output messages M_PONE or M_PTWO respectively. If the square receiving the messages is occupied, no legal move to the square is possible and a M_NONE is output; if the square is empty a legal move to it is possible, and on receiving a M_PONE or M_PTWO message the square will send a M_NONE or M_PONE message respectively to the next square in the row. When the direction is forward diagonal, the move capability will be C_PDIAG and a M_PDIAG message is generated. If the diagonal square is occupied by an opposing man a capture is possible, but if there is no piece to be captured, the M_PDIAG message is treated like a M_NONE message.

In Appendix A a table of the various combinations of square capabilities and incoming and outgoing messages is presented. The capability indicated in the Appendix is generated by using the colour and type of piece occupying the square and the global side to move, direction of motion, and type of move information.

A *square machine* implementing the above algorithm would consist of an input port, an occupant port, global control inputs, and an output port (Figure 4.4). The occupant port specifies the colour (black or white) and type of occupant (king, queen, bishop, knight, rook, P1 pawn, P2 pawn, or empty). The global control inputs supply the side to move (white or black), which of 8 directions, and the type of move (normal or knight) to be made. There are 6 different types of messages that are propagated between *square machines* from the output port of one machine to the input port of its neighbour. Besides the input information, we need a means of extracting the moves via a *legal move* output port.

Using such *square machines*, we can construct a *Legal Move Matrix Machine*, or *LM**. One configuration for the *LM** would be 16 boards (one for each direction and move type), 64 *square machines* each (Figure 4.5) viewed as a 3 dimensional box, 16 x 8 x 8. Each board has its *square machines* interconnected in the manner appropriate for in the assigned direction. Also, *square machines* are aligned so that each column of the *LM** matrix comprises machines working on the same square. The global control ports for each board are wired in common, and the input values are set to generate moves for a particular direction. The occupant ports in each column of the *LM** matrix are also wired in common. To use the *LM** to generate

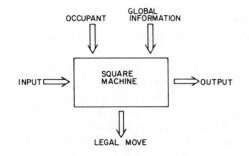

Figure 4.4 Square Machine Block Diagram

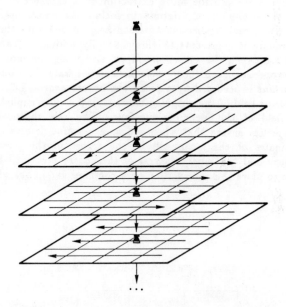

Figure 4.5 LM∗ Configuration

legal moves, the board position is broadcast to all the *square machines* through the occupant port inputs causing each *square machine* in the matrix to update its output message port and legal move output based on the occupant and the input port messages. The time taken to find a legal move depends on the propagation time of messages through a *square machine* and the number of machines through which messages must pass (at most 7 for any particular row).

So far, we have deliberately avoided discussing the important issue of how to translate legal moves into sets of from-square/to-square pairs and output them.

Although the *square machines* in the *LM** detect the legal move condition, it must also indicate what the move is. There are two ways of doing so: a) decoding logic external to the *square machines* b) or by additional structures in the *square machines* themselves. There is the additional problem of whether to output the moves sequentially or in parallel. If the legal move from/to pairs were output sequentially, then only 12 output lines would be needed, compared to a much larger number for parallel output.

If the translation were done externally we would need at least one bit of information from each *square machine* to signal the existance of a legal move. This indicates the to-square of the move. To get the from-square, we observe that because of *row independence* it must be from the first preceding occupied *square machine* in that row. Extensive logic is required to extract this information since in the worst case, the contents of the entire row must be known in order to discover the from-square.

If we allow readout of legal moves in a sequential order, then we can enhance the *square machines* to provide move extraction in a simple and effective manner. By expanding the occupant port inputs to include an additional 6 bits of *square address* information, each *square machine* can be provided with the location on the board of the square it represents (Figure 4.6). By adding a *from address* to the messages passed between the *square machines*, and using the move uniqueness and independence properties, each *square machine* can generate the from/to information. If a *square machine* is occupied by a piece, the *square address* information is placed in the *from address* field of the outgoing message; if it is unoccupied, it passes on the *from address* field of the incoming message. The *row independence* and *move uniqueness* properties ensure that when a *square machine* detects a legal move, the from/to coordinates of the move will be the values of the *from address* of the incoming message and the *square address* of the *square machine*. By allowing each *square machine* to place this information on a common *output bus* when enabled by a

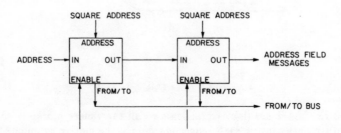

Figure 4.6 Internal From/To Pair Generation

simple output enable signal, readout of the legal moves is easy to perform.

In its final form, the *LM** would consist of circuits to do the legal move generation, and a sequencer to look for *square machines* with a legal move and read out the from/to information.

5. VLSI Algorithms

Each implementation technology has its own set of constraints and benefits. The benefits of VLSI are fairly clear: fewer parts, greater speed, cheaper reproduction, etc. The constraints are placed on the form of a design, to make it both feasible and tractable:

(1) Regularity, both at the system and logic level. The design should be modular in nature, consisting of a small number of similar, reproduced cells.

(2) Simple interconnection. One of the main restrictions of VLSI is that while logical functions are relatively cheap (take up a small area), interconnections are expensive (take up a large area).

(3) Planarity. VLSI designs are flat - that is, most functional blocks are implemented as areas of circuitry which have connections only along the periphery.

These restrictions have motivated design methodologies such as the Mead-Conway [7] and Bristle Blocks [8] methods, where design and implementation are intimately related. The design process consists of structuring a design so that it can be implemented (at the circuit level) as a hierarchically structured, easily interconnected set of rectangular blocks of logic.

The *LM** is a highly regular architecture consisting of 16 boards of 64 *square machines*, arranged as a 3 dimensional matrix. However, it is neither planar nor simply connected, as there are many communication paths between each *square machine* and its neighbours. In its current form, the *LM** is a totally combinatorial network, with a separate sequencer finite state machine providing sequential readout of from/to pairs. Let us re-examine the design, and relax the restriction that legal move generation must be combinatorial, adding a sequential element to the legal move generation algorithm.

Each of the 16 boards of *LM** is devoted to generating legal moves in one direction. We can view this as 8 pairs of boards, each handling a particular direction and its inverse (Figure 5.1). If we make the input and output message ports *bidirectional,* and switch the direction with a global direction input signal (left/right, up/down etc.), we need only 8 boards. Legal moves are now extracted by a two phase procedure. With more connectivity, even fewer boards are needed. If each *square machine* communicated with four other machines, and each port were bidirectional, legal moves could be generated with 4 boards and 4 phases. A connectivity of 8 would give 2 boards and 8 phases; the *full square machine* would have 1 general purpose board capable of handling all 16 directions and requiring 16 phases.

We can construct a version of *LM** using 64 *full square machines (FM*)* in MSI technology. However, implementing *FM** on a single chip in VLSI is quite difficult because of the many communication signals needed between each *full square machine.* For example, each of the 16 message ports would require a minimum of 3

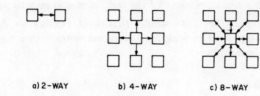

a) 2-WAY   b) 4-WAY   c) 8-WAY

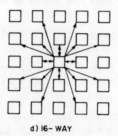

d) 16-WAY

Figure 5.1 Reconfigurations of LM*

signals, for a total of 48 input/output signals; together with global signals, power, ground, clocks, etc. this total would exceed 64. This large number of communication paths makes it impractical to split the algorithm over several chips.

Let us re-examine the *LM** from a different perspective, and see if there is another way to reduce complexity. Given row independence and move uniqueness we can regard the legal move computation as a computational *wavefront* moving along the rows, like the wavefront of computations in a systolic array [9]. By using the inverse of a systolic algorithm transformation, we can construct a sequential or finite state machine to perform the same algorithm (Figure 5.2).

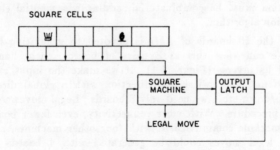

Figure 5.2 Systolic Transformation

Consider a single row of *square machines*. Each machine will generate an output message based on its current occupant, the global information, and its input message. If we store the current occupant information in a sequence of *square cells*, we can replace the *square machines* with a single *square machine* and a network to read the information in the *square cells* and present it to the single *square machine*. The *square machine* is provided a *output latch* which can store the value of the *square machine's* last output message. Legal move generation would consist of the following steps.

(1) Initialize the *output latch* with the M_NONE message. This will simulate the end of the row condition for the *square machine*. The input message is selected from the *square cell* corresponding to the end of the row. The output message of the *square machine* can now be saved in the *output latch*.

(2) The occupant input for the *square machine* is now selected from the *square cell* corresponding to the second square in the row. The output message of the *square machine* is then stored in the memory, and the operation repeats for the next cell in the row.

(3) This repeats until all the squares in the row have been done.

Using this technique, legal moves in the left to right direction can be generated in a straight forward manner. To do the opposite direction, the *square cells* are read out from right to left. If we duplicate this arrangement 8 times, we can do left/right and right/left legal move generation in parallel for a "board." Following this lead, we will construct the *VM** (VLSI Move Machine) in Figure 5.3.

To generate legal moves in the vertical (up/down) direction, we could add another set of *square machines* and a method of reading out the *square cells* along a row. However, this may be done by using a *folded* cell matrix (Figure 5.4) in which each cell has two output ports, H_out and V_out (Horizontal out and Vertical out), which are enabled by separate H_enable and V_enable inputs respectively. All the

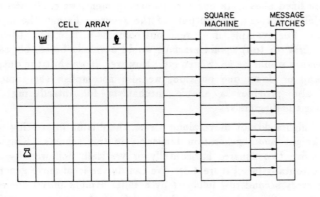

Figure 5.3 VM*

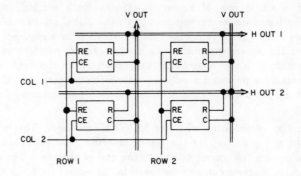

Figure 5.4 Folded Cell Matrix

H_out signals of the cells in a column are attached to a *column bus* and all the V_out signals of the cells in a row, to a *row bus*. All of the H_enable signals of the cells in row x are connected to the Row_x signal, and all of the V_enable signals of the cells in column x are connected to the Col_x signal. When the Row_x signal is activated, the cells in row x will place their contents on the *column busses,* and similarly when the Col_x signal is activated, the cells in column x will place their contents on the *row busses*. If we now connect the row and column busses at the cells on the diagonal of the matrix, the contents of *column bus z* will also appear on *row bus z,* allowing us to read out the cells in either rows or columns. By adding this capability to the *VM** we can now generate horizontal and vertical moves using the same basic structure.

To generate legal moves along a diagonal, it is necessary to present the sequence of diagonal squares to a *square machine,* and to perform the same type of legal move generation as outlined above. In our current arrangement, we could accomplish this by placing a shifter between the outputs of the *square cells* and the inputs of the *square machines*. For example, if we read out columns a, b, c, etc., then the input sequence to the first (or top) *square machine* would normally be a8, b8, c8, etc., and the shifted sequence would be a8, b7, c6 etc. However, if we shift the outputs of the *output latches* up or down one position, we also accomplish the same effect; in addition this also provides a simple mechanism for introducing "boundary conditions" along the diagonals.

Although knights do not move along a row, there is an interesting observation which makes knight move generation realizable in this scheme. Knights can be considered to move two squares horizontally or vertically followed by one square vertically or horizontally (see Figure 4.3). We can do the legal move generation then by reading out every second row followed by a shift. Knight moves then essentially become the same as diagonal moves, differing only in the order in which rows are read.

The from-square/to-square pair generation of *LM** can be transformed in the same manner (Figure 5.5). In the *LM**, the *square machines* have a *square address* input which provides the 6 bit location of the square. If a square were occupied by a

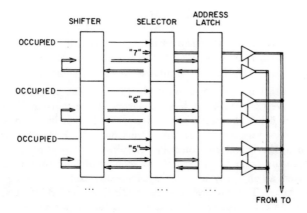

Figure 5.5 From/To Pair Generation

piece, the value of the *square address* is placed in the output message *from address* field, otherwise the *from address* of the *input message* is placed in the output message. In *VM**, when columns of the *square cells* are being read out, the locations on the chess board of the cells read out have the form (column, row_y), where the row_y is fixed and the column value corresponds to the column being read out. The *To Generator* logic block inputs are the number of the row or column that is being read out; the outputs are *to addresses* of the form (number, row_y). Note that if the *square machines* are being read out in row order, the corresponding coordinate read out of the *To Generator* will have the form (row_y, number), or the transpose of the actual value; this requires an *exchanger* in the output stage to generate the correct order. The *Address Multiplexer* is used to control the values the *Address Latches* are updated with. Each *square machine* in *VM** supplies a legal move and a square occupied output signal. If the square is occupied, the *Address Multiplexer* logic will update the corresponding latch with the value from the To Generator logic block. The *Address Multiplexer* also shifts the output values from the *Address Latches* so that they will remain in correspondence with the correct *square machines*.

To read out the from/to pair information, if there is a legal move, the from information will be the value in the *Address latch* and the to information will be the value generated by the To Generator logic. These values are multiplexed onto an *From/To Output Bus*.

6. Implementation

The choice of algorithm for implementation depends on the constraints imposed. For this project, the considerations, in decreasing order of priority, were:

(1) that it be implementable as a single VLSI chip,

(2) that it require minimum effort to layout due to an unavailability of sophisticated tools, and

(3)   that it give satisfactory performance.

LM* clearly violates (1) and *FM**, (2). *VM** represents a compromise in that it does not give the performance of the other 2, but does yield a relatively straight-forward implementation. It turns out that the entire algorithm can be implemented in terms of PLAs, Finite State Machines (FSM), RAM cells, and Shift Registers. A PLA compiler was written to help further minimize the layout effort.

The implementation of *VM**[10] was split into two functional subsections: move generation and move extraction. A block diagram of the move generation hardware is shown in Figure 6.1. The move generation procedure is initiated when the host computer asserts the start bit on the Direction Controller FSM. For each of the 16 possible directions, the Direction Controller enables the Row/Column Counter FSM to release information from the Board RAM sequentially by rows or columns. The Board is organized as an array of 4 bit *square cells*, each containing the occupant colour and piece type. The data released from the Board is fed into Capability Logic which translates the colour/piece information into the proper move capability. The resulting data is then fed into the Legal Move Logic. This logic is arranged as an array of finite state machines which get their state either from themselves or the machine directly above or below them via the Shifter. If the new data causes a legal move to be generated, the "hit" bit for that particular row will be set signaling the move extraction circuitry to begin the extraction sequence.

A block diagram of the move extraction hardware is shown in Figure 6.2. If a legal move is present at the output of the Legal Move Logic (i.e. "hit" bit is on), the Output Controller FSM will generate a 12 bit "move vector" that has 6 bits representing the from-square and 6 bits representing the to-square. The from-square coordinates are extracted from the Shift Register (must be able to shift up, down, in, and out) and the to-square coordinates are extracted from the current states of the Row/Column Counter and the Output Controller. Depending on the direction, the row and column information may have to be exchanged by the Row/Column

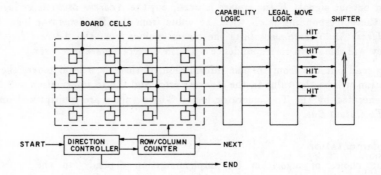

Figure 6.1 Move Generation Block Diagram

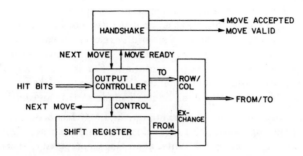

Figure 6.2 Move Extraction Block Diagram

Exchanger. Simultaneously, the Output Controller signals the Handshake Controller FSM to initiate the I/O protocol with the outside world. When the data transfer is completed the Output Controller is signaled and begins extracting the next move. When all the extractions have been completed, new data will be clocked into the Legal Move Logic and the move generation/extraction sequence repeats.

Besides the obvious parallelism achieved by processing 8 rows at a time, pipelining is also used to further increase the throughput of the system. For example, while the Output Controller is extracting the legal moves, it signals the Row/Column Counter to release the next row or column of data from the Board. Thus, the next set of data is present at the input of the Legal Move Logic and can be clocked in as soon as move extraction is completed. Also, to increase speed, finite state machines which communicate with each other are oppositely phased. Thus, one machine will accept inputs on $\Phi 1$ and produce outputs on $\Phi 2$ while the other machine will accept inputs on $\Phi 2$ and produce outputs on $\Phi 1$. This technique allows a speedup by a factor of 2 in inter-machine communication and eliminates the need for intermediate storage registers.

The performance of *VM** is measured in terms of the number of clock cycles required to generate all the legal moves for a given board configuration. For the purposes of this evaluation, it is assumed that the host computer can accept and acknowledge a legal move within one clock cycle. The "cost" (C) of legal move generation in clock cycles is [10]

$$C = 2 \qquad \{ \text{For } n = 0 \}$$
$$= 2n + 1 \qquad \{ \text{For } n > 0 \}$$

where n is the number of hits. For a typical board configuration there are approximately 40 legal moves. In the best case, 296 cycles are required to generate and extract the moves while 331 are necessary in the worst case. It is expected that the initial hardware design will operate at approximately 3 MHz thus generating moves at the rate of about 360,000 per second. By comparison, a software version of the bit-vector algorithm running on a VAX 11/780 generates moves at the rate of approximately 20,000 per second.

7. Further Developments

The Legal Move Generator machine was born of a desire to explore the capabilities of VLSI in a practical application. In general, the project successfully demonstrated that a single chip legal move generator not only was feasible, but also performed quite acceptably. The next step in developing a hardware computer chess contender is to consider the other algorithms used in move evaluation and tree searching for possible hardware implementation. For instance, an additional set of LMGs in a tree organization could generate a full two ply of moves in a very simple manner (Figure 7.1). Other VLSI candidates include sorting and hash table lookups.

The most interesting part of working on the LMG has been the development of an awareness of the capabilities of a technology for implementing algorithms, and the interactions between technological limits, algorithms, complexity, space, and time.

At the time of this writing, the functional blocks of *VM** have been sent for fabrication and the chips are expected in April.

8. Acknowledgements

The *VM** was designed and implemented as part of the EE755 VLSI Design course at the University of Waterloo, taught by Prof. M.I. Elmasry in the summer of 1982. The Chess Chip team consisted of Greg Bakker, Jim Jonkman, Jonathan Schaeffer, and Tom Schultz. Jim Barby made many valuable contributions to the project. Thanks to Amy Goldwater for her careful proof-reading.

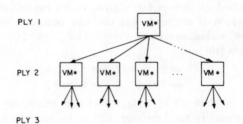

Figure 7.1 Legal Move Tree

9. References

[1]   R.P. Gurd, *User's Reference to B*, University of Waterloo, 1982.

[2]   Joe Condon and Ken Thompson, Belle Chess Hardware, *Advances in Computer Chess* **3**, 1982, Pergammon Press, M.R.B. Clarke (ed.).

[3]   Andy Soltis, Four Roads to Success, *Chess Life and Review*, February, 1981.

[4]   Peter Frey, (ed.), *Chess Skill in Man and Machine*, Springer-Verlag, 1975.

[5]   John Mick and Jim Brick, *Bit-Slice Microprocessor Design*, McGraw-Hill, 1980.

[6]   Leonard S. Haynes, Richard L. Lau, Daniel P. Siewiorek, and Daniel Mizell, A survey of highly parallel computing, *Computer* **Vol. 15**(No. 1), January, 1982.

[7]   Carver Mead and Lynne Conway, *Introduction to VLSI Systems*, Addison-Wesley, 1980.

[8]   Dave Johannsen, Bristle Blocks: A Silicon Compiler, *Proceedings of the 16th Design Automation Conference*, 1979.

[9]   Uri Weiser and Al Davis, A wavefront notation tool for VLSI array design, *CMU Conference on VLSI Systems and Computations*, October 19-21, 1981, Computer Science Press, H.T. Kung, Bob Sproull, Guy Steel (eds.).

[10] Greg Bakker, Jim Jonkman, Jonathan Schaeffer, and Tom Schultz, VLSI Implementation of a Chess Legal Move Generator, EE755S-1, University of Waterloo, August, 1982.

10. Appendix A - Square Machine

Moving	Colour	Piece	Direction	Capability
B/W	W/B	-	-	C_NONE
W/B	W/B	King	Not Knight	C_ONE
W/B	W/B	King	Knight	C_OURS
W/B	W/B	Queen	Not Knight	C_MANY
W/B	W/B	Queen	Knight	C_OURS
W/B	W/B	Bishop	Diagonal	C_MANY
W/B	W/B	Bishop	Not Diagonal	C_OURS
W/B	W/B	Rook	Horizontal and Vertical	C_MANY
W/B	W/B	Rook	Diagonal and Knight	C_OURS
W/B	W/B	Knight	Knight	C_ONE
W/B	W/B	Knight	Not Knight	C_OURS
W/B	W/B	P1 Pawn	Forward	C_PONE
W/B	W/B	P2 Pawn	Forward	C_PTWO
W/B	W/B	P1 or P2 Pawn	Diagonal Forward	C_PDIAG
W/B	W/B	P1 or P2 Pawn	Not Forward	C_PDIAG

a) Move Capabilities

Capability	Input Message					
	MNONE	MONE	MMANY	MPONE	MPTWO	MPDIAG
CNONE	MNONE	MNONE*	MMANY*	MNONE*	MPONE*	MNONE
CONE	MONE	MONE	MONE	MONE	MONE	MONE
CMANY	MMANY	MMANY	MMANY	MMANY	MMANY	MMANY
CPONE	MPONE	MPONE	MPONE	MPONE	MPONE	X
CPTWO	MPTWO	MPTWO	MPTWO	X	X	X
CPDIAG	MPDIAG	MPDIAG	MPDIAG	X	X	MPDIAG
CTHEIRS	MNONE	MNONE*	MNONE*	MNONE	MNONE	MNONE*
COURS	MNONE	MNONE	MNONE	MNONE	MNONE	MNONE

b) Input/Output Message Relationships

Figure A.1 Square Machine Move Capabilities and Messages

New VLSI Architectures with Reduced Hardware*

Joseph Ja'Ja' and Robert Michael Owens

Abstract

This paper explores the possibility of designing special-purpose chips with limited processing area. We propose a new architecture which allows many problems to be solved quite efficiently
on chips with very small processing areas. We consider in detail
the sorting problem and show how it can be solved quickly and elegantly on our model. We show that sorting n numbers can be done
on a chip with processing area $A = o(n)$ in time $T = O(\frac{n\log^2 n}{A} + \sqrt{n}\log^2 n)$ in a network with mesh-connected interconnections, which
is optimal within a $\log n$ factor. The control is shown to be
simple and easily implementable in VLSI. Several other examples,
such as the Fast Fourier Transform and integer multiplication, are
considered and are shown to possess simple and fast algorithms in
our model.

1. Introduction

The operation of a network is defined by specifying the sequence of operations to be performed by each processor in the network and the movement of operands over data paths between the pro-

* Supported by the U.S. Army Research Office, Contract No.
  DAAG29-82-K-0110 and U.S. Office of Naval Research, Contract No.
  N00014-80-C-0517.

cessors. For a parallel algorithm P to be executed by a network, the concurrently executable set of operations of each step must be mapped to a set of processors in the network such that, when a processor performs an operation, all the operands of that operation must reside at that processor. The ability to reduce the execution time of parallel algorithms is limited by having to insure that data flow and execution conflicts are avoided. A data flow conflict would arise if an attempt is made to transfer more information over a data path in some period of time than is possible. An execution conflict would arise if an attempt is made to have a processor perform more operations in some unit of time than is possible.

Suppose we know we can solve a problem of size n by a network with $q(n)$ processors in time $t(n)$. An interesting question is how long it would take to solve this same problem by a network with only $r(n) = o(q(n))$ processors. One approach to answer this question in general has been to describe a method to map a virtual network of $q(n)$ onto a physical network of $r(n)$ processors. This mapping can be defined by associating a physical processor with each virtual processor, associating a physical data path with each virtual data path, and mapping each possible execution step of the virtual network to one or more execution steps of the physical network. For example a virtual square mesh of 2^{2n} processors can be mapped to physical mesh of $2^{2(n-m)}$ processors by folding the mesh $2m$ times. Hence, if a suitable mapping can be found, a problem of size n can be solved in time $\frac{t(n)q(n)}{r(n)}$.

There exists a concern when the results of such work are ex·-

trapolated to VLSI. Note that the previous result is expressed in terms of the time it takes to solve a given problem on a network and the number of processors which are connected within the network. However, VLSI designers are interested in chip area and time. The simplistic assumption of course is to assume that each processor requires only some constant area and that the required chip area is the sum of areas required by each processor, i.e. $A = r(n)$. Unfortunately, this assumption omits at least one important factor.

As more and more virtual processors are mapped to a single processor, it is well understood that a physical processor can not execute simultaneously all of the instructions of the virtual processors associated with it. Hence, the smaller physical network requires more time (by a factor of $\frac{q(n)}{r(n)}$) to solve a given problem than the larger virtual network. On the other hand, it is often overlooked that the numbers which were once stored at several virtual processors are associated with a single physical processor, the physical processor must be allocated more chip surface (again by a factor of $\frac{q(n)}{r(n)}$) so that all the necessary numbers can be retained. If it is assumed that the chip surface occupied by a processor has a regular shape and an area which is proportional to the amount of data stored at the processor and that the time required to exchange data between any two processors is proportional to the distance between them, we find that each step of the physical network takes $\sqrt{\frac{q(n)}{r(n)}}$ times longer than each step of the virtual network. Hence, the time required to solve a problem of size n is $\frac{t(n)q(n)}{r(n)} \sqrt{\frac{q(n)}{r(n)}}$ instead of $\frac{t(n)q(n)}{r(n)}$.

In the case of sorting n numbers on a $\sqrt{n} \times \sqrt{n}$ mesh, we find $t(n) = \sqrt{n}$, $q(n) = n$, and $r(n) = A$. Hence, sorting n numbers on a virtual mesh of area A takes time $\dfrac{n^2}{A\sqrt{A}}$. For a chip of small area $A = o(n)$, a significant loss in performance can occur. We explore, in this paper, the effects of mapping a network of basically unlimited size onto a physical network of limited size. Our results show that such a mapping can be done without a significant loss in performance.

2. A New Architecture

The model proposed by this paper differs from the earlier homogeneous models in that it assumes that the network contains two different types of basic elements. One type, called the processing element, is capable of performing instructions and has a very limited storage capability. Processing elements can be interconnected by wires to form a processing component. The other type, called the storage element, is capable of storing some small bounded amount of information but has no instruction execution capability. Storage elements can be interconnected by wires to form a storage component. Also storage and processing elements can be interconnected by wires to form a network. In this way, a storage component is used to retain the intermediate results needed by the processing component. We propose this model to study the possibility of designing a chip of a certain area to solve problems of arbitrary size using a separate storage region which can be enlarged in a regular fashion whenever needed.

We now describe the main assumptions about the model more rigorously. We start by assuming that the storage and processing components of a network can be segregated into two regions. The

processing region contains the processing components of a network.
We define the area of the chip to be the area of just the processing region. The storage region contains all the storage components
of a network.

We assume that wires have minimal width $\geq \lambda$ and gates occupy a minimal area $\geq \lambda^2$. We allow a constant number of layers
and, hence, a constant number of wires can intersect at any point.
The area of a processing region is determined by processing elements connected by wires. We assume that a wire which connects a
processing element to a storage element must pass through a port
which lies on the boundary of the processing region. Note that we
make no assumption as to how the processing elements themselves are
connected together.

We assume that each storage element is capable of storing
some constant number of bits. The area of the storage region is
determined by storage elements connected by wires. We assume that
all the elements of the network are synchronized by a clock with
period T. At least one unit of information can be transferred
across any wire in time T. A unit of information can consist of a
single bit in some cases (e.g. integer multiplication) or many bits
in other cases (e.g. sorting n bit integers, matrix multiplication.) The format of the information being transferred into and
out of the network as well as the times when the information into
the network are made available are fixed and independent of the
value of input. The input can be read into either the processing
region or the storage region or both.

We now introduce a new parameter B which is related to the propagation characteristics of data paths. Let $p(e)$ be the time necessary for one unit of information to be transferred across a wire of length e. Then, let $B = \max_e \frac{e}{p(e)}$, where the maximum is taken over all the wires in the chip. Hence, given our assumptions, it must be that all the wires of the network must have length $\leq BT$. Note that without loss of generality, we assume $T = 1$. Note that if B is a constant, we obtain the model of Chazelle and Monier [CM] and, if B is unbounded, we obtain the usual "constant model". We feel that dealing with B rather than the propagation delay directly leads to more practical insights.

Thompson [T2] developed a different model where he basically used RAMS's for the storage component. In this case, no two words can be accessed simultaneously unless they are stored in different RAM's. We will later show that many problems can be solved much faster in our model which we will call the network model. The other model will be called the RAM model.

3.   A Sorting Chip with Reduced Processing Area

Let $\{x_1, x_2, \ldots, x_n\}$ be a set of n numbers to be sorted on a chip whose processing component has a given area $A = o(n)$.

A sequence $X = \{x_1, x_2, \ldots, x_n\}$ is called bitonic if there exists an index i such that either $x_1 \leq x_2 \leq \ldots \leq x_i \geq \ldots \geq x_{n-1} \geq x_n$ or $x_1 \geq x_2 \geq \ldots \geq x_i \leq \ldots \leq x_{n-1} \leq x_n$. To sort a bitonic sequence $X = \{x_1, x_2, \ldots, x_n\}$, the sequence is divided into two bitonic subsequences $\{x_1, x_3, x_5, \ldots\}$ and $\{x_2, x_4, x_6, \ldots\}$ which are recursively sorted and then merged together by applying pairwise a comparison-exchange operation to corresponding elements of the now sorted subsequences. Our algorithm is based on an adaptation of the bitonic sort algorithm for a mesh interconnected network given by D. Nassimi and S. Sahni [NS]. However, there are significant dif-

ferences between the two algorithms due to the characteristics of
the networks on which they run.

The algorithm given by [NS] runs on a mesh interconnected array of n identical processors. The processor array is organized
as $\sqrt{n}$ rows by $\sqrt{n}$ columns (the assumption is made that n is an
even power of 2). Each processor has several registers, an arithmetic unit, a control unit, and a stored program. The mesh interconnections afford the transfer of data in global shifing like
movements. In contrast, our algorithm runs on a mesh interconnected
array of n nonidentical processors. That is, while every proces-
sor $p_{i,j}$, $0 \leq i, j \leq \sqrt{n} - 1$, contains one value register v_{ij} and
four flag registers $r_{i,j}$, $s_{i,j}$, $c_{i,j}$, and $d_{i,j}$, only the processors in the first 2 rows by λ columns (without loss of generality,
we assume that λ is even and divides $\sqrt{n}$) have an arithmetic
unit, a nontrivial control unit, and a stored program. Note that
these 2λ processors and their interconnections form the processing
component, while the remainder of the mesh forms the storage component.

$p_{0,0}$	$p_{0,1}$	...	$p_{0,\lambda-1}$	$p_{0,\lambda}$	...	...	$p_{0,\sqrt{n}-1}$
$p_{1,0}$	$p_{1,1}$	...	$p_{1,\lambda-1}$	$p_{1,\lambda}$	...	...	$p_{1,\sqrt{n}-1}$
$p_{2,0}$	$p_{2,1}$	...	$p_{2,\lambda-1}$	$p_{2,\lambda}$	...	...	$p_{2,\sqrt{n}-1}$
...	...	...	...	...	...	...	...
...	...	...	...	...	...	...	...
$p_{\sqrt{n}-1,0}$	$p_{\sqrt{n}-1,1}$	...	$p_{\sqrt{n}-1,\lambda-1}$	$p_{\sqrt{n}-1,\lambda}$	...	...	$p_{\sqrt{n}-1,\sqrt{n}-1}$

Figure 1. Mesh Architecture

The movement of data between adjacent processors is controlled
by six clocks. We call these clocks the row shuffle, row unshuffle,
column shuffle, column unshuffle, row rotate, and column rotate

clocks. The clocks are issued by one of the processors in the control region to all processors. Only one clock is issued at any
time. Note that the existence of six clocks need not imply that six
separate control lines must be utilized. We use six clocks to simplify our description. The effect of each clock on the contents of
each processor is described in the following paragraphs.

When a row shuffle clock is issued, the contents of $v_{i,j}$,
$r_{i,j}$, $s_{i,j}$, $c_{i,j}$, and $d_{i,j}$ are exchanged with the contents of
$v_{i,j+1}$, $r_{i,j+1}$, $s_{i,j+1}$, $c_{i,j+1}$, and $d_{i,j+1}$ respectively, if
$\bar{r}_{i,j} \cdot r_{i,j+1} \cdot \bar{s}_{i,j} \cdot \bar{s}_{i,j+1}$. Note that it is not possible to set
$r_{i,j}$ and $s_{i,j}$, $0 \le i$, $j \le \sqrt{n} - 1$ in such a way that the attempt
would be made to exchange the contents of a register with the contents of more than one other register. It is also possible that a
register may retain its original contents. The following figure
illustrates the effect of issuing a row shuffle clock.

clock	$v\,r\,s$	$v\,r\,s$	$v\,r\,s$	$v\,r\,s$	$v\,r\,s$	$v\,r\,s$	$v\,r\,s$	$v\,r\,s$
0	$1\,F\,T$	$4\,F\,F$	$6\,F\,F$	$2\,F\,F$	$8\,T\,F$	$3\,T\,F$	$5\,T\,F$	$9\,T\,T$
1	$1\,F\,T$	$4\,F\,F$	$6\,F\,F$	$8\,T\,F$	$2\,F\,F$	$3\,T\,F$	$5\,T\,F$	$9\,T\,T$
2	$1\,F\,T$	$4\,F\,F$	$8\,T\,F$	$6\,F\,F$	$3\,T\,F$	$2\,F\,F$	$5\,T\,F$	$9\,T\,T$
3	$1\,F\,T$	$8\,T\,F$	$4\,F\,F$	$3\,T\,F$	$6\,F\,F$	$5\,T\,F$	$2\,F\,F$	$9\,T\,T$

Figure 2. Row Shuffling

Note by properly setting the flags r and s of 2k row adjacent
processors, it is possible to shuffle together the original contents
of their value registers, using k-1 row shuffle clocks. Observe
that the processors for which s is true delineate the boundarys
of the segment of row adjacent processors.

When a row unshuffle clock is issued, the contents of $v_{i,j}$,
$r_{i,j}$, $s_{i,j}$, $c_{i,j}$, and $d_{i,j}$ are exchanged with the contents of

$v_{i,j+1}$, $r_{i,j+1}$, $s_{i,j+1}$, $c_{i,j+1}$, and $d_{i,j+1}$ respectively, if $r_{i,j} \cdot \bar{r}_{i,j+1} \cdot \bar{s}_{i,j} \cdot \bar{s}_{i,j+1}$. The following figure illustrates the effect of issuing a row unshuffle clock.

clock	v r s	v r s	v r s	v r s	v r s	v r s	v r s	v r s
0	1 F T	8 T F	4 F F	3 T F	6 F F	5 T F	2 F F	9 T T
1	1 F T	4 F F	8 T F	6 F F	3 T F	2 F F	5 T F	9 T T
2	1 F T	4 F F	6 F F	8 T F	2 F F	3 T F	5 T F	9 T T
3	1 F T	4 F F	6 F F	2 F F	8 T F	3 T F	5 T F	9 T T

Figure 3.   Row Unshuffling

Note by properly setting the flag values r and s, the issuing of the row unshuffle clock can be used (as its name implies) to undo the shuffling effect produced by issuing the shuffle clock.

When a row rotate clock is issued, the contents of $v_{i,j}$, $r_{i,j}$, $s_{i,j}$, $c_{i,j}$, and $d_{i,j}$ are transferred to $v_{i,j-1}$, $r_{i,j-1}$, $s_{i,j-1}$, $c_{i,j-1}$, and $d_{i,j-1}$ respectively (the contents of $v_{i,0}$, $r_{i,0}$, $s_{i,0}$, $c_{i,0}$, and $d_{i,0}$ are transferred to $v_{i,\sqrt{n}-1}$, $r_{i,\sqrt{n}-1}$, $s_{i,\sqrt{n}-1}$, $c_{i,\sqrt{n}-1}$, and $d_{i,\sqrt{n}-1}$ respectively). The definition for the column shuffle, column unshuffle, and column rotate clocks are similar to the definitions for the row shuffle, row unshuffle, and row rotate clocks respectively.

Procedure Sort begins with each of the value registers $v_{i,j}$, $0 \leq i, j \leq \sqrt{n} -1$ containing one the numbers to be sorted and terminates with these numbers sorted in row major order form. That is at termination the value register $v_{i,j}$ contains the $(i \times \sqrt{n} + j)$'th largest number. The procedure is based on two main procedures Horizontal_Merge(k) and Vertical_Merge(k). Horizontal_Merge(k) simultaneously merges pairwise $\dfrac{n}{k^2}$ horizontally aligned submeshes of k rows by k columns into $\dfrac{n}{2k^2}$ submeshes of k rows by 2k columns. Vertical_Merge(k) simultaneously merges pairwise

$\dfrac{n}{2k^2}$ vertically aligned submeshes of k rows by $2k$ columns into $\dfrac{n}{4k^2}$ submeshes of $2k$ rows by $2k$ columns. The overall procedure can be briefly described as follows.

procedure Sort

    $k = 1$

    while $k < \sqrt{n}$ do

        Horizontal_Merge(k)

        Vertical_Merge(k)

        $k = 2 \times k$

    endwhile

  endprocedure

Note that the loop inside procedure Sort is executed $\dfrac{\log_2 n}{2}$ times. Figure 4 illustrates the structure of the submeshes when procedure Vertical_Merge(k) is called.

$p_{0,0}$	...	$p_{0,2k-1}$	$p_{0,2k}$	...	$p_{0,4k-1}$	$p_{0,4k}$	...	...	$p_{0,\sqrt{n}-1}$
...	...	...	...	...	...	...	...	...	...
$p_{k-1,0}$	...	$p_{k-1,2k-1}$	$p_{k-1,2k}$	...	$p_{k-1,4k-1}$	$p_{k-1,4k}$	...	...	$p_{k-1,\sqrt{n}-1}$
$p_{k,0}$	...	$p_{k,2k-1}$	$p_{k,2k}$	...	$p_{k,4k-1}$	$p_{k,4k}$	...	...	$p_{k,\sqrt{n}-1}$
...	...	...	...	...	...	...	...	...	...
$p_{2k-1,0}$	...	$p_{2k-1,2k-1}$	$p_{2k-1,2k}$	...	$p_{2k-1,4k-1}$	$p_{2k-1,4k}$	...	...	$p_{2k-1,\sqrt{n}-1}$
$p_{2k,0}$	...	$p_{2k,2k-1}$	$p_{2k,2k}$	...	$p_{2k,4k-1}$	$p_{2k,4k}$	...	...	$p_{2k,\sqrt{n}-1}$
...	...	...	...	...	...	...	...	...	...
...	...	...	...	...	...	...	...	...	...
$p_{\sqrt{n}-1,0}$	...	$p_{\sqrt{n}-1,2k-1}$	$p_{\sqrt{n}-1,2k}$	...	$p_{\sqrt{n}-1,4k-1}$	$p_{\sqrt{n}-1,4k}$	...	...	$p_{\sqrt{n}-1,\sqrt{n}-1}$

Figure 4.   Vertical_Merge(k) Submeshes

For each merged pair of processor submeshes, the contents of the value registers of one subarray are assumed to be sorted in nondecreasing row major order and the contents of the value registers of the other subarray are assumed to be sorted in nonincreasing row major order. The contents of the value registers of the merged processor subarray are sorted into either nondecreasing or nonincreasing order by procedure Vertical_Merge(k).

Figure 5 illustrates the structure of the submeshes when the procedure Horizontal_Merge(k) is called.

$P_{0,0}$	...	$P_{0,b-1}$	$P_{0,b}$	...	$P_{0,2b-1}$	$P_{0,2b}$	...	...	$P_{0,\sqrt{n}-1}$
...	...	...	...	...	...	...	...	...	...
$P_{b-1,0}$	...	$P_{b-1,b-1}$	$P_{b-1,b}$	...	$P_{b-1,2b-1}$	$P_{b-1,2b}$	...	...	$P_{b-1,\sqrt{n}-1}$
$P_{b,0}$	...	$P_{b,b-1}$	$P_{b,b}$	...	$P_{b,2b-1}$	$P_{b,2b}$	...	...	$P_{b,\sqrt{n}-1}$
...	...	...	...	...	...	...	...	...	...
$P_{2b-1,0}$	...	$P_{2b-1,b-1}$	$P_{2b-1,b}$	...	$P_{2b-1,2b-1}$	$P_{2b-1,2b}$	...	...	$P_{2b-1,\sqrt{n}-1}$
$P_{2b,0}$	...	$P_{2b,b-1}$	$P_{2b,b}$	...	$P_{2b,2b-1}$	$P_{2b,2b}$	...	...	$P_{2b,\sqrt{n}-1}$
...	...	...	...	...	...	...	...	...	...
...	...	...	...	...	...	...	...	...	...
$P_{\sqrt{n}-1,0}$	...	$P_{\sqrt{n}-1,b-1}$	$P_{\sqrt{n}-1,b}$	...	$P_{\sqrt{n}-1,2b-1}$	$P_{\sqrt{n}-1,2b}$	...	...	$P_{\sqrt{n}-1,\sqrt{n}-1}$

Figure 5. Horizontal_Merge(k) Submeshes

For each merged pair of processor submeshes, the contents of the value registers of one subarray are assumed to be sorted in nondecreasing row major order and the contents of the value registers of the other subarray are assumed to be sorted in nonincreasing row major order. The contents of the value registers of the merged processor subarray are sorted into either nondecreasing or nonincreasing order by procedure Horizontal_Merge(k). We now explain in some detail how to execute Vertical_Merge(k) and Horizontal_Merge(k).

Procedure Vertical_Merge(k) can be described as follows.

Procedure Vertical_Merge(k)

Column_Merge(k)

Row_Merge(k, even)

endprocedure

Procedure Row_Merge(k, s) simultaneously sorts $\frac{n}{2k}$ bitonic sequences of length 2k. The numbers in each such sequence are contained in the value registers of 2k row adjacent set of processors. Figure 6 illustrates the structure of the submeshes when procedure Row_Merge(k, s) is called.

$P_{0,0}$	...	$P_{0,2k-1}$	$P_{0,2k}$	...	$P_{0,4k-1}$	$P_{0,4k}$	...	...	$P_{0,\sqrt{n}-1}$
$P_{1,0}$	...	$P_{1,2k-1}$	$P_{1,2k}$	...	$P_{1,4k-1}$	$P_{1,4k}$	...	...	$P_{1,\sqrt{n}-1}$
$P_{2,0}$	...	$P_{2,2k-1}$	$P_{2,2k}$	...	$P_{2,4k-1}$	$P_{2,4k}$	...	...	$P_{2,\sqrt{n}-1}$
...	...	...	...	...	...	...	...	...	...
...	...	...	...	...	...	...	...	...	...
$P_{\sqrt{n}-1,0}$	...	$P_{\sqrt{n}-1,2k-1}$	$P_{\sqrt{n}-1,2k}$	...	$P_{\sqrt{n}-1,4k-1}$	$P_{\sqrt{n}-1,4k}$	...	...	$P_{\sqrt{n}-1,\sqrt{n}-1}$

Figure 6. Row_Merge Submeshes

Depending on s and the row and columns in which a bitonic sequence resides, Row_Merge(k, s) sorts the sequence into either a nonincreasing or nondecreasing order.

```
procedure Row_Merge(k, s)

    ℓ = k

    while ℓ > 0   do

        set r_{i,j} and s_{i,j} as indicated in the figure following
            this procedure

        issue ℓ - 1 row shuffle clocks

        Row_Exchange(k, s)

        issue ℓ - 1 row unshuffle clocks

        ℓ = ℓ / 2

    endwhile

endprocedure
```

r s	r s	...	r s	r s	...	r s	r s	r s	r s	...	...	r s	r s
F T	F F	...	F F	T F	...	T F	T T	F T	F F	...	...	T F	T T
F T	F F	...	F F	T F	...	T F	T T	F T	F F	...	...	T F	T T
F T	F F	...	F F	T F	...	T F	T T	F T	F F	...	...	T F	T T
...	...	...	...	...	...	...	...	...	...	...	...	...	...
...	...	...	...	...	...	...	...	...	...	...	...	...	...
F T	F F	...	F F	T F	...	T F	T T	F T	F F	...	...	T F	T T

Figure 7.   Row Flag Bits

Note that when procedure Row_Merge(k, s) is invoked, $\log_2 k$
calls to Row_Exchange(k, s) are made and at most 2k row shuffle
and row unshuffle clocks are issued. Also except for the calls to
Row_Exchange(k, s), all the operations performed when Row_Merge(k, s)
is invoked are executed synchronously by every processor in the
original mesh. However, the calls to procedure Row_Exchange(k, s)
are made to do the comparison/exchange operations. Once such operation is required for each pair of processors in a particular group-

ing of row adjacent processors. Unfortunately, given our earlier assumptions about the processors in the original mesh, most of these operations can not be performed at the processors where the operands are actually stored. Instead the operands must be brought to the $2 \times \lambda$ processors located in the processing region. The following procedure shows how this can be done.

```
procedure Row_Exchange(k, s)
    for j = 0 to √n - λ by λ
        for i = 0 to √n - 2 by 2
            Row_Exchange_{ℓ,m}(i,j,k,s), 0≤ℓ≤1, 0≤m≤λ-2 and m even.
            issue 2 column rotate clocks
        endfor
        issue λ row rotate clocks
    endfor
endprocedure
```

Each of procedures Row_Exchange $_{\ell,m}$ is defined in the following manner. Let the processors in the processing region be grouped in the following way.

$p_{0,0}$ $p_{0,1}$	$p_{0,2}$ $p_{0,3}$	$p_{0,4}$ $p_{0,5}$	$\cdots$	$p_{0,\lambda-2}$ $p_{0,\lambda-1}$
$p_{1,0}$ $p_{1,1}$	$p_{1,2}$ $p_{1,3}$	$p_{1,4}$ $p_{1,5}$	$\cdots$	$p_{1,\lambda-2}$ $p_{1,\lambda-1}$

Figure 8.   Row_Exchange Processor Grouping

Then Row_Exchange $_{\ell,m}(i,j)$ is performed on the numbers contained on the processor pair $p_{\ell,m}$ and $p_{\ell,m+1}$ as follows.

procedure Row_Exchange$_{\ell,m}(i,j,k,s)$

$$\text{let } t = \begin{cases} 1 & s \text{ odd and } \lfloor \frac{\ell+i}{k} \rfloor \text{ odd} \\ 1 & s \text{ even and } \lfloor \frac{m+j}{2k} \rfloor \text{ odd} \\ -1 & s \text{ odd and } \lfloor \frac{\ell+i}{k} \rfloor \text{ even} \\ -1 & s \text{ even and } \lfloor \frac{m+j}{2k} \rfloor \text{ even} \end{cases}$$

if $t\,(v_{\ell,m} - v_{\ell,m+1}) < 0$ then

    exchange the numbers contained in value registers

    $v_{\ell,m}$ and $v_{\ell,m+1}$

        (Note the contents of the flag registers are not

        changed)

  endif

 endprocedure

Note that, since k is a power of 2, the expression $\lfloor \frac{i}{2k} \rfloor$ odd
can be evaluated by examining only the $(1 + \log_2 k)$'th bit of i.
Hence, division is not actually required. Also, note that when
procedure Row_Exchange(k, s) is invoked, $\frac{n}{\lambda}$ column rotate clocks
and $\sqrt{n}$ row rotate clocks are issued and $\frac{n}{2\lambda}$ parallel compare-exchange operations are performed.

Procedure Column_Merge(k) simultaneously sorts $\frac{n}{2k}$ bitonic
sequences of length 2k. The numbers in each such sequences are
contained in the value registers of 2k column adjacent set of
processors. The following figure illustrates how the $\sqrt{n}$ by $\sqrt{n}$
processors of the original mesh are organized into these sets.

$p_{0,0}$	$p_{0,1}$	$p_{0,2}$	...	...	$p_{0,\sqrt{n}-1}$
...	...	...	...	...	...
$p_{2k-1,0}$	$p_{2k-1,1}$	$p_{2k-1,2}$	...	...	$p_{2k-1,\sqrt{n}-1}$
$p_{2k,0}$	$p_{2k,1}$	$p_{2k,2}$	...	...	$p_{2k,\sqrt{n}-1}$
...	...	...	...	...	...
$p_{4k-1,0}$	$p_{4k-1,1}$	$p_{4k-1,2}$	...	...	$p_{4k-1,\sqrt{n}-1}$
$p_{4k,0}$	$p_{4k,1}$	$p_{4k,2}$	...	...	$p_{4k,\sqrt{n}-1}$
...	...	...	...	...	...
...	...	...	...	...	...
$p_{\sqrt{n}-1,0}$	$p_{\sqrt{n}-1,1}$	$p_{\sqrt{n}-1,2}$	...	...	$p_{\sqrt{n}-1,\sqrt{n}-1}$

Figure 9. Column_Merge Submeshes

Depending on column in which a bitonic sequence resides,
Column_Merge(k) sorts the sequence into either a nonincreasing or
nondecreasing order.

    procedure Column_Merge(k)

        ℓ = k

        while ℓ > 0 do

            set $c_{i,j}$ and $d_{i,j}$ as indicated in the figure following
                this procedure

            issue ℓ - 1 column shuffle clocks

            Column_Exchange(k)

            issue ℓ - 1 column unshuffle clocks

            ℓ = ℓ / 2

        endwhile

    endprocedure

	r s	*F T*	*F T*	*F T*	...	...	*F T*
	r s	*F F*	*F F*	*F F*	...	...	*F F*
k processors	...	...	...	...	...	...	...
	r s	*F F*	*F F*	*F F*	...	...	*F F*
	r s	*T F*	*T F*	*T F*	...	...	*T F*
	...	...	...	...	...	...	...
k processors	*r s*	*T F*	*T F*	*T F*	...	...	*T F*
	r s	*T T*	*T T*	*T T*	...	...	*T T*
	r s	*F T*	*F T*	*F T*	...	...	*F T*
	r s	*F F*	*F F*	*F F*	...	...	*F F*
	...	...	...	...	...	...	...
	...	...	...	...	...	...	...
	r s	*T F*	*T F*	*T F*	...	...	*T F*
	r s	*T T*	*T T*	*T T*	...	...	*T T*

Figure 10.   Column Flag Bits

Note that when procedure Column_Merge(k) is invoked, $\log_2 k$ calls to Column_Exchange(k) are made and at most $2k$ column shuffle and column unshuffle clocks are issued. Also except for the call to Column_Exchange(k), all the operations performed when Column_Merge(k) is invoked are executed synchronously by every processor in the original mesh. However, the calls to procedure Column_Exchange(k) are made to do the comparison/exchange operations. One such operation is required for each pair of processors in a particular grouping of column adjacent processors. Again the numbers must be brought to the $2 \times \lambda$ processors located in the processing region. This can be expressed in the following manner.

```
procedure Column_Exchange(k)
```

for $j = 0$ to $\sqrt{n} - \lambda$ by λ

for $i = 0$ to $\sqrt{n} - 2$ by 2

Column_Exchange$_{\ell,m}$ (i,j,k), $\ell = 0$, $0 \leq m \leq \lambda-1$

issue 2 column rotate clocks

endfor

issue λ row rotate clocks

endfor

endprocedure

Each of the procedure Column_Exchange$_{\ell,m}$ is defined in the following manner. Let the processors in the processing region be grouped in the following way.

$p_{0,0}$	$p_{0,1}$	$p_{0,2}$	$p_{0,3}$	$p_{0,4}$	$p_{0,5}$	$\cdots$	$p_{0,\lambda-2}$	$p_{0,\lambda-1}$
$p_{1,0}$	$p_{1,1}$	$p_{1,2}$	$p_{1,3}$	$p_{1,4}$	$p_{1,5}$	$\cdots$	$p_{1,\lambda-2}$	$p_{1,\lambda-1}$

Figure 11.   Column_Merge Processor Grouping

Then Column_Exchange$_{\ell,m}$ (i,j,k), is performed on the numbers contained in the processor pair $p_{0,m}$ and $p_{1,m}$ as follows.

procedure Column_Exchange$_{\ell,m}$ (i,j,k)

$$\text{let } t = \begin{cases} 1 & \lfloor \frac{m+j}{2k} \rfloor \text{ odd} \\ -1 & \lfloor \frac{m+j}{2k} \rfloor \text{ even} \end{cases}$$

if $t \times (v_{\ell,m} - v_{\ell+1,m}) < 0$ then

exchange the numbers contained in value regesters

$v_{\ell,m}$ and $v_{\ell+1,m}$

(note the contents of the flags registers are not changed)

endif

endprocedure

Note that when procedure Column_Exchange(k,s) is invoked, $\frac{n}{\lambda}$ column rotate clocks and $\sqrt{n}$ row rotate clocks are issued and $\frac{n}{2\lambda}$ parallel compare-exchange operations are performed.

We now briefly show that procedure Vertical_Merge(k) works correctly. We start by establishing the correctness of the two procedures Column_Merge(k) and Row_Merge(k,s).

Lemma 1: Procedure Column_Merge(k) correctly sorts $\frac{n}{2k}$ bitonic sequences each of which consists of 2k elements as shown in figure 9.

Proof: Let $k = 2^t$ (note k is always a power of 2). Once can easily check that, after the i'th iteration of the loop, elements within a submesh which are 2^{t-i} apart will be compared and possibly interchanged. But this is precisely how bitonic sort works. □

Lemma 2: Procedure Row_Merge(k) correctly sorts $\frac{n}{2k}$ bitonic sequences each of which consists of 2k elements as shown in figure 6.

Proof: The proof is similar to that for Lemma 1 and will be omitted. □

Lemma 3: Procedure Vertical_Merge(k) correctly sorts $\frac{n}{2k^2}$ vertically aligned submeshes of k rows by 2k columns.

Proof: One can easily check that the result follows by inspection and Lemmas 1 and 2. □

We now turn our attention to procedure Horizontal_Merge(k) which can be expressed as follows.

    procedure Horizontal_Merge(k)

        Two_Column_Merge(k)

        Row_Merge(k, odd)

    endprocedure

Procedure Row_Merge(k, s) has been previously described. Procedure Two_Column_Merge(k) simultaneously sorts $\frac{n}{2k}$ bitonic sequences of length 2k. The numbers in each such sequence are contained in the value registers of a pair of k row adjacent set of processors. Furthermore the two row adjacent sets in each pair are k processors apart. Figure 12 illustrates the structure of the submeshes when procedure Two-Column_Merge(k) is called.

$p_{0,0}$	...	$p_{0,k}$	...	...	$p_{0,\sqrt{n}-1}$
...	...	...	...	...	...
$p_{2k-1,0}$	...	$p_{2k-1,k}$	...	...	$p_{2k-1,\sqrt{n}-1}$
$p_{2k,0}$	...	$p_{2k,k}$	...	...	$p_{2k,\sqrt{n}-1}$
...	...	...	...	...	...
$p_{4k-1,0}$	...	$p_{4k-1,k}$	...	...	$p_{4k-1,\sqrt{n}-1}$
$p_{4k,0}$	...	$p_{4k,k}$	...	...	$p_{4k,\sqrt{n}-1}$
...	...	...	...	...	...
...	...	...	...	...	...
$p_{\sqrt{n}-1,0}$	...	$p_{\sqrt{n}-1,k}$	...	...	$p_{\sqrt{n}-1,\sqrt{n}-1}$

Figure 12.   Two_Column_Merge Submeshes

Two_Column_Merge(k) sorts each sequence into either a nonincreasing on nondecreasing row major order (as viewed when the two sets of each pair are adjacent). The precise algorithm is given below.

```
procedure Two_Column_Merge(k)

    set r_{i,j} and s_{i,j} as indicated in figure 7
    issue k - 1 row shuffle clocks
    ℓ = k
    while ℓ > 1 do
        Row_Exchange (k, odd)
        set c_{i,j} and d_{i.j} as indicated in figure 10
        issue ℓ - 1 column shuffle clocks
        Two_Column_Exchange()
        issue ℓ - 1 column unshuffle clocks
        ℓ = ℓ / 2
    endwhile
    Row_Exchange(k, odd)
    issue k - 1 row unshuffle clocks
endprocesure
```

Note that when procedure Two-Column_Merge(k) is invoked, $\log_2 k$ calls to both Two_Column_Exchange() and Row_Exchange(k) are made and $2k$ column shuffle and column unshuffle clocks are issued. Also except for the calls to Row_Exchange(k) and Two_Column_Exchange() all the operations performed when Two_Column_Merge(k) is invoked are executed synchronously by every processor in the original mesh. However, the calls to Two_Column_Exchange and Row_Exchange are made to do the comparison/exchange operations. One such operation is required for each pair of processors in a particular grouping of adjacent processors. Again the numbers must be brought to the $2 \times \lambda$ processors located in the processing region. Two_Column_Exchange() can be expressed as follows.

```
procedure Two_Column_Exchange()

    for j = 0 to √n - λ by λ

        for i = 0 to √n - 2 by 2

            Two_Column_Exchange_{ℓ,m}(i,j),  ℓ=0, 0≤m≤λ-1, m even

            issue 2 column rotate clocks

        endfor

        issue λ row rotate clocks

    endfor

endprocedure
```

Each of the procedures Two_Column_Exchange$_{\ell,m}$ is defined in the following manner. Let the processors in the processing region now be grouped in the following way.

$p_{0,0}$ $p_{0,1}$	$p_{0,2}$ $p_{0,3}$	$p_{0,4}$ $p_{0,5}$	$\cdots$	$p_{0,\lambda-2}$ $p_{0,\lambda-1}$
$p_{1,0}$ $p_{1,1}$	$p_{1,2}$ $p_{1,3}$	$p_{1,4}$ $p_{1,5}$	$\cdots$	$p_{1,\lambda-2}$ $p_{1,\lambda-1}$

Then Two_Column_Exchange$_{\ell,m}$(i,j) is performed on the numbers continued on the processor pair $p_{\ell+1,m}$ and $p_{\ell,m+1}$ as follows

```
procedure Two_Column_Exchange_{ℓ,m}(i,j)

    exchange the numbers contained in value registers v_{ℓ+1,m}

    and v_{ℓ,m+1} (note the contents of the flag registers are

    not changed).

endprocedure
```

Note that when procedure Two_Column_Exchange is invoked, $\frac{n}{\lambda}$ column rotate clocks and $\sqrt{n}$ row rotate clocks are issued and $\frac{n}{2\lambda}$ parallel compare/exchange operations are performed.

Lemma 4: Procedure Two_Column_Merge(k) correctly sorts pairs of columns which are k columns apart such that the resulting sequence in each pair is in row major order form.

<u>Proof</u>: The first two lines of procedure Two_Column_Merge(k) cause

the proper columns in each pair to be moved so that they are adja-

cent. Let's now consider the first iteration of the while loop. If

we view the elements of every pair as belonging to one list, the

first line of the loop will introduce a compare-exchange operation

between elements which are k elements apart. The next 4 lines

will interchange the lower of the left column with the upper half

of the right column in each pair. In the next iteration, the

Row_Exchange(k,s) will compare and exchange the elements which are

$\frac{k}{2}$ elements apart and so on. Note that the last line of the proce-

dure will cause the columns to be returned to their original posi-

tions. ☐

We now consider the problem of setting the flag registers.

The flag registers can of course be set by rotating the contents of

the value and flag registers of each processor (as is done for each

of the parallel comparison/exchange operations) through the proces-

sing component. At which time, they could be correctly set. The

running time of <u>Sort</u> at most doubles since the flag registers are

set at most once preceding any parallel comparison/exchange opera-

tion. In this way, the running time of <u>Sort</u> would not be increased.

<u>Theorem 1</u>: Procedure <u>Sort</u> correctly sorts n number into row major

order form.

<u>Proof</u>: The previous lemmas establish essentially the correctness of

the procedures Horizontal_Merge(k) and Vertical_Merge(k). All that

is left to be shown is that, whenever either of these procedures is

called, for each pair of submeshes to be merged, one is in nonin-

creasing row major order form and the other is in nondecreasing row

major order form.

Let's consider a call to Vertical_Merge(k) which merges pairs of vertically aligned $k \times 2k$ submeshes. We will show that, when this procedure terminates, successive $2k \times 2k$ horizontally aligned submeshes alternate in their sorting orders. Notice that what determines the order of the sorting is the value of t in Column_Merge(k) and Row_Merge(k). It is easy to check by inspection that $j + m$ is the column index of the corresponding entry and t changes value only when we move from one submesh to another (if $k = 2^i$, then $(i+1)$'st bits of the column indices of the entries withing a submesh are identical). Hence, when Vertical_Merge(k) terminates, horizontally aligned $2k \times 2k$ submeshes will be sorted in different orders. A similar proof holds for Horizontal_Merge(k). □

Theorem 2: It is possible to do sorting on a chip with processing area $A = o(n)$ in time $T = O(\frac{n\log^2 n}{A} + \sqrt{n}\, \log^2 n)$.

Proof: Assume $A \leq \sqrt{n}$. We study the running time of procedure Row_Merge(k,*) in some detail. Procedure Row_Exchange(k,s) takes $O(\frac{\sqrt{n}}{\lambda} \times \sqrt{n}) = O(\frac{n}{A})$ time regardless of the value of k. The while loop in procedure Row_Merge(k) runs for $O(\log k)$ iterations and the execution time of each iteration is dominated by that of the Row_Exchange(k,s) procedure. Hence, the running time of Row_Merge(k,*) is $O(\frac{n\log k}{A})$. Similarly, the running time of Column_Merge(k) and Two_Column_Merge(k) is dominated by $\frac{n\log k}{A}$. Hence the same holds true for Vertical_Merge(k) and Horizontal_Merge(k). Note that the while loop in procedure Sort runs for $\log n$ iterations and that $k < n$. Therefore, the running time of procedure Sort is given by $O(\frac{n\log^2 n}{A})$. If $A > \sqrt{n}$, it is clear that we can use $\lambda = \sqrt{n}$ and obtain an algorithm with a running time $O(\sqrt{n}\, \log^2 n)$. □

We now show that the above algorithm is almost optimal and that our model is superior to the RAM model.

Theorem 3:   For B   constant (as introduced in section 2), sorting n numbers requires time   T   such that

   (1)   Network model

$$T = \Omega(\frac{n\log n}{A} + \sqrt{n})$$

   (2)   RAM model

$$T = \Omega(\frac{n\sqrt{n}}{A} \div \sqrt{n})$$

Proof: Any sorting algorithm (decision tree model) uses $\Omega(n\log n)$ comparisons. Hence, the time required is at least $\Omega(\frac{n\log n}{A})$. On the other hand, the $\Omega(\sqrt{n})$ is a lower bound regardless of the size of the area available. Therefore, (1) follows. For (2), notice that the total number of data movements required for sorting is at least $O(n\sqrt{n})$ (this can be shown by using well-known techniques in VLSI complexity theory). These data movements can only be performed through the processing component. Hence, (2) follows. $\square$

We now turn to the case where B could be unbounded. It is clear that if $A = O(\sqrt{n})$ then the above algorithm is still optimal to within a logn factor, since the obvious lower bound of $\Omega(\frac{n\log n}{A})$ still holds. However for the case when A is of order greater than $\sqrt{n}$, an improvement is possible. The above algorithm will be implemented to run in time $T = O(\frac{n\log^2 n}{A} + \log^3 n)$. This can be done as follows. We use a cube-connected-cycles [PV] for the storage component. This can be laid out in an area $A = O(\frac{n}{\log n}) \times O(\frac{n}{\log n})$. If we examine the layout described in [PV], it is not hard to see that the contents of the n storage elements can be moved into the processing region in time $T = O(\frac{n}{A} + \log n)$.

Note that for the comparison-exchange operation, we first move the proper data elements so that any two elements which must be compared are in the same storage element. We then move the contents of the storage elements into the processing region and perform the comparisons there.

Hence, we conclude:

<u>Theorem 4</u>: Sorting n number on a chip with processing area $A = o(n)$ with arbitrary long wires allowed can be done in time
$$T = O(\frac{n\log^2 n}{A} + \log^3 n).$$

4. <u>Other Problems</u>

The approach described in the previous section is quite general and can be applied to many problems. Consider, for example, the discrete Fourier transform and integer multiplication mesh-connected algorithms described in [P]. Then we can easily implement these algorithms in the network model with processing area $A = o(n)$ to run in time $T = O(\frac{n\log n}{A} + \sqrt{n}\log n)$. If we allow arbitrary long wires, then the corresponding running time is given by
$T = O(\frac{n\log n}{A} + \log^2 n)$ using the cube-connected-cycles as given in [PV]. Below we give a table of some of the corresponding algorithms.

Problem	*Long Wires*	*Bounded Wires*
Sorting	$O(\frac{n\log^2 n}{A} + \log^3 n)$	$O(\frac{n\log^2 n}{A} + \sqrt{n}\log^2 n)$
Merging	$O(\frac{n\log n}{A} + \log^2 n)$	$O(\frac{n\log n}{A} + \sqrt{n}\log n)$
Integer Multiplication	$O(\frac{n\log n}{A} + \log^2 n)$	$O(\frac{n\log n}{A} + \sqrt{n}\log n)$
FFT	$O(\frac{n\log n}{A} + \log^2 n)$	$O(\frac{n\log n}{A} + \sqrt{n}\log n)$
Convolution	$O(\frac{n\log n}{A} + \log^2 n)$	$O(\frac{n\log n}{A} + \sqrt{n}\log n)$
Cyclic Shift	$O(\frac{n\log n}{A} + \log^2 n)$	$O(\frac{n\log n}{A} + \sqrt{n}\log n)$

Note that the sorting problem is somewhat the hardest problem in the above list.

5. Conclusions

We have presented a new architecture which allows many problems to be solved quite efficiently on VLSI chips with very limited processing areas. We have considered in detail the sorting problem and shown how it can be done on a chip with reduced processing area. Furthermore, the control was shown to be simple and easily implementable in VLSI. We feel the same architecture can be used to solve other problems. As a matter of fact we show in a forthcoming paper how the FFT could be implemented optimally on a chip with a very simple control and few multipliers.

References

[CM] B. Chazelle and L. Monier, "A Model of Computation for VLSI with Related Complexity Results", 13th Annual Symp. on Theory of Computing, ACM (May 1981), pp. 318-325.

[MC] C. Mead and L. Conway, Introduction to VLSI Systems, Addison-Wesley, 1980.

[NS] D. Nassimi and S. Sahni, "Bitonic Sort on a Mesh-Connected Parallel Computer", IEEE Trans. on Computers, C-27(1), Jan. 1979, pp. 2-7.

[P] F. Preparata, "A Mesh-Connected Area-Time Optimal VLSI Integer Multiplier", Proceedings of the CMU conference on VLSI Systems and Computations, Oct. 1981, pp. 311-316.

[PV] F. Preparata and J. Vuillemin, "The Cube-Connected-Cycles: A Versatile Network for Parallel Computation", CACM 24(5), pp. 300-309, May 1981.

[T1] C. Thompson, "A Complexity Theory for VLSI", Ph.D. Thesis, Department of Computer Science, CMU, August 1980.

[T2] C. Thompson, "The VLSI Complexity of Sorting", Tech. Report, U.C. Berkeley.

[TK] C. Thompson and H. Kung, "Sorting on a Mesh-Connected Parallel Computer", CACM, 20, pp. 263-271, April 1977.

Silicon Compilation

Dumbo, A Schematic-to-Layout Compiler
W. Wolf, J. Newkirk, R. Mathews and R. Dutton

Macrocell Design for Concurrent Signal Processing
S.P. Pope and R.W. Broderson

A Case Study of the F.I.R.S.T. Silicon Compiler
N. Bergmann

Dumbo, A Schematic-to-Layout Compiler

Wayne Wolf, John Newkirk, Robert Mathews and Robert Dutton
Center for Integrated Systems
Stanford University

Abstract

This paper describes a technique for compiling logic circuit descriptions into layout and a program, Dumbo, to demonstrate the technique. Cells are described to the compiler as structural primitives, which describe the general form of the layout, and electrical primitives, which describe the circuit connectivity in terms of Boolean and pass gates. Dumbo maps these primitives into stick figures; it then solves a placement and wiring problem to produce a stick diagram that can be compacted into a layout. Implementation of Dumbo has focused on nMOS technology. Experimental results show that the resulting system gives promising results for a large class of cells with much less effort on the designer's part than required for traditional design techniques.

Introduction

The advantages of a VLSI design methodology that emphasizes a structured approach to both the logical and physical design [9] are clear. In order to make such an approach feasible for large chips, we must allow the designers to push detail onto subordinates, preferably computers, so that they can concentrate on problems critical to the design. This paper describes a designer's assistant, Dumbo, that compiles circuits and abstract physical structures into layouts.

We conjectured that it would be possible to compile stick diagrams from simple descriptions because not all cells are equally hard to design. Many cells in a chip are not very area-critical and some not at all. These cells tend to be straightforward applications of common physical structures to implement logical elements. Abstractions of these structures can be used to describe the desired cell, letting the compiler worry about their details.

To test this hypothesis we built a cell compiler, Dumbo. Our implementation works with nMOS circuits, but the techniques used are not restricted to that

This research was supported by ARO contract DAAG29-80-K-0046 and by DARPA contract MDA903-79-C-0680

technology. Dumbo takes as input a connectivity list of the circuit and a description of its external connections. It then solves a placement and routing problem to synthesize a planar topology (topography) for the cell and wires it. Dumbo's output is a stick diagram that can be turned into a layout by a sticks compactor; Figure 1 shows a compiled stick diagram and its compacted layout.

The compilation process is extremely flexible: if physical structures are not given for parts of the circuit the compiler will use simple structures as defaults. If the result is inadequate, the designer can improve the compiled cell as necessary by adding structural detail to the input. In this way the designer can match the effort in designing the cell to the cell's importance in the overall design.

A stick diagram compiler is a logical next step given the existence of sticks compactors, and the compaction problem is well understood. Some typical sticks compactors are SLIM [12] and FLOSS [7], which are shearing compactors, and CABBAGE [5] and Lava [8], which are critical path compactors. Lava was used as the back end compactor in this work.

The topography synthesis problem has attracted much attention in the past several years. Related work includes single-layer block layout [2], automatic generation of schematic artwork [3] [4], one-dimensional gate layout, including Weinberger's work [14] [1] [13], and the miniature gate array cells produced by SLAP [11].

The next step in automatic layout synthesis is full topography synthesis, which is the subject of this paper. To describe our solution, we first enumerate the goals for Dumbo. We then describe the algorithms used to design cells and give the results of quality measurements on Dumbo-designed cells.

Goals

Our ultimate objective is to study the relationship of logical and physical design. The canonical silicon compiler allows great flexibility in logic design by focusing on a very limited set of allowable layouts. We are interested in exploring the properties of silicon compilers that offer flexibility in both the logical and physical architectures. A leaf cell compiler is a good first step for two reasons: it would be a useful subsystem in such a chip compiler, and it lets us explore structural description on a manageable scale.

Given our emphasis on structural plus logical description of cells, we decided that these properties were necessary for a practical cell design system:

1. The program should be able to work with arbitrary digital logic circuits, including pass transistor designs. (We do not insist that parasitic circuit elements--capacitance, resistance--be precisely controllable.)
2. Cells should not be forced into a limited set of topographies. The program should be able to synthesize arbitrary layouts.
3. The program should design an electrically correct layout from an absolute minimum of information.
4. The program should be controllable and predictable: the algorithms should be able to use information in addition to the minimal description, and the designer should be able to add information and correctly predict the result.

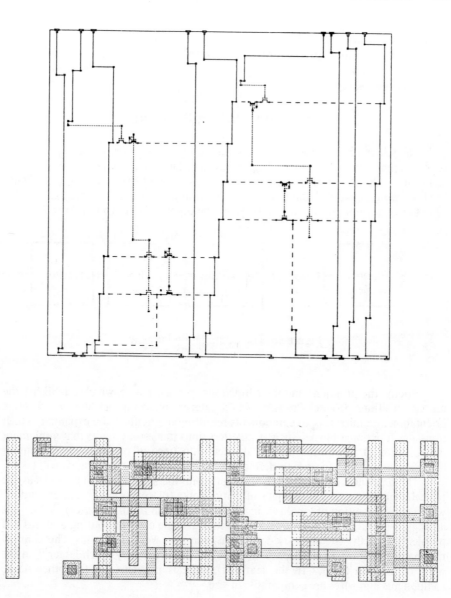

Figure 1: An automatically generated stick diagram and its layout

5. Execution time should be no more than a few CPU minutes. This restriction limits the extent of the program's search. We will trade a slightly more detailed input description for computation time in order to achieve area-efficient results.

The Compilation Process

The schematic-to-layout process occurs in two steps: first Dumbo produces a stick diagram; then the stick diagram is translated into a layout by a sticks compactor. Dumbo itself is implemented as a pipeline of five programs (see Figure 2). Each program adds detail to the description until a unique stick diagram results. The state of a simple cell at various stages in the pipeline is shown in Figure 2.

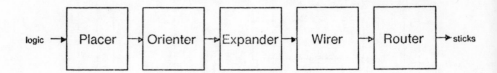

Figure 2:   The Compilation Pipeline

Briefly, the programs and their functions are: the placer, which establishes the relative positions of gates, transistors and external pins; the orienter, which gives orientation and mirroring to components requiring orientation; the expander, which translates functional blocks, such as inverters, into transistor circuits that implement the functions; the wirer, which breaks nets into trees of pairwise connections or branches; and the router, which defines the precise path that each branch will take.

The pipeline can be partitioned roughly into the same functional blocks as the typical channel router. placement of components (placer, orienter and expander); wiring tree generation or loose routing (wirer); and final routing (router). While this analogy is useful, it is important to remember that the channel routing and leaf cell problems, and the methods used to solve them, are very different. The channel routing problem works with components (blocks) that are very much larger than the wires connecting them. In the leaf cell synthesis problem, on the other hand, components and wires are comparable in size.

The Input Description

The input to Dumbo has two sections: an electrical description of the circuit, and structural information suggesting the desired layout. The level of detail in the electrical description is fixed. In contrast, the amount of structural information given depends on the degree to which the designer wants to control the form of the output. It is by varying the size and content of the structural description that the designer matches the cell to the requirements.

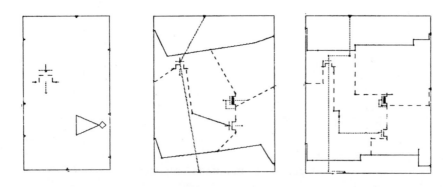

| After Placement and Orientation | After Expansion and Wiring | After Routing |

Figure 3:   Partial Designs of a Cell

Electrical description

The electrical description lists the components in the circuit and the connections among these components. Each electrical node (net) is described by a list of the component pins connected to the net. A component can be an external pin, a transistor or a compound component made up of several transistors. Compound components are used to describe common functional blocks. Figure 4 shows two compound components, an inverting superbuffer and a multiplexer.

Structural Description

The minimum structural information required is a description of the external connections into the cell. Each pin is placed on one of four cell edges--north, south, east or west--and the pins on each edge are partially ordered. The pin ordering is required because it defines the cell's communication and helps define the cell structure; it is a partial ordering so that the placement process can help define the communication plan.

The designer can optionally specify additional structural information about both nets and components. These structures specify the form that the layout is to take; the compiler fills in the details.

There are two wiring structures, the spanning tree and the track. Examples of these primitives are shown in Figure 5.

If the pins on the net being wired are considered as nodes on a net, then the spanning tree is a set of arcs that directly connect all the nodes. Tracks differ in that connections can also be made indirectly through intermediate nodes that are added to the problem. These intermediate nodes are roughly collinear and the wires between them form the spine of the track. Spurs connect the points on the spine to the component pins that are to be connected. Tracks are modeled after the long sets of control lines commonly used in layout.

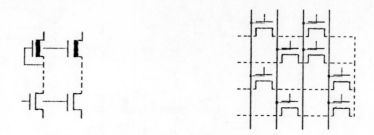

Figure 4: Compound Components

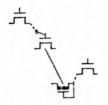

Spanning Tree   *Track*

Figure 5: Wiring Structures

A net can be implemented as a single wiring structure, or it can be composed of several structures each wiring a subset of the pins on the net. For instance, a net may be wired as two separate tracks connected by a spanning tree. Thus the designer can build arbitrary structures for a net from combinations of the basic wiring structures.

Compound components are used to supply structural information, in addition to their role in the electrical description. Functional units usually have one or a few best layouts. The best layout for a function may vary with the application, but generally a particular layout for a function is used many times in a design. The designer chooses a topography for a function by specifying a template to be used. Each template has a different placement of the internals and a different set of prewired connections. Templates supply information on the relative placement of elements of the function and on device sizing; the internal electrical connections can be made by the template designer or by the wiring program in any combination. Figure 6 shows two templates for an inverter that differ in the placement of their internal transistors.

The final method of structural specification is to give implementation hints to the programs in the pipeline. The hints mechanism is discussed in the next section.

Figure 6:   Component Templates

Algorithm Summary

We now discuss the algorithms used to implement each phase of the compiler. The phases will take as input, in addition to the information described above, hints that specify a part of the solution. The final solution will be in general changed by the existence of the solution seed supplied by the hint. For instance, the wiring phase will take as a hint a branch to be included in the spanning tree of a net; the existence of this first branch may be enough to produce a different spanning tree. Hints allow the designer finer control of the cell design than do structural elements; a hint is a rough equivalent to a block of assembly language code inserted into a high-level language program. While hints are not intended for extensive use, they are useful and add to the controllability of the compiler. Each algorithm summary below describes the hints that the algorithm will recognize.

Placer

The placer produces two total orderings of the transistors, logic elements and external pins in the cell, one in x and one in y. The algorithm used is a modified form of the force-directed placement method of Quinn and Breuer [10]. The force-directed model uses springs to guide placement. For each net a complete graph of springs is built so that every pair of pins on the net is connected by a spring. The system is then relaxed to find the minimum energy state, under the restriction that external pins are allowed to move in only one dimension along the cell edge. The ordering of components in the stable state is taken as the placement.

Hints take the form of added springs. The user specifies a pair of components that are to be attached by a spring and the stiffness of the attached spring. Figure 7 shows how user-defined springs may modify the placement. The original placement put the topmost transistor to the left of both tracks connected to its source and drain. To move it between the tracks the designer stiffened the springs between the source and drain and the two tracks.

Orienter

The orienter uses a torque-directed algorithm to compute the orientation of each component. Consider the component to be oriented as a pivot point around which rotate a set of arms, one for each component pin; then a wire connecting to a

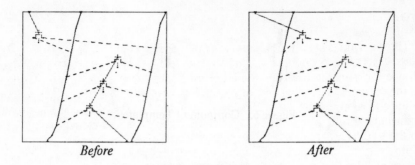

Before   *After*

Figure 7: Placement Hints

pin exerts a torque on the arm that drives the component to a preferred orientation, as shown in Figure 8.

Figure 8: Torque-directed Orientation

The torques exerted by each connection to the component can be summed to find the torque that rotates the component to its minimum energy orientation. Since there may be two stable states, one with wires crossed and one with uncrossed wires, mirroring may also be necessary. Mirroring is done so that the minimum number of wires cross over the component to get to their destination. The solution is actually calculated by a simple voting algorithm. Each branch votes on the orientations it prefers and the orientation with the most votes wins. If the orientation is not satisfactory the user can override it with a hint.

Expander

The expander converts a description of compound components and transistors into a description containing only transistors. The template specified by the designer for each compound component is used as a macro for the transistor implementation of the function; if no template is specified a default template is used. The template contains information on device sizes and on placement of internal components, which may be transistors or vias. It may also specify parts of the internal wiring of the function. For this stage the hint supplied by the designer is the template to be used for an instance of a compound component.

Wirer

The wirer translates each net into a set of pairwise connections (branches) with assigned layers and introduces vias where they are required. Two algorithms are used, one for spanning trees and the other for tracks. The spanning tree algorithm finds the minimum spanning tree of a complete graph of branches, which contains all connections on the net. Branches to be included in the tree can be supplied by the designer as hints. or by earlier stages of the compiler, and they are guaranteed to be in the tree. Vias are inserted into the cell on a branch-by-branch basis: a via is put into every branch that connects pins on different layers. The wirer determines the placement for these intermediate vias with a very simple heuristic.

Track nets are implemented as a particular form of Steiner tree. The spine of the net is built at a location specified in the placement, and spurs connect the Steiner points on the spine to components on the net. The Steiner points can be simple points or vias, depending on the layers of the spine and the component connection.

Router

The router translates the branches created by the wirer into a Manhattan stick diagram, one whose line segments all intersect at right angles. The cell routing problem is more general than the channel routing problem. In a channel routing problem the channels are formed by the four edges of the blocks that are being wired; but in the cell routing problem the transistors to be connected can be considered as point objects, so there are no natural wiring channels.

Our area routing algorithm, called L-routing, uses a very simple form of interconnect: all branches are drawn with two line segments in the form of either an L or an inverted L (see Figure 9).

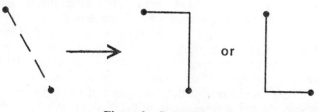

Figure 9:   L-routing

If the problem is constructed such that no two pins share a common centerline then no branches will be left unrouted: all undesired shorts can be removed by adding a jumper to a shorted branch that temporarily changes the wire to a safe layer. The routing problem is therefore to find a set of orientations for the Ls that minimizes the number of shorts. The optimal L-routing problem is NP-complete. However, typical cells do not require an exhaustive search to find a reasonable solution. A straightforward rip-up algorithm is sufficient to give good results. The algorithm also allows the designer to specify a routing for a branch as a hint.

Measurements

It would be desirable to test Dumbo's effectiveness by directly comparing compiled stick diagrams to hand-drawn stick diagrams. However, a stick diagram is only a means toward the desired end of a manufacturable layout, and there is no reasonable quality measures for stick diagrams. Therefore, Dumbo's performance must be measured through the filter of a sticks compactor. Figure 10 shows our method for measuring the quality of compiled layouts.

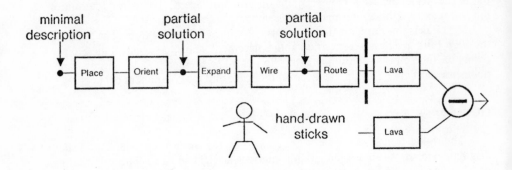

Figure 10:   Comparing Hand-drawn Stick Diagrams to Compiled Sticks

First, a test cell is rendered as a stick diagram by hand and compacted. Then a Dumbo description of the cell is written and compiled into a stick diagram, and that stick diagram is also compacted. Finally the areas of the two layouts are compared.

We can also measure the layout quality as a function of the work expended on the input. We do this by inserting a partially complete description of the cell into the middle of the compilation pipeline. For instance, to measure the penalty incurred with automatic placement we prepare a description of the cell that has the same placement and orientation of the components as does the hand-drawn stick diagram, and we start compilation of this description at the expander. We have taken measurements starting at three stages of the pipeline: starting at the placer, which gives a totally automatic cell; starting at the expander, which is equivalent to generating a layout from a schematic drawing (taking the component placement from the drawing); and starting at the router, which measures routing efficiency.

The cells used for comparison are taken from actual designs. Primary sources are Stanford chip design projects and the Stanford Cell Library [6]. Seven cells have been tested to date; they range in complexity from six to 24 transistors and cover a variety of physical design styles.

Table 1 summarizes the area penalties incurred by Dumbo, rounded to 1.5 significant digits. The variance of area penalty for designs introduced at the router is small, while the variance for designs introduced at later stages is larger. The mean and variance are large because the compilation algorithms compile some structures more efficiently than others. We have identified two problems that we think

Starting stage	Area penalty (% larger than hand design)
placer	120%
expander	60%
router	15%

Table 1: Average Area Penalty Compared to Hand-drawn Sticks

account for the inefficiencies: the placer is insufficiently attentive to the planarity of the wiring implied by the placement; and the wirer sometimes chooses very bad locations for the vias it adds to the cells. We are currently conducting experiments to test major improvements to the placement and wiring algorithms. Some of the rest of the area penalty is due to differences in aspect ratio on the tested cells: very thin cells suffer a large area penalty for a small increase in width. If the two thinnest cells are thrown out of the test suite the average compilation penalty starting at the expander is about 30%.

We have investigated the relationship between area penalty and the number of jumpers in the routing to help quantify the cost of suboptimal compilation in the early phases. Remember that a single jumper adds two vias to the component count. We correlated the area penalty to the number of jumper vias added during routing (see Figure 11). The percentage of components added is calculated as the number of jumper vias added divided by the number of transistors and vias input to the router.

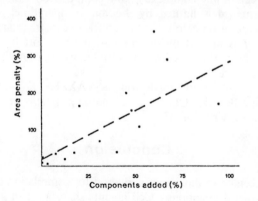

Figure 11: Increase in component count versus area penalty

The percentage increase in area of a cell is roughly 2.5 times the increase in component count caused by jumpering. The area penalty per jumper is large because jumpers add unusable white space to the cell as well as the area of their vias. A jumper via standing between two wires will leave a strip of white space the width of the via (plus design rule spacing) next to the via and between the wires.

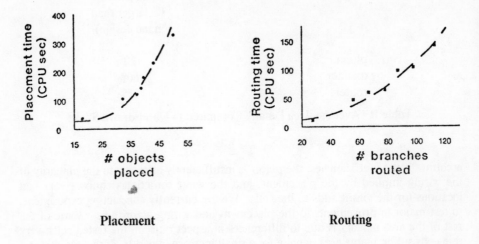

Placement   Routing

Figure 12:   Computation time versus cell complexity

The vast majority of the CPU time required to run Dumbo is spent in the placer and router. Figure 12 shows the average time complexity of these programs; in the placement time graph transistors, compound components, external pins and tracks are counted as placed components. Both programs show a quadratic relationship between input complexity and computation time. The placement algorithm's performance is limited by the matrix operations performed; sparse matrix techniques are of no help since the coefficient matrices are not sparse. The speed of the router is limited by the method used to sort branches. The time complexity of routing could be reduced to $O(n \log n)$ from $O(n^2)$ if more care were taken in implementation.

The measurements were made on a VAX/11-780 with floating point acceleration running Berkeley Unix[2]. The Dumbo programs were written in Pascal and the compactor was written in C.

Conclusion

We have presented in this paper a method for compilation of leaf cell layouts from simple, abstract descriptions. Cell layouts share a set of physical primitives that can be succinctly described and reconstructed using straightforward techniques. Structural compilation allows the designer to vary the design effort expended on a cell, with a commensurate payoff in output layout quality. These properties of the compiler encourage the designer to concentrate on the most critical areas of the design, so the designer can be more productive.

---

[2]Unix is a Trademark of Bell Laboratories.

Our future work will concentrate on improvements in the compilation techniques used, notably the placement and wiring algorithms. We expect to be able to greatly enhance the placer with the addition of an improvement phase that concentrates on maximizing the planarity of the wiring. We are also working on a wiring algorithm that searches for good locations for vias. These enhancements should significantly improve the quality of compiled cells.

References

1. Isao Shirakawa, Noboru Okuda, Takashi Harada, Sadahiro Tani and Hiroshi Ozaki. "A Layout System for the Random Logic Portion of an MOS LSI Chip." *IEEE Transactions on Computers C-30*, 8 (1981), 572-581.

2. S.B. Akers, J.M. Geyer and D.L. Roberts. IC Mask Layout With a Single Conductor Layer. Seventh Design Automation Workshop Proceedings, ACM/IEEE, 1970, pp. 7-16.

3. James A. Smith. *Automated Generation of Logic Diagrams.* Ph.D. Th., University of Waterloo, 1975.

4. R. J. Brennan. An Algorithm for Automatic Line Routing on Schematic Drawings. Design Automation Conference Proceedings, ACM/IEEE, 1975, pp. 324-330.

5. Min-Yu Hsueh. *Symbolic Layout and Compaction of Integrated Circuits.* Ph.D. Th., University of California, Berkeley, December 1979.

6. John Newkirk, Robert Mathews, Charlie Burns and John Redford. The Stanford nMOS Cell Library. Dept. of Electrical Engineering, Stanford University, 1981.

7. Y. E. Cho, A. J. Korenjak and D. E. Stockton. FLOSS: An Approach to Automated Layout for High-Volume Designs. 14th Design Automation Conference Proceedings, ACM/IEEE, 1977, pp. 138-141.

8. Robert Mathews, John Newkirk and Peter Eichenberger. A Target Language for Silicon Compilers. Compcon Proceedings, IEEE Computer Society, Spring, 1982, pp. 349-353.

9. Carver Mead and Lynn Conway. *Introduction to VLSI Systems.* Addison-Wesley, 1979.

10. Neil R. Quinn, Jr. and Melvin A. Breuer. "A Force Directed Component Placement Procedure for Printed Circuit Boards." *IEEE Transactions on Circuits and Systems CAS-26*, 6 (1979), 377-388.

11. Steven P. Reiss and John E. Savage. SLAP - A Methodology for Silicon Layout. International Conference on Circuits and Computers Proceedings, IEEE Computer and Circuits and Systems Societies, 1982, pp. 281-284.

12. A. E. Dunlop. "SLIM-The Translation of Symbolic Layouts into Mask Data." *Journal of Digital Systems V*, 4 (1981), 429-451.

13. J. Luhukay and W. J. Kublitz. A Layout Synthesis System for nMOS Gate-cells. 19th Design Automation Conference Proceedings, ACM/IEEE, 1982, pp. 307-314.

14. A. Weinberger. "Large Scale Integration of MOS Complex Logic: A Layout Method." *IEEE Journal of Solid-State Circuits SC-2* (1967), 182-196.

Macrocell Design for Concurrent Signal Processing

Stephen P. Pope and R.W. Brodersen
Department of Electrical Engineering and Computer Science
University of California, Berkeley

1. Introduction: What Is a Macrocell?

The term "macrocell", used in the context of large-scale integrated circuit design, has had a variety of meanings in the past. There is now general agreement that macrocells have the following properties:

(1) A macrocell is large, larger than for example a cell from a typical standard cell library. The macrocells to be described here range in complexity from several hundred to ten thousand transistors.

(2) Macrocells have sufficient internal functionality, and a sufficiently small need for external communication, that they may be assembled into a circuit using general placement and routing techniques with acceptable area efficiency.

(3) When a macrocell, stored in a library, is called into use by the designer, there will in general be parameters and options. This optimizes the macrocell to its particular application.

Thus, macrocells serve as subsystems at the top level of the hardware design hierarchy. The second property listed above is important since it implies that no predetermined floorplan for the circuit is required. This distinguishes the macrocell approach from the conventional standard-cell approach, in which the cells are typically arranged in rows separated by wiring channels. The macrocells themselves will have a compact, well-defined floorplan, such as a bit-slice organization.

Presented here is the macrocell design technique as applied to the implementation of real-time, sampled-data signal processing functions. The design of such circuits is particularly challenging due to the computationally intensive nature of signal-processing algorithms and the constraints of real-time operation. The most efficient designs make use of a high degree of concurrency -- a property facilitated by the macrocell approach.

The following functions have been identified as appropriate for macrocells in signal processing circuits:

(1) Processors, consisting of data memory and a pipelined arithmetic unit.

(2) Control sequencers, consisting of counters, ROM and PLA.

(3) Control modification units, such as address arithmetic units (AAU's) and conditional logic operators.

(4) Buffer memories and external I/O interfaces.

The way in which these macrocells are organized into a signal processing circuit will be briefly outlined. A control sequencer is used to generate a sequence of horizontal control words (typically periodic over a sample interval, or some other interval of interest). The control word feeds the control points of a processor, after possible modification by control modification units. In multiprocessor organizations, a single control sequencer may control more than one processor. Buffer memories may be used to synchronize communications between processors or with external devices.

Two circuit design projects whose development resulted largely from the macrocell methodology described above will be used as examples throughout this report: a linear-predictive vocoder circuit, and a front-end filter-bank chip for a speech recognition system. Both are monolithic multiprocessor implementations: the LPC vocoder circuit contains three processors, the filter-bank chip two processors.

In the design of multiprocessor systems such as these the individual processors must have properties that facilitate the multiprocessor organization. The following properties may be identified:

(1) The individual processors must be compact, so that several of them may reasonably be used in one monolithic circuit.

(2) The processors must have sufficient arithmetic capability and internal state memory that the need for inter-processor communication is small, preferably implemented in a bit-serial fashion to avoid parallel busing of signals between processors.

(3) The design must be applicable to any intended algorithm without involving a large amount of manual effort. This would imply either that the processor is not algorithm-

specific, or that the algorithm-specific portions may be realized in a fashion that is not labor-intensive.

The construction of algorithm-specific processors using silicon compiler techniques has been reported by Denyer [1]. In the following section an architecture suitable for the non-algorithm-specific alternative is presented.

2. Signal Processor Architectures using Parallel/Serial Multiplication

At the computational level signal processing algorithms primarily involve multiply-accumulate operations. Samples of data are multiplied by coefficients and the resulting terms are accumulated. If the coefficients are constant, fixed-frequency filters are the result; the coefficients are variable for adaptive or programmable filters. In either case, the sampled data represents a signal of interest -- speech, for example -- while the coefficients are dimensionless values.

In implementing filters, it is also necessary to delay the data samples by one or more sample intervals. Assuming the data is stored in an addressable memory, implementing such delays requires generating the desired sequence of addresses. Address generation is straightforward and is performed concurrently with the processing of the signal data. Address processing is discussed in section 4 of this report.

Most signal processing algorithms also involve some form of non-linear processing. Common varieties of non-linearities are: absolute-value operations, used for estimating average amplitudes; division operations, used for normalization; and multiplication of two signals, used for estimating second moments and for modulation.

The requirement that the individual processors in our multi-processor designs be physically compact argues against the use of parallel array multipliers, even in an efficient pipelined form. The alternative presented here is parallel-serial multiplication: the sequence of partial products is generated in a bit-parallel, word-serial form, and accumulated by a single accumulator. An arithmetic unit using this technique can use the same accumulator to sum the partial products in a single term and to sum several terms. This leads to a "single accumulator" architecture, an example of which is shown in Fig.1.

This architecture is used for the three processors in the LPC vocoder circuit.

This processor consists of a data memory, an arithmetic unit, and an I/O section that provides parallel-serial and serial-parallel conversion for the bit-serial I/O ports. Provision for parallel I/O is also included. Signals are represented in the data memory as fixed-point two's complement values. Constant values may also be stored in read-only locations in the memory.

The arithmetic unit consists of four pipeline registers, a complementer, an adder, and a certain amount of multiplexing and gating. Pipeline segmentation allows concurrent memory access, addition, and invert or shift operation.

The memory output register (MOR) is a master-slave register which is loaded each cycle with the output of the data memory. Its output feeds the complementer, which under program control will give the true, complement, or absolute value of its input.

The shift register (SR) is a master-slave register which, under program control will either load or shift right (arithmetic). In the first case this register is loaded with the output of the complementer divided by two; in the second case it is loaded with its previous contents divided by two. Thus the possible values that may be loaded into this register are summarized as follows:

    sr := mor/2
    sr := -mor/2
    sr := |mor/2|
    sr := sr/2

When performing a parallel-serial multiply or divide operation, this register is used to shift right repeatedly on successive cycles. This presents a sequence of partial products to the adder inputs. The procedures for these operations will be discussed in more detail shortly.

The adder is saturating: if an overflow is detected, the output will be the maximum or minimum value that can be represented, for positive or negative overflow respectively. The output of the adder is loaded into the accumulator (ACC). Given the possible values for the adder inputs, the following summarizes the values that may be loaded into the accumulator:

acc := 0
acc := mor
acc := acc
acc := sr
acc := sr + mor
acc := sr + acc

One additional feature of the accumulator, included to allow for a parallel-serial division operation, is a "accumulate if positive" control; when this feature is used, the output of the adder is loaded into the accumulator only if it is positive.

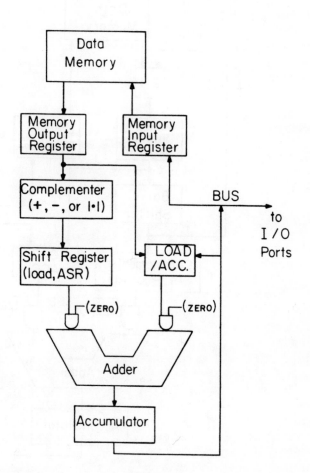

Fig.1 Processor Data Path (Sign-Magnitude Coefficients)

The output of the accumulator drives the Bus, which feeds the memory input register (MIR). This register is a transparent latch, which may either load or hold data under program control. All memory write operations store the output of this register; by holding the latch transparent the accumulator is written directly into memory.

All I/O transfers are through the Bus as well. Provided are parallel input and output ports, two serial output ports, and a serial input port.

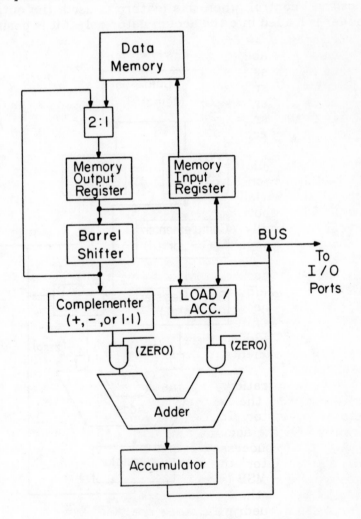

Fig.2 Processor Data Path (Signed-digit Coefficients)

The fashion in which the arithmetic unit is used to multiply a word in data memory by a signed constant is straightforward. For example, consider an eight bit constant in sign-magnitude format such as:

-.1100001

First the data is read from memory into the MOR. The multiply proceeds as the following microcode fragment:

```
sr := -mor/2
acc := sr; sr := sr/2
acc := sr + acc; sr := sr/2
acc := acc; sr := sr/2
acc := acc; sr := sr/2
acc := acc; sr := sr/2
acc := acc; sr := sr/2
acc := sr + acc
```

The accumulator now contains the original data multiplied by the above constant. Note that the eight bits of the constant are needed serially on successive cycles, MSB first, in order to control the above sequence of instructions. Thus, a multiply by an externally available coefficient in serial form can be accomplished by routing the coefficient into the appropriate control inputs for the processor. In the vocoder circuit, two of the three processors always multiply by constant coefficients, thus these coefficients are generated directly by the control sequencer. The third processor, which implements an adaptive analysis filter and a programmable synthesis filter, requires the addition of a multiplexer circuit to direct these variable coefficients into the appropriate arithmetic unit control points.

Divide operations, by means of long division, are implemented using the "accumulate if positive" control option for the accumulator. To find the quotient N/D, $D > N \geq 0$, N is loaded into the accumulator and -D/2 is loaded into the shift register. On successive cycles $D/2$, $D/4$, ... is subtracted from the accumulator, the result being accumulated only if it is positive. The MSB (sign bit) of the adder then represents the magnitude of the quotient in a bit serial, MSB-first format. Two or four quadrant divides are synthesized using the absolute value operator.

The parallel/serial architecture takes advantage of the fact that many of the multiplies encountered in signal processing algorithms are by low-precision coefficients. A variation of this approach, represented by the architecture shown in Fig.2, is optimized for performing multiplies by high-precision constant coefficients by representing these coefficients in the canonical signed-digit form [6]. In this scheme, a constant of the form

$$\pm \, k_0 *2^{-1} + k_1 *2^{-2} + k_2 *2^{-3} + \ldots$$

where the k's have values of 0 or 1 is represented as

$$g_0 *2^{n} + g_1 *2^{m} + \ldots$$

where the g's have values of ± 1, and the exponents n,m,... are chosen so as to minimize the total number of digits. This representation typically has one third the number of digits as does a binary representation, with no loss in precision.

It should be kept in mind that this is very effective for fixed coefficients, where the canonical form can be calculated before the processor's code is written; however, this device cannot be readily applied to the case of variable coefficients.

The processor of Fig.2 contains a parallel shifter array which is used to present terms in the above canonical expansion to the accumulator. A path is included to recirculate the output of the shifter to the register that feeds the shifter's input. This serves to relax the requirement that the depth of the shifter be determined by the number of bits of precision of the coefficients; instead, the depth need only equal the the maximum difference of the exponents of adjacent terms of the canonical expansion. Even this requirement may be relaxed at the expense of occasional additional clock cycles. In practice a shifter depth of four or six provides near-optimal performance. This organization, using a depth of six, was chosen for the two processors in the front-end filter-bank circuit.

It should also be noted that if the shifter depth is one, with the only available shift field being a shift-right of one place, the architecture of Fig.2 reduces essentially to that of Fig.1.

A few words should be said about the arithmetic precision of architectures such as these. Since the arithmetic shift-right operations involve truncation of the LSB's, it is necessary to take this effect into account when determining the required number of bits in the data word, for a given algorithm and given performance criteria.

R. Apfel [10] found that the cumulative round-off error can be quite large when implementing frequency selective FIR filters. He therefore modified the basic parallel-serial architecture to a form which he calls the "modified shift and add" architecture. Here the barrel shifter follows the accumulator rather than preceding it; when accumulating the partial products, the least-significant partial product is accumulated first, and quantity in the accumulator is shifted right as each additional partial product is added. Thus the relative weight of the quantity in the accumlator increases as the multiply proceeds. A decrease in data-word width of two to four bits is achieved.

3. Control Flow and Conditional Operations

To a large extent signal processing algorithms involve the unconditional execution of the same sequence of operations every sample interval. If functional multiplexing (wherein repeated sections of the algorithm are realized with the same hardware elements) is used, the control sequencer must loop through a section of code several times. Still, the control flow is data-independent; the number of iterations is predetermined and does not vary from sample to sample. It is very desirable to maintain this data-independent control flow in pipelined systems.

Many signal processing algorithms also require some sort of conditional operation. This is often the case when decision-making is required, or when heuristics have been applied to improve the performance of a purely analytical algorithm. In conventional, general purpose computer architectures, the only type of conditional instruction available is the conditional branch instruction. This practice has been carried over to programmable signal processor architectures such as those by Bell Laboratories [7], N.E.C. [8], or Texas Instruments [9]. To implement a conditional branch it is necessary to imbed the branch instruction in the same instruction stream that controls the data operations, which is inefficient. Unconditional control flow, on the other hand, can be generated by the control sequencer concurrently with the processing of signal data.

A well-known result from the theory of computer languages [4,5] states that it is never necessary to have branch instructions, other than iterations, when representing an algorithm by a computer program. This may be accomplished at the expense of more instructions being executed, and additional Boolean variables. While this sort of transformation is impractical in a general-purpose environment (where about one-third of the instructions executed are branches), it may make sense in the case of a pipelined signal processor where the density of conditional operations is relatively low. This transformation is facilitated by the following:

(1) Boolean state variables used as condition-code bits. These bits may be set as the result of comparisons or by external events.

(2) A conditional write instruction, wherein an assignment is made to a variable in data memory only if a condition-code bit is set.

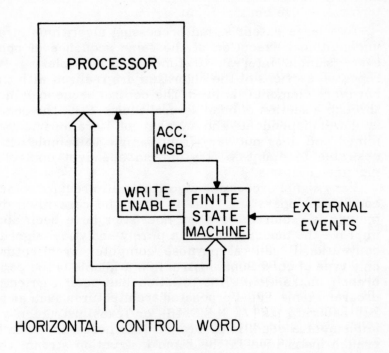

Fig.3 Use of Conditional Logic Operator

The general hardware implementation of the above is shown in Fig.3. The main hardware feature is the inclusion of a finite state machine, whose single output controls the data memory write cycle, and whose inputs are of three types:

(1) The sign bit of the accumulator, which is the result of a comparison operation;
(2) Control signals generated by the control sequencer;
(3) External events.

This organization was used in one of the three processors of the LPC vocoder circuit to facilitate the implementation of a pitch detection and voicing decision algorithm. The type of conditional operations needed were: peak detection; incrementing and resetting counters; thresholding of signals; and selecting the maximum of a set of variables. The finite state machine in this case has three bits of internal state, 18 minterms, and five control inputs from the sequencer. Thus with the inclusion of a modest hardware circuit it was possible to avoid conditional branches. Since the control word is horizontal, instructions which change the state of the FSM are processed in parallel with data path operations.

The above organization is quite general and expands considerably the set of algorithms that may be implemented with data-independent control flow. It should also be noted that when control flow is data-independent, execution time is also data-independent, simplifying design of real-time systems.

4. Address Processing

Due to the pipeline segmentation in the processor data paths, there is no possibility of performing address arithmetic with the processor arithmetic unit. Since the address field of the horizontal control word is relativley small (around six bits), and the type of address processing needed is varied but generally simple, address processing is implemented by individually designed macrocells for each processor.

In the simplest case (direct addressing) the address field of the control word may be used as the data memory address without modification so that no address modification macrocell is needed. Two common situations where modification is required are when some form of indexing is needed to support functional multiplexing, and when delays in the signal flow are most efficiently implemented by making the memory appear as a circular shift register, an arrangement referred to as "precessing".

Functional multiplexing involves looping through a section of microcode to implement a repeated section of the algorithm. Indexing is required to access the state variables for the individual sections. A counter serves as the index register. The unmodified address may be added to the index register; or the index register may be concatenated with a subfield of the unmodified address. In either case the control word contains a bit which enables the indexing operation. More generally, more than one indexing counter may be needed.

In some cases the hardware needed for indexing can be reduced by feeding the indexing counter directly into the memory address lines, and programming the address decoder such that the these inputs are "don't care" terms for non-indexed locations. If the indexing counter is included in the control sequencer the need to have an address arithmetic macrocell is avoided. This is the case in one of the three processors in the LPC vocoder circuit.

Direct or indexed addressing as described above generates the same address sequence each time the control sequencer goes through its cycle (sample interval). This is reasonably effective when only single-sample delays are needed in the signal flow; when longer delays are needed, it is more efficient to precess the addresses: each sample interval the sequence of addresses has been incremented (or decremented) in each of its terms. The effect of this is that the entire address space appears to shift from sample to sample, allowing delays of many samples without the need to move data each sample. This arrangement is advantageous when implementing direct-form filters. Interestingly, it was found to be advantageous for an all-zero lattice topology, despite the fact that all delays are single sample.

Precessing can be implemented by providing a base address counter, incremented once each sample, and a modulo N adder, where N is the number of words in data memory. It follows that N is most conveniently a power of two, and that the memory address decoder uses a straight binary code.

Another implementation of precessing addresses is to accumulate modulo N the unmodified addresses. In this approach each unmodified address is relative to the previous address, and the sum of the terms of the sequence of unmodified addresses is one modulo N. Indexing, if used, must follow this accumulation.

Along with the conditional logic units described in the previous section, address arithmetic units tend to be the smallest macrocells in the circuit.

5. Comparison with bit-serial architectures

A number of researchers have recently advocated the use of bit-serial architectures for monolithic implementations of signal processing functions. These approaches use as their basic computational element the pipelined two's-complement bit serial multiplier of Lyon [3]. These architectures have the attractive fundamental property that the combinatorial elements are heavily exercised due to the bit-level pipelining. Multipliers, adders, and other primitives are combined into arithmetic elements which are in general algorithm specific.

Bit-serial architectures were considered at an early stage for portions of the LPC vocoder circuit. The decision not to use them was made not for reasons of performance or efficiency, but primarily because of the manual effort involved in generating an algorithm dependent layout from the bit-serial primitives. Recent work by Denyer [1], wherein silicon compilation eliminates the need for manual layout (except for the primitive set), suggests that these two alternate approaches should be compared more carefully.

Before discussing the relative performance of the two methodologies there are several basic similarities that may be identified. In both cases, the system is divided into local memory/computation subsystems; these subsystems contain the state memory and the computational elements of some portion of the algorithm. Functional multiplexing is used to implement repeated portions of the algorithm with the same subsystem. The subsystems communicate via bit-serial signals. Horizontal control words are generated by essentially similar arrangements using counters, ROM's, and PLA's.

Another similarity that is not immediately evident is that signal data is stored in bit-parallel form in an addressable memory. Conceptually, in a bit-serial architecture the primitive element for data storage is a shift register. However, shift registers are quite inefficient with respect to area and power when compared to random-access memories. Thus, Lyon [2] proposes the use of RAM's for signal data storage by converting bit-serial data to bit-parallel form for storage in RAM, and converting back to bit-serial form after retrieving data from RAM. Thus both approaches store the data in bit-parallel form, and use parallel to serial conversion to interface with the bit-serial inputs and outputs. The difference lies in whether the computational elements operate on the bit-serial data or the bit-parallel data. In general the two systems will require data memories of the same size and bandwidth, if each implements the same algorithm.

Denyer [1] uses a somewhat different arrangement wherein FIFO's are used for signal data storage. Large FIFO's are most effectively implemented using random-access memory and the appropriate addressing logic, in which case the system resembles those just described.

With these basic similarities understood, the performance differences of the two approaches can be discussed. The discussion will be limited to a few of the most important considerations. First, some disadvantages of using the bit-parallel arithmetic unit relative to a bit-serial architecture:

(1) Maximum clock rate is limited by ripple carry delays. In designing a bit-parallel ALU, much care must be taken to make the carry chain as fast as possible. For the arithmetic units described in section (2), ripple-carry time does not become the limiting factor in the maximum clock rate until the word size is greater than 32 bits. Other factors tend to limit maximum clock rate, such as clock generation and distribution, driving long signal lines, and control word generation. Also, once a fast carry circuit is designed it may be applied in different situations without further effort, whereas optimizing other limiting factors tends to require design effort for each application. Still, given careful circuit design one should expect that significantly higher clock rates are possible in the bit-serial architectures.

(2) There is no direct way to multiply two signals, since signals and coefficients are handled separately. One approach is to convert one of the signals to the required format for coefficients (bit-serial, MSB first, sign-magnitude); for high-precision multiplies this is expensive in terms of time and area. By contrast, bit serial architectures which use the same representation for signals and coefficients can readily form the product of two signals with full precision.

(3) Multiplies must be microcoded, resulting in a more complex control structure. This is the price paid for using a more general arithmetic element as opposed to a dedicated multiplier.

The following describes some of the disadvantages of bit-serial architectures relative to the bit-parallel arithmetic unit:

(1) To avoid intractable timing problems, a "system word length" must be established. Shorter and longer words may be represented only with loss of efficiency. Thus the design cannot take advantage of the fact that different types of data require different precision. In the LPC vocoder circuit, the three bit-parallel processors have word widths of 16, 18, and 26 bits, the widest words being needed for second moment, (cross-correlation) estimation. Dimensionless coefficients are represented with eight or fewer bits.

(2) More clock cycles are needed to perform multiplies. Using Lyon's pipelined serial multiplier [3] and a serial adder, the number of cycles required for a multiply-accumulate operation is equal to the system word length. The architecture of Fig.1 can perform multiply-accumulate operations with n-bit variable coefficients at the rate of one every n-1 clock cycles. For the filters required in the LPC vocoder circuit, the number of cycles required may be estimated to be three times greater for the bit serial architecture. The amount of circuitry involved is roughly the same.

(3) There is no easy way to divide two signals. Long division methods by necessity yield the MSB of the quotient first, the opposite of the LSB-first format required for serial adds and multiplies. Division is needed for normalization in most signal processing algorithms. For example, in the LPC vocoder algorithm normalized cross-correlations are calculated. In the filter-bank circuit, the amplitude estimates of the filter outputs are normalized by the output of the largest channel prior to further processing by speech-recognition hardware. In general, the only way to derive a dimensionless value from a set of signals is by dividing two equally-dimensioned signals. Denyer [1] does not list a division operator as part of his primitive set.

To summarize, the pipelined bit-serial architecture is more efficient for multiplies by high precision variable coefficients, but cannot take advantage of the fact that many coefficients are low precision. Bit-parallel arithmetic units of the type described in this paper are more versatile than bit-serial elements of similar complexity, but require more complex control sequencing to achieve this functionality. If the system word length for the bit-serial architecture is N, and the average number of bits needed to represent a coefficient is K, the two systems will achieve equal multiply-accumulate throughput if the bit-serial architecture allows a faster clock rate by a factor of $N/(K-1)$.

6. Summary and conclusions

In Section (1) a very brief outline was given describing how this particular set of macrocells can be organized into a circuit. Now that the macrocells themselves have been described in some detail, it seems appropriate to describe the high-level design procedure more thoroughly.

As in any top-down design of a dedicated system, the algorithm must be developed first. The capabilities of the hardware should be considered during this phase. One important example is found in the case of linear filter implementations. Here, the lattice topology is often the best choice since this topology requires less coefficient precision than other topologies, especially for higher order filters. This is advantageous with the parallel- serial multiplier scheme. For the same reason, non-critical multipliers should be selected to be some simple value such as a power of two. A second example is that the use of adaptive algorithms, which avoid the need to buffer a large block of input data, is generally preferred.

A critical step in the design is the mapping of portions of the algorithm onto specific macrocells. Considerations here are to minimize inter-processor communication; to partition the algorithm according to the precision required for signal representation; and to group together those computations that would require similar sorts of address processing and other auxillary functions.

Configuration of the individual macrocells can now proceed. For the processors, parameters that can be specified by the designer are width of the data word; size of data memory; programming of constants into read-only memory locations; and programming the memory address decoder, which is essentially the AND-plane of a PLA. Similar parameters and options apply to the other macrocells in the circuit.

A very important part of the design is generation of the microcode, from which ROM's and PLA's in the control sequencers are created.

Part of our definition of a macrocell is that the macrocells may be assembled into a chip using general placement and routing techniques, with acceptable area efficiency. This procedure is the final step in the design process.

CAD software is used to facilitate most of the steps outlined above. Algorithm development makes use of a general-purpose simulator suitable for signal-processing applications. The basic tool of the circuit designer is a graphics editor, augmented by routers and ROM/PLA generation software. Microprogram development is aided by a symbolic assembler which accepts a register-transfer style description of the code, and is adaptable to the various hardware configurations encountered by different processors. Verification of both hardware configuration and microcode is possible using high-level functional simulation.

A design methodology based on macrocells has been presented which satisfies the specific characteristics of signal processing systems. These systems in general require real-time operation that is dominated by multiply-accumulate computation. We have designed our macrocells to provide the high performance required for these applications through the use of parallelism and pipelining. Our processor macrocells, which implement the bulk of the computation, have an extremely high degree of functionality for their small area; a single bit slice of the processor's arithmetic unit contains only 65 transistors. Perhaps even more important is that our macrocells are clearly defined, easy to apply and can be cast into a near-optimal configuration for a given section of an algorithm. The macrocell technique simultaneously provides the best aspects of a general purpose approach and a full custom design. Extreme flexibility results from both the programmability of a microcoded architecture, and the building-block approach to system organization, while efficiencies approach those of a dedicated design. The system is currently seeing a wide variety of applications, demonstrating the ease with which this methodology can be used.

Acknowledgements

Much circuit design for the macrocell library was done by Bjorn Solberg and Peter Ruetz. This research was sponsored by DARPA Contract No. MDA903-79-C-0429.

References

1. Denyer, Peter B., "An Introduction to Bit-serial Architectures for VLSI Signal Processing", University of Bristol, U.K. July 1982.

2. Lyon, Richard F., "A Bit-serial VLSI Architecture Methodology for Signal Processing", in Gray, J. P., (ed.), "VLSI 81", Academic press.

3. Lyon, Richard F., "Two's Complement Pipeline Multipliers", IEEE Trans. Comm., v. COM-24, 1976, pp. 418-425.

4. Bohm, C., and Jacopini, G., "Flow-diagrams, Turing Machines, and languages with only two formulation rules", Comm. ACM 9,5 May 1976, pp. 366-371.

5. Cooper, D. C., "Bohm and Jacopini's reduction of flow charts", Comm. ACM 10,8, August 1967, p. 463, p. 473.

6. Hwang, K., "Computer Arithmetic", Wiley, New York, 1979, p. 149.

7. Boddie, J.R., et al., "A digital signal processor for telecommunications applications", Int. Solid State Circuits Conf. Digest, San Francisco, 1980, pp. 44-45.

8. Kawakami, Y., et al., "A single chip signal processor for voiceband applications", Int. Solid State Circuits Conf. Digest, San Francisco, 1980, pp. 40-41.

9. Magar, S.S., Caudel, E.R., Leigh, A.W., "A Microcomputer with Digital Signal Processing Capability", Int. Solid State Circuits Conf. Digest, San Francisco, 1982, pp. 32-33.

10. R. Apfel, private communication.

A Case Study of the F.I.R.S.T. Silicon Compiler

Neil Bergmann
Department of Computer Science
University of Edinburgh

1: INTRODUCTION

This paper presents a case study of a particular
silicon compiler and the development of the software
environment necessary to support it. The FIRST silicon
compiler (Fast Implementation of Real-time Signal
Transforms) has been developed as a cooperative project
between the departments of Electrical Engineering and
Computer Science at the University of Edinburgh, in
order to allow the rapid investigation and
implementation of VLSI digital signal processing
systems.

The FIRST system is built around an underlying
bit-serial signal representation as proposed by Lyon
[Lyon 81], and systems are implemented as hard-wired
networks of pipelined bit-serial operators. A typical
flow diagram for a system suitable for implementation by
the FIRST compiler is shown in figure 1.

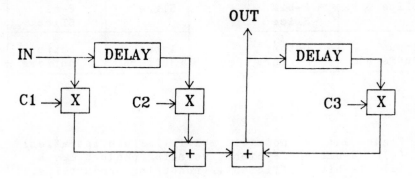

Figure 1 : A Signal Processing System
Suitable for FIRST Implementation

2: AN INNOVATIVE ARCHITECTURE

The hardware implementation of a FIRST circuit consists of a network of interconnected bit-serial operators, laid out according to a relatively fixed floorplan. Each bit-serial operator is implemented as a seperate function block, which is, in turn, assembled from a library of hand-designed leaf cells. A typical leaf cell might comprise, say, a single bit-slice of a given function, and the complete operator would then be arranged, both logically and physically, as a linear systolic array of these individual bit-slices. In this way, the logical size and exact function of each function block can be easily varied by selecting different numbers and types of leaf cells. For example, figure 2 shows two possible configurations for a bit serial multiplier - one which uses an 8-bit coefficient and rounds the least significant product bit, and the other which uses a 12-bit coefficient and truncates. (Note: the multiplier design uses 2-bit systolic array elements). The FIRST system includes a library containing a relatively small number of such predefined operators, as listed in appendix 1.

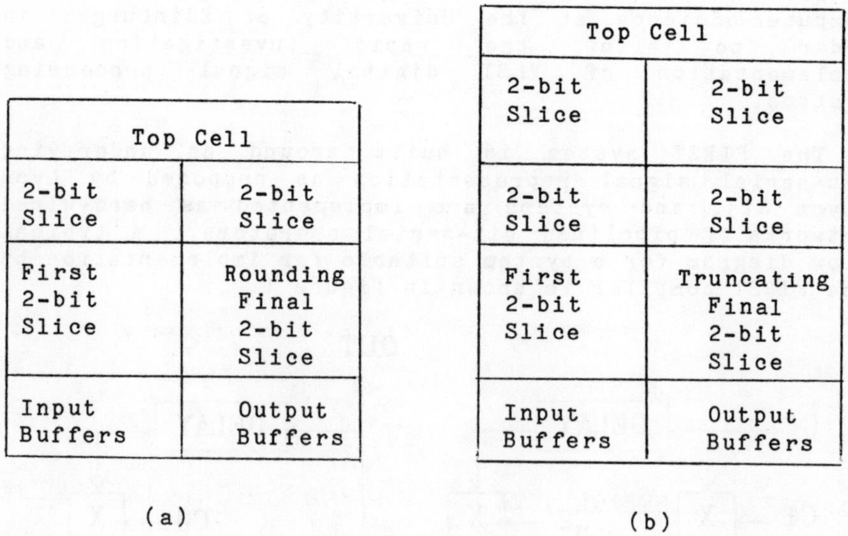

Top Cell	
2-bit Slice	2-bit Slice
First 2-bit Slice	Rounding Final 2-bit Slice
Input Buffers	Output Buffers

(a)

Top Cell	
2-bit Slice	2-bit Slice
2-bit Slice	2-bit Slice
First 2-bit Slice	Truncating Final 2-bit Slice
Input Buffers	Output Buffers

(b)

Figure 2 : Two possible multiplier configurations
(a)   8-bit coefficient, rounding product l.s.b.
(b)   12-bit coefficient, truncating product l.s.b.

The function blocks on a chip are arranged in two rows along either side of a single, central communications channel.   Interconnections between function blocks and connections to bonding pads are all made within this channel. A typical floorplan is shown in figure 3.   Some silicon area is wasted by this approach, since function blocks may differ in height. Typically, this area is about 20% of the total chip area, which, since it is not active area, has only a linear effect on good die/wafer yield.

Bonding pads are arranged more or less evenly around the chip periphery.   After some thought it was decided to allow the pad order to be user controlled, in order to improve PCB level wiring management.

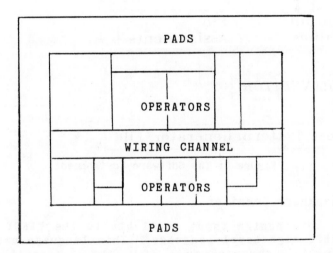

Figure 3 : Typical Floorplan

3: SOFTWARE SUPPORT FOR AN INTEGRATED DESIGN ENVIRONMENT

The software support for the FIRST system consists of a small suite of programs which are able to provide the designer with a complete, specialised design environment. The structure of this environment is shown in figure 4, and each of the major components is described below.

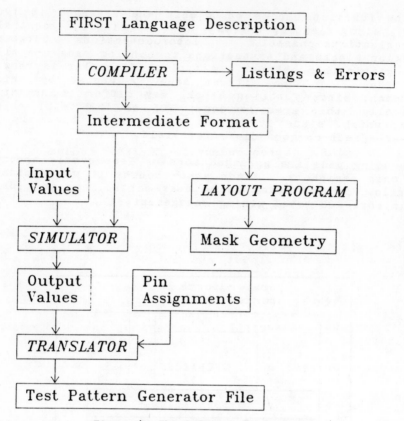

Figure 4: The Software Environment

3.1: Language Compiler.

The sole design input available to the FIRST user is the FIRST high level language. This language provides a structural description of the circuit under consideration, in that it describes the function blocks (i.e. bit-serial operators) which are present in a circuit, and their interconnections. In addition, because the structural design primitives have been chosen to correspond to functional design primitives, it also serves as a de facto functional description of a circuit. The language is able to capture the designer's intent in a form which closely matches the designer's logical conception of a system. This is a major advantage over designing with proprietry parts, where the designer must translate from his or her logical conception of a system into a quite different structural realisation, often in terms of ill-matched and inconvenient functional units.

The FIRST language identifies four distinct data types needed to build a circuit description:

(1) OPERATORS, corresponding to arithmetic and logical functions.

(2) CONSTANTS, which are integer expressions corresponding to the values of parameterisable operator attributes.

(3) SIGNALS, corresponding to network nodes which carry sampled-signal data.

(4) CONTROLS, corresponding to network nodes which carry timing information such as 'START OF WORD' and 'START OF FRAME'.

A circuit is a collection of parameterised operators, each with a set of input and output ports which are connected to control or signal nodes. Interconnections between different operator ports are made implicitly by connecting several ports to the same node. Each operator invocation, corresponding to a line of the FIRST language description, contains the following information:

- the operator name.

- the values of any parameterisable attributes.

- ordered lists of the names of the nodes to which input/output ports are connected.

These different data types are syntactically seperated in an operator invocation, which is in essence a mathematical expression in prefix notation. The general form is:

NAME [params] (ctls in -> ctls out) sigs in -> sigs out

e.g.
ADD [DELAY+2,0,0] (LSBTIME) A,B,CARRYIN -> SUM,CARRYOUT

The FIRST language also allows the definition of common, parameterisable sub-circuits as user-defined operators. These operators may then be invoked any number of times in exactly the same manner as primitive operators. Operator calls are grouped into CHIP definitions, corresponding to physical integrated circuits. These CHIP definitions may then be grouped into SUBSYSTEM definitions, and finally into a single SYSTEM definition. Thus the design language spans the whole hierarchy from structural primitives, to complete multi-chip signal processing systems.

The language compiler reduces this hierarchy into a list of primitive operator invocations, with node names replaced by unique node numbers. The resultant desription is called the FIRST intermediate format, and it is this description which is used by later phases of the design software.

A high level language design interface allows considerable error checking to be performed, e.g. type checking, undefined names, incorrect number of operator port connections etc. The relatively constricted syntax of the language helps to avoid many of these errors in the first place.

The form of the language has been derived from the structural design language for the MODEL gate-array design system [Gray 82]. The language compiler was built using the lexical and syntax analysis phases of an existing teaching compiler, SKIMP [Rees 80], with a custom "code generation" section added. The result is a very simple single pass, recursive descent compiler. Since FIRST circuit descriptions are typically only about one page in length, the simplicity and ease with which the compiler can be altered have far outweighed any considerations about run-time efficiency.

3.2: Simulation.

Two different simulators have been produced in the development of the FIRST system. The earlier simulator was clock driven. On each clock cycle, every operator in the system would be simulated in turn, using the present binary values on its input nodes along with any stored internal state to produce new values on its output nodes at the next clock cycle. Once all operators had been so invoked, the clock would advance by one cycle, and the process repeated. External inputs to the simulator were entered via an external data file. Similarly the values on any nodes could be output to another data file for subsequent inspection.

The operation of this simulator was a direct algorithmic interpretation of the physical circuit - so much so in fact that the functional definitions of operators could often be written such that each logic equation in the functional description would have a direct hardware counterpart in the physical realisation. This proved useful in determining if a proposed hardware realisation of an operator did, in fact, implement the desired function, and also gave a great deal of confidence that the simulator provided an accurate model of operator behaviour.

However, because this simulator was using a sequential, word oriented machine (i.e. the computer) to directly simulate the operation of a highly concurrent, bit-oriented architecture, the computational effort required for a thorough simulation of a large system was unexceptable.

For this reason, another simulator has been designed which simulates a system at a higher level of abstraction. This simulator is event driven, and simulates the operation of individual operators on a word by word basis. The values on nodes, which in reality consist of serial bit streams, are modelled as discrete words of data occuring at discrete time intervals. When a new word of data reaches a node, an event is said to have occurred. Events are described in terms of the node to which they refer, the new value of that node as a result of the event, and the time at which the event occurs.

The scheduling of events is handled by keeping all pending events on an event queue. All events due to occur at a given time are removed from the queue, and the values of the associated nodes are updated to reflect their new values. All operators which have any of these nodes as their inputs are activated. The new values which have arrived at their inputs will generate new values at their outputs. These new values are modelled as further events scheduled to occur at some later time, as determined by the latency of the particular operator.

In general, new values should arrive at all inputs of a particular operator at the same time. If not, this usually implies that the designer has made an error in matching the latency of the various signal paths leading to this operator. When such mismatching occurs, the simulator issues a warning, and continues the simulation. The operation of the simulator in such cases will not truly echo that of the physical system, which is itself unlikely to be a valid circuit. Once all such timing "bugs" have been eliminated (and this simulator provides a powerful tool for such debugging) the designer can have a high degree of confidence in the simulation results.

Inputs and outputs to the system are again via external data files. These data files are essentially

lists of events relevant to the system input or output
nodes. Since the system is being modelled at a higher
level than in the clock driven simulator, the computing
effort required for a given simulation has been reduced
by about an order of magnitude.

A program also exists within the FIRST suite to
convert the simulator's output into a form suitable for
use with an automatic test pattern generation system.

3.3: Physical Layout

The task of the FIRST layout program is to produce a
physical realisation of the system implied by a FIRST
language description. This process proceeds in distinct
phases, according to a strict layout strategy.

For every invocation of a predefined operator in a
FIRST language description of a system, a corresponding
function block appears in the physical layout.   Each
function block is assembled from the appropraite leaf
cells as described above.   Once constructed, operator
blocks are placed along "waterfronts" at the top and
bottom edges of a central wiring channel.   The layout
program uses the criteria of minimum chip area to decide
on a suitable arrangement of the blocks.   The size of
operator bounding boxes typically covers quite a range
of sizes.   The overall size of the chip, however,
depends on four main factors:-

- the height of the tallest block on the top row
- the height of the tallest block on the bottom row
- the width of the wider of the two rows
- the size of the wiring channel

The first three factors depend only on which blocks
are placed on the top row, and which are placed on the
bottom, and the placement algorithm first decides on
this subdivision.   Without attempting to rigorously
explain the algorithm used, the basic idea is to place
all the tall blocks on the top row, and the short blocks
on the bottom row, where the division between tall and
short is chosen so that the total area is minimized.
Perturbations are then made to this arrangement to
reduce the area further by making the rows more equal in
width.   The size of the wiring channel is considered
constant during these calculations.

Next, the arrangement of the blocks within each row
is decided.   The width of the wiring channel, and hence
total chip area will differ with different arrangements.
Rather than use an algorithmic method for determining a

good arrangement, the blocks are placed left to right
within each row in the same relative order that they
were invoked in the FIRST language description of the
system. Since a designer may be expected to write this
description such that the general flow of information is
from one operator to the next down the page, such a
strategy should lead to closely coupled operators being
placed close together on the chip. Wiring channel area
is only a relatively small part of the overall chip
area, and so it was not considered worth the effort,
both in terms of programming time, and run time, to
calculate any more optimal arrangement.

All routing between operators is done in the single
central wiring channel. Inputs and outputs of all
operators are available along the channel waterfront. A
very simple two layer router is used, with metal lines
running horizontally, and diffusion lines running
vertically. The input and output ports of operators are
restricted to points on a fixed grid, with the grids for
the two sides being offset. This ensures that
connections can be made between any two ports with a
single horizontal wire and two vertical wires, i.e.
without dog-legs. Figure 5 shows a section of a typical
wiring channel.

Finally bonding pads are arranged around the chip,
and ancillaries such as power, ground and a global
two-phase non-overlapping clock are added.

The layout program has been wriiten using an imbedded
IC design language called ILAP [Hughes 82], based on LAP
[Locanthi 80], but using IMP as a host language. IMP
[Robertson 80] is a high level Algol-like programming
language used extensively within the Edinburgh computing
environment.

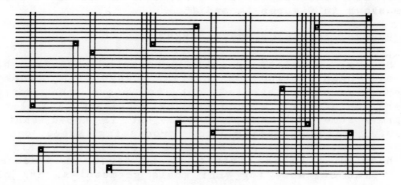

Figure 5: Wiring Channel Section

4: EXAMPLE SYSTEMS

As an illustrative, if perhaps not very practical example of a FIRST circuit, consider a chip to implement a simple four-stage, cascadable FIR filter section. A flow diagram of such a section is shown in figure 6. For illustrative purposes, consider it divided into two "TWO STAGE" sections as shown in figures 7 and 8. From these flow diagrams, an implementation in terms of FIRST operators could be derived to give the circuits shown in figures 9 and 10. Note that the final delay element in TWO STAGE has been made parameterisable in order to allow the final TWO STAGE section on the chip to have a slightly shorter inter-stage delay which compensates for the delay in going off-chip to the next stage in a multi-chip system. Also note that a network of timing signals, shown as broken lines in figures 9 and 10, has been added in order to give 'start of word' information to the various bit-serial operators.

The FIRST language description of a single section is shown in figure 11, while the resultant layout is shown in figures 12 and 13, in outline and detailed form respectively. If such a chip was fabricated using a 5 micron Nmos process, it would be approximately 5mm x 5mm, and have a clock speed of 8 MHz, corresponding to a sample rate for 10 bit samples of 800 kHz.

A more realistic example of a FIRST circuit is a chip, designed by Dr. Peter Denyer, to implement a multiplexed section of an adaptive transversal filter function. By suitable use of multiplexing, this chip is able to match the 8 MHz processing speed of a 5 micron Nmos realisation to the desired sample rate of 12 kHz for 14 bit samples, and so implement 47 stages of the filter function on a single chip approximately 5mm x 7mm. The FIRST language description and detailed layout are shown in figures 14 and 15.

5: PRESENT AND FUTURE DEVELOPMENT

Several complete signal processing systems have already been designed using FIRST [Denyer 83]. Work is currently being carried out to prove the correct operation of the 5 micron Nmos cell library, and also to improve some of the operators. It is hoped that by careful circuit design, the processing speed can be increased to 32 Mhz so that real time video signals can be handled. In the future it is hoped to produce a floorplan suitable for VLSI scale geometries, and also to produce cell libraries for other technologies, most notably CMOS.

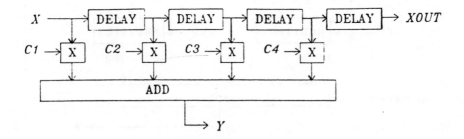

Figure 6 : Flow Diagram

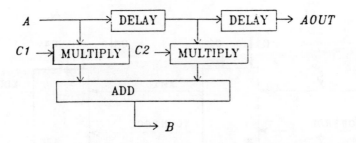

Figure 7 : TWOSTAGE sub-circuit

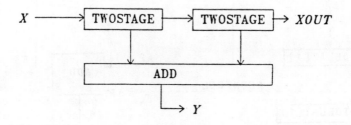

Figure 8 : Simplified Flow Diagram

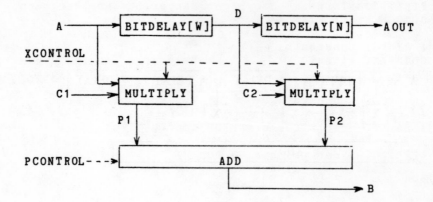

Figure 9 : Flow Diagram for TWOSTAGE[N]

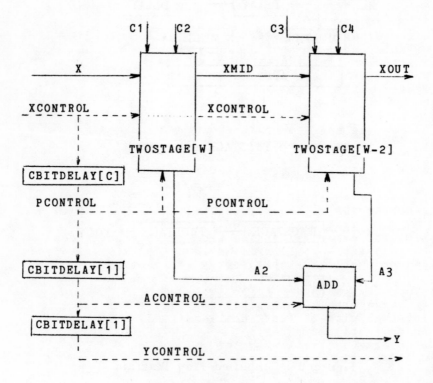

Figure 10 : Complete Circuit Flow Diagram

```
FIRST COMPILER    - Copyright Denyer,Renshaw,Bergmann
SOURCE FILE: FIR

! Global Constants
CONSTANT wlth=10,           ! Data word length
         cbits =10,         ! Coefficient word length
         d = cbits/2,       ! Multiplier latency
         truncate = 0       ! Type of multiplier

! Define simple forms of ADD and MULTIPLY

OPERATOR Adder (c) a,b -> sum
  Add[1,0,0,0] (c) a,b,gnd -> sum,nc
END
OPERATOR Multiplier (c) a,b -> p
  Multiply[truncate,cbits,0,0] (c->nc) a,b -> p,nc
END

! Two stages of an FIR filter, n=2nd stage delay

OPERATOR TwoStage[n] (xctl,pctl) a,c1,c2 -> aout,b
  SIGNAL d,p1,p2
  Bitdelay[wlth] a -> d
  Bitdelay[n] d -> aout
  Multiplier (xctl) a,c1 -> p1
  Multiplier (xctl) d,c2 -> p2
  Adder (pctl) p1,p2 -> b
END

! Define the whole chip,including pad  ordering
CHIP FIR (xxctl -> yyctl) xx,d1,d2,d3,d4 -> xxout,yy
  CONTROL pctl, actl, xctl, yctl
  SIGNAL xmid,a2,a3,x,y,c1,c2,c3,c4,xout

!!!!Specify order of bonding pads
  PADORDER VDD,xx,xxout,yy,GND,
           xxctl,yyctl,CLOCK,d1,d2,d3,d4
!!!!Equate external bonding pads to internal nodes
  Padin (xxctl->xctl) xx,d1,d2,d3,d4 -> x,c1,c2,c3,c4
  Padout (yctl -> yyctl) xout,y -> xxout,yy

!!!!Specify operators to be included
  TwoStage[wlth] (xctl,pctl) x,c1,c2 -> xmid,a2
  TwoStage[wlth-2] (xctl,pctl) xmid,c3,c4 -> xout,a3
  Adder (actl) a2,a3 -> y
  Cbitdelay[d] (xctl -> pctl)
  Cbitdelay[1] (pctl -> actl)
  Cbitdelay[1] (actl -> yctl)
END
ENDOFPROGRAM
```

Figure 11 : FIR Filter Listing

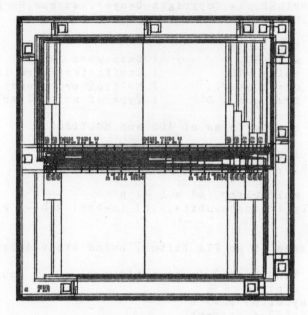

Figure 12: FIR Filter - Outline of Layout

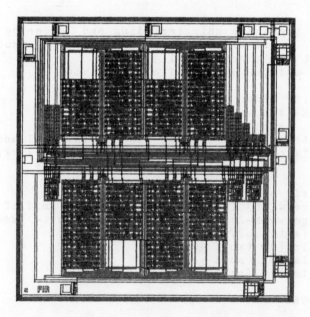

Figure 13: FIR Filter - Detailed Layout

```
        FIRST COMPILER    - Copyright Denyer,Rensnaw,Bergmann - 1982

SOURCE FILE: LMSFIR1

!Adaptive(LMS)Transversal(FIR)Filter
!filter stage of multiplexed filter sections

CONSTANT round=1,
          wordlength = 14,
          idles = 10,
          stages = 47

CHIP LMSFIR -
(cc0,cc1,initt -> cc1t0,cc1t1,cc1t2,cc1t3,cc1t5,cc1t6,cc2t0) -
xxin,tmeinn,wwmsbin,wwlsbin -> xxout,yymsbout,yylsbout

    SIGNAL Xin,TMEIN,WMSBin,WLSBin,Xout,YMSBout,YISBout,YLSBout
    CONTROL c0,c1,init,c1t0,c1t1,c1t2,c1t3,c1t5,c1t6,c2t0

    PADORDER VDD,xxin,tmeinn,cc0,cc1,initt,wwmsbin,
             GND,cc1t0,xxout,yymsbout,yylsbout,
             CLOCK,cc1t1,cc1t2,cc1t3,cc1t5,cc1t6,cc2t0,wwlsbin

    PADIN (cc0,cc1,initt -> c0,c1,init)
    PADIN xxin,tmeinn,wwmsbin,wwlsbin -> xin,tmein,wmsbin,wlsbin
    PADOUT (c1t0,c1t1,c1t2,c1t3,c1t5 -> cc1t0,cc1t1,cc1t2,cc1t3,cc1t5)
    PADOUT (c1t5,c1t6,c2t0 -> cc1t5,cc1t6,cc2t0)
    PADOUT xout,ymsbout,ylsbout -> xxout,yymsbout,yylsbout

    SIGNAL s1,s2,s3,s4,s5,s6,s7,s8,s9,s10,s11,s12,s13,s14,s15,
           s16,s19,s20,s21,s22,ss1,ss2,ss3,ss4,ss5,ss6

        Multiplex[1,0,0] (c2t0) s3,Xin -> s1
        Multiplex[1,0,0] (INIT) s15,WMSBin -> s7
        Multiplex[1,0,0] (INIT) s16,WLSBin -> s8
        Multiplex[1,0,0] (c1t2) s6,s5 -> s6
        Multiplex[1,0,0] (c1t6) YMSBout,YISBout -> YMSBout

        Multiply[round,wordlength-2,0,0] (c1t0->nc) s2,TMEin -> s5,nc
        Multiply[round,wordlength-2,0,0] (c1t5->nc) s22,s15 -> YLSBout,YISBout
        Add[1,0,0,0] (c1t3) s6,s11,gnd -> s15,nc
        Add[1,0,0,0] (c1t3) s5,s12,gnd -> s16,nc

        constant p1=(idles+1)*(wordlength)

        Bitdelay[18] s1 -> ss1
        Bitdelay[18] ss1 -> ss2
        Bitdelay[18] ss2 -> ss3
        Bitdelay[18] ss3 -> ss4
        Bitdelay[18] ss4 -> ss5

        constant p2 = p1-90 , p3 = p2/2 , p4 = p2-p3

        Bitdelay[p3] ss5 -> ss6
        Bitdelay[p4] ss6 -> Xout

        Worddelay[idles+stages-1,wordlength-2,0] (c1t1) s7 -> s19

        Worddelay[idles+stages-1,wordlength,0] (c1t1) s8 -> s10

        Constant half = stages/2,
                 rest = (stages - 1) - half
        Worddelay[half,wordlength-2,0] (c1t0) Xout -> s4
        Worddelay[rest,wordlength-2,0] (c1t0) s4 -> s2

        Bitdelay[wordlength] s1 -> s19
        BItdelay[wordlength] s19 -> s20
        Bitdelay[wordlength] s20 -> s21

        Bitdelay[wordlength-1] s2 -> s3
        Bitdelay[wordlength-4] s9 -> s11
        Bitdelay[wordlength-4] s10 -> s12
        Bitdelay[wordlength/2+1] s21 -> s22
    END
ENDOFPROGRAM
```

Figure 14: Adaptive Transversal Filter Listing

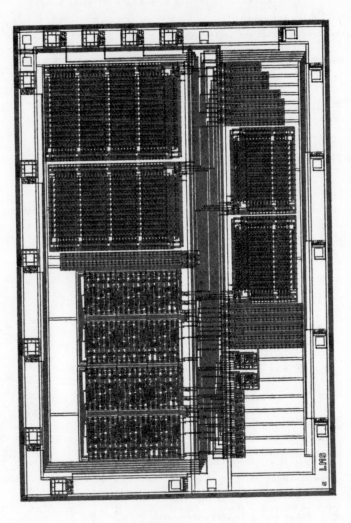

Figure 15: Adaptive Transversal Filter Layout